Sampling Methods
Exercises and Solutions

Pascal Ardilly

Yves Tillé

Translated from French by Leon Jang

Sampling Methods
Exercises and Solutions

 Springer

Pascal Ardilly
INSEE Direction générale
Unité des Méthodes Statistiques,
    Timbre F410
18 boulevard Adolphe Pinard
75675 Paris Cedex 14
France
Email: pascal.ardilly@insee.fr

Yves Tillé
Institut de Statistique,
Université de Neuchâtel
Espace de l'Europe 4, CP 805,
2002 Neuchâtel
Switzerland
Email: yves.tille@unine.ch

Library of Congress Control Number: 2005927380

ISBN-10: 0-387-26127-3
ISBN-13: 978-0387-26127-0

Printed on acid-free paper.

Printed in the United States of America. (MVY)

9 8 7 6 5 4 3 2 1

springeronline.com

Preface

When we agreed to share all of our preparation of exercises in sampling theory to create a book, we were not aware of the scope of the work. It was indeed necessary to compose the information, type out the compilations, standardise the notations and correct the drafts. It is fortunate that we have not yet measured the importance of this project, for this work probably would never have been attempted!

In making available this collection of exercises, we hope to promote the teaching of sampling theory for which we wanted to emphasise its diversity. The exercises are at times purely theoretical while others are originally from real problems, enabling us to approach the sensitive matter of passing from theory to practice that so enriches survey statistics.

The exercises that we present were used as educational material at the *École Nationale de la Statistique et de l'Analyse de l'Information* (ENSAI), where we had successively taught sampling theory. We are not the authors of all the exercises. In fact, some of them are due to Jean-Claude Deville and Laurent Wilms. We thank them for allowing us to reproduce their exercises. It is also possible that certain exercises had been initially conceived by an author that we have not identified. Beyond the contribution of our colleagues, and in all cases, we do not consider ourselves to be the lone authors of these exercises: they actually form part of a common heritage from ENSAI that has been enriched and improved due to questions from students and the work of all the demonstrators of the sampling course at ENSAI.

We would like to thank Laurent Wilms, who is most influential in the organisation of this practical undertaking, and Sylvie Rousseau for her multiple corrections of a preliminary version of this manuscript. Inès Pasini, Yves-Alain Gerber and Anne-Catherine Favre helped us over and over again with typing and composition. We also thank ENSAI, who supported part of the scientific typing. Finally, we particularly express our gratitude to Marjolaine Girin for her meticulous work with typing, layout and composition.

Pascal Ardilly and Yves Tillé

Contents

5 Multi-stage Sampling 159

6 Calibration with an Auxiliary Variable 209

1

Introduction

1.1 References

This book presents a collection of sampling exercises covering the major chapters of this branch of statistics. We do not have as an objective here to present the necessary theory for solving these exercises. Nevertheless, each chapter contains a brief review that clarifies the notation used. The reader can consult more theoretical works. Let us first of all cite the books that can be considered as classics: Yates (1949), Deming (1950), Hansen et al. (1993a), Hansen et al. (1993b), Deming (1960), Kish (1965), Raj (1968), Sukhatme and Sukhatme (1970), Konijn (1973), Cochran (1977), a simple and clear work that is very often cited as a reference, and Jessen (1978). The *post-mortem* work of Hájek (1981) remains a masterpiece but is unfortunately difficult to understand. Kish (1989) offered a practical and interesting work which largely transcends the agricultural domain. The book by Thompson (1992) is an excellent presentation of spatial sampling. The work devoted to the basics of sampling theory has been recently republished by Cassel et al. (1993). The modern reference book for the past 10 years remains the famous Särndal et al. (1992), even if other interesting works have been published like Hedayat and Sinha (1991), Krishnaiah and Rao (1994), or the book Valliant et al. (2000), dedicated to the model-based approach. The recent book by Lohr (1999) is a very pedagogical work which largely covers the field. We recommend it to discover the subject. We also cite two works exclusively established in sampling with unequal probabilities: Brewer and Hanif (1983) and Gabler (1990), and the book by Wolter (1985) being established in variance estimation.

In French, we can suggest in chronological order the books by Thionet (1953) and by Zarkovich (1966) as well as that by Desabie (1966), which are now classics. Then, we can cite the more recent books by Deroo and Dussaix (1980), Gouriéroux (1981), Grosbras (1987), the collective work edited by Droesbeke et al. (1987), the small book by Morin (1993) and finally the manual of exercises published by Dussaix and Grosbras (1992). The 'Que Sais-je?' by Dussaix and Grosbras (1996) expresses an appreciable translation of the theory. Obviously, the two theoretical works proposed by the authors Ardilly (1994) and Tillé (2001) are fully adapted to go into detail on the subject. Finally, a very complete work is suggested, in Italian, by Cicchitelli et al. (1992) and, in Chinese, by Ren and Ma (1996).

1.2 Population, variable and function of interest

Consider a finite population composed of N observation units; each of the units can be identified by a *label*, of which the set is denoted

$$U = \{1, ..., N\}.$$

We are interested in a variable y which takes the value y_k on unit k. These values are not random. The objective is to estimate the value of a function of interest

$$\theta = f(y_1, ..., y_k, ..., y_N).$$

The most frequent functions are the total

$$Y = \sum_{k \in U} y_k,$$

the mean

$$\overline{Y} = \frac{1}{N} \sum_{k \in U} y_k = \frac{Y}{N},$$

the population variance

$$\sigma_y^2 = \frac{1}{N} \sum_{k \in U} (y_k - \overline{Y})^2,$$

and the corrected population variance

$$S_y^2 = \frac{1}{N-1} \sum_{k \in U} (y_k - \overline{Y})^2.$$

The size of the population is not necessarily known and can therefore be considered as a total to estimate. In fact, we can write

$$N = \sum_{k \in U} 1.$$

1.3 Sample and sampling design

A sample without replacement s is a subset of U. A sampling design $p(.)$ is a probability distribution for the set of all possible samples such that

$$p(s) \geq 0, \text{ for all } s \subset U \text{ and } \sum_{s \subset U} p(s) = 1.$$

The random sample S is a random set of labels for which the probability distribution is

$$\Pr(S = s) = p(s), \text{ for all } s \subset U.$$

The sample size $n(S)$ can be random. If the sample is of fixed size, we denote the size simply as n. The indicator variable for the presence of units in the sample is defined by

$$I_k = \begin{cases} 1 & \text{if } k \in S \\ 0 & \text{if } k \notin S. \end{cases}$$

The inclusion probability is the probability that unit k is in the sample

$$\pi_k = \Pr(k \in S) = \mathrm{E}(I_k) = \sum_{s \ni k} p(s).$$

This probability can (in theory) be deduced from the sampling design. The second-order inclusion probability is

$$\pi_{k\ell} = \Pr(k \in S \text{ and } \ell \in S) = \mathrm{E}(I_k I_\ell) = \sum_{s \ni k, \ell} p(s).$$

Finally, the covariance of the indicators is

$$\Delta_{k\ell} = \mathrm{cov}(I_k, I_\ell) = \begin{cases} \pi_k(1 - \pi_k) & \text{if } \ell = k \\ \pi_{k\ell} - \pi_k \pi_\ell & \text{if } \ell \neq k. \end{cases} \tag{1.1}$$

If the design is of fixed size n, we have

$$\sum_{k \in U} \pi_k = n, \quad \sum_{k \in U} \pi_{k\ell} = n\pi_\ell, \quad \text{and} \quad \sum_{k \in U} \Delta_{k\ell} = 0.$$

1.4 Horvitz-Thompson estimator

The Horvitz-Thompson estimator of the total is defined by

$$\widehat{Y}_\pi = \sum_{k \in S} \frac{y_k}{\pi_k}.$$

This estimator is unbiased if all the first-order inclusion probabilities are strictly positive. If the population size is known, we can estimate the mean with the Horvitz-Thompson estimator:

$$\widehat{\overline{Y}}_\pi = \frac{1}{N} \sum_{k \in S} \frac{y_k}{\pi_k}.$$

The variance of $\widehat{Y}_\pi$ is

$$\mathrm{var}(\widehat{Y}_\pi) = \sum_{k \in U} \sum_{\ell \in U} \frac{y_k y_\ell}{\pi_k \pi_\ell} \Delta_{k\ell}.$$

If the sample is of fixed size ($\mathrm{var}(\#S) = 0$), then Sen (1953) and Yates and Grundy (1953) showed that the variance can also be written

$$\mathrm{var}(\widehat{Y}_\pi) = -\frac{1}{2} \sum_{k \in U} \sum_{\ell \in U} \left(\frac{y_k}{\pi_k} - \frac{y_\ell}{\pi_\ell} \right)^2 \Delta_{k\ell}.$$

The variance can be estimated by:

$$\widehat{\mathrm{var}}(\widehat{Y}_\pi) = \sum_{k \in S} \sum_{\ell \in S} \frac{y_k y_\ell}{\pi_k \pi_\ell} \frac{\Delta_{k\ell}}{\pi_{k\ell}},$$

where $\pi_{kk} = \pi_k$. If the design is of fixed size, we can construct another estimator from the Sen-Yates-Grundy expression:

$$\widehat{\mathrm{var}}(\widehat{Y}_\pi) = -\frac{1}{2} \sum_{k \in S} \sum_{\substack{\ell \in S, \\ \ell \neq k}} \left(\frac{y_k}{\pi_k} - \frac{y_\ell}{\pi_\ell} \right)^2 \frac{\Delta_{k\ell}}{\pi_{k\ell}}.$$

These two estimators are unbiased if all the second-order inclusion probabilities are strictly positive. When the sample size is 'sufficiently large' (in practice, a few dozen most often suffices), we can construct confidence intervals with a confidence level of $(1 - \alpha)$ for Y according to:

$$\mathrm{CI}(1 - \alpha) = \left[\widehat{Y}_\pi - z_{1-\alpha/2} \sqrt{\mathrm{var}(\widehat{Y}_\pi)}, \widehat{Y}_\pi + z_{1-\alpha/2} \sqrt{\mathrm{var}(\widehat{Y}_\pi)} \right],$$

where $z_{1-\alpha/2}$ is the $(1 - \alpha/2)$-quantile of a standard normal random variable (see Tables 10.1, 10.2, and 10.3). These intervals are estimated by replacing $\mathrm{var}(\widehat{Y}_\pi)$ with $\widehat{\mathrm{var}}(\widehat{Y}_\pi)$.

2

Simple Random Sampling

2.1 Simple random sampling without replacement

A design is simple without replacement of fixed size n if and only if, for all s,

$$p(s) = \begin{cases} \binom{N}{n}^{-1} & \text{if } \#s = n \\ 0 & \text{otherwise,} \end{cases}$$

or

$$\binom{N}{n} = \frac{N!}{n!(N-n)!}.$$

We can derive the inclusion probabilities

$$\pi_k = \frac{n}{N}, \quad \text{and} \quad \pi_{k\ell} = \frac{n(n-1)}{N(N-1)}.$$

Finally,

$$\Delta_{k\ell} = \frac{n(N-n)}{N^2} \times \begin{cases} 1 & \text{if } k = \ell \\ \dfrac{-1}{N-1} & \text{if } k \neq \ell. \end{cases}$$

The Horvitz-Thompson estimator of the total becomes

$$\widehat{Y}_\pi = \frac{N}{n} \sum_{k \in S} y_k.$$

That for the mean is written as

$$\widehat{\overline{Y}}_\pi = \frac{1}{n} \sum_{k \in S} y_k.$$

The variance of $\widehat{Y}_\pi$ is

$$\text{var}(\widehat{Y}_\pi) = N^2 \left(1 - \frac{n}{N}\right) \frac{S_y^2}{n},$$

and its unbiased estimator

$$\widehat{\mathrm{var}}(\widehat{Y}_\pi) = N^2(1 - \frac{n}{N})\frac{s_y^2}{n},$$

where

$$s_y^2 = \frac{1}{n-1}\sum_{k\in S}\left(y_k - \widehat{\overline{Y}}_\pi\right)^2.$$

The Horvitz-Thompson estimator of the proportion P_D that represents a sub-population D in the total population is

$$p = \frac{n_D}{n},$$

where $n_D = \#(S \cap D)$, and p is the proportion of individuals of D in S. We verify:

$$\mathrm{var}(p) = \left(1 - \frac{n}{N}\right)\frac{P_D(1 - P_D)}{n}\frac{N}{N-1},$$

and we estimate without bias this variance by

$$\widehat{\mathrm{var}}(p) = \left(1 - \frac{n}{N}\right)\frac{p(1 - p)}{n-1}.$$

2.2 Simple random sampling with replacement

If m units are selected with replacement and with equal probabilities at each trial in the population U, then we define $\tilde{y}_i$ as the value of the variable y for the i-th selected unit in the sample. We can select the same unit many times in the sample. The mean estimator

$$\widehat{\overline{Y}}_{WR} = \frac{1}{m}\sum_{i=1}^{m}\tilde{y}_i,$$

is unbiased, and its variance is

$$\mathrm{var}(\widehat{\overline{Y}}_{WR}) = \frac{\sigma_y^2}{m}.$$

In a simple design with replacement, the sample variance

$$\tilde{s}_y^2 = \frac{1}{m-1}\sum_{i=1}^{m}(\tilde{y}_i - \widehat{\overline{Y}}_{WR})^2,$$

estimates σ_y^2 without bias. It is possible however to show that if we are interested in n_S units of sample $\tilde{S}$ for distinct units, then the estimator

$$\widehat{\overline{Y}}_{DU} = \frac{1}{n_S}\sum_{k\in\tilde{S}}y_k,$$

is unbiased for the mean and has a smaller variance than that of $\widehat{\overline{Y}}_{WR}$. Table 2.1 presents a summary of the main results under simple designs.

Table 2.1. Simple designs : summary table

Simple sampling design	Without replacement	With replacement
Sample size	n	m
Mean estimator	$\widehat{\overline{Y}} = \dfrac{1}{n} \sum_{k \in S} y_k$	$\widehat{\overline{Y}}_{WR} = \dfrac{1}{m} \sum_{i=1}^{m} \tilde{y}_i$
Variance of the mean estimator	$\mathrm{var}\left(\widehat{\overline{Y}}\right) = \dfrac{(N-n)}{nN} S_y^2$	$\mathrm{var}\left(\widehat{\overline{Y}}_{WR}\right) = \dfrac{\sigma_y^2}{m}$
Expected sample variance	$\mathrm{E}\left(s_y^2\right) = S_y^2$	$\mathrm{E}\left(\tilde{s}_y^2\right) = \sigma_y^2$
Variance estimator of the mean estimator	$\widehat{\mathrm{var}}\left(\widehat{\overline{Y}}\right) = \dfrac{(N-n)}{nN} s_y^2$	$\widehat{\mathrm{var}}\left(\widehat{\overline{Y}}_{WR}\right) = \dfrac{\tilde{s}_y^2}{m}$

EXERCISES

Exercise 2.1 *Cultivated surface area*

We want to estimate the surface area cultivated on the farms of a rural township. Of the $N = 2010$ farms that comprise the township, we select 100 using simple random sampling. We measure y_k, the surface area cultivated on the farm k in hectares, and we find

$$\sum_{k \in S} y_k = 2907 \text{ ha and } \sum_{k \in S} y_k^2 = 154593 \text{ ha}^2.$$

1. Give the value of the standard unbiased estimator of the mean

$$\overline{Y} = \frac{1}{N} \sum_{k \in U} y_k.$$

2. Give a 95 % confidence interval for $\overline{Y}$.

Solution

In a simple design, the unbiased estimator of $\overline{Y}$ is

$$\widehat{\overline{Y}} = \frac{1}{n} \sum_{k \in S} y_k = \frac{2907}{100} = 29.07 \text{ ha}.$$

The estimator of the dispersion S_y^2 is

$$s_y^2 = \frac{n}{n-1} \left(\frac{1}{n} \sum_{k \in S} y_k^2 - \widehat{\overline{Y}}^2 \right) = \frac{100}{99} \left(\frac{154593}{100} - 29.07^2 \right) = 707.945.$$

The sample size n being 'sufficiently large', the 95% confidence interval is estimated in hectares as follows:

$$\left[\widehat{\overline{Y}} \pm 1.96 \sqrt{\frac{N-n}{N} \frac{s_y^2}{n}} \right] = \left[29.07 \pm 1.96 \sqrt{\frac{2010 - 100}{2010} \times \frac{707.45}{100}} \right]$$

$$= [23.99; 34.15].$$

Exercise 2.2 *Occupational sickness*

We are interested in estimating the proportion of men P affected by an occupational sickness in a business of 1500 workers. In addition, we know that three out of 10 workers are usually affected by this sickness in businesses of the same type. We propose to select a sample by means of a simple random sample.

1. What sample size must be selected so that the total length of a confidence interval with a 0.95 confidence level is less than 0.02 for simple designs with replacement and without replacement ?
2. What should we do if we do not know the proportion of men usually affected by the sickness (for the case of a design without replacement) ?

To avoid confusions in notation, we will use the subscript WR for estimators with replacement, and the subscript WOR for estimators without replacement.

Solution

1. a) Design with replacement.
    If the design is of size m, the length of the (estimated) confidence interval at a level $(1 - \alpha)$ for a mean is given by

    $$\text{CI}(1 - \alpha) = \left[\widehat{\overline{Y}} - z_{1-\alpha/2} \sqrt{\frac{\tilde{s}_y^2}{m}}, \widehat{\overline{Y}} + z_{1-\alpha/2} \sqrt{\frac{\tilde{s}_y^2}{m}} \right],$$

    where $z_{1-\alpha/2}$ is the quantile of order $1 - \alpha/2$ of a random normal standardised variate. If we denote $\widehat{P}_{WR}$ as the estimator of the proportion for the design with replacement, we can write

    $$\text{CI}(1 - \alpha) = \left[\widehat{P}_{WR} - z_{1-\alpha/2} \sqrt{\frac{\widehat{P}_{WR}(1 - \widehat{P}_{WR})}{m - 1}}, \right.$$

    $$\left. \widehat{P}_{WR} + z_{1-\alpha/2} \sqrt{\frac{\widehat{P}_{WR}(1 - \widehat{P}_{WR})}{m - 1}} \right].$$

Indeed, in this case,

$$\widehat{\text{var}}(\widehat{P}_{WR}) = \frac{\widehat{P}_{WR}(1 - \widehat{P}_{WR})}{(m - 1)}.$$

So that the total length of the confidence interval does not exceed 0.02, it is necessary and sufficient that

$$2z_{1-\alpha/2}\sqrt{\frac{\widehat{P}_{WR}(1 - \widehat{P}_{WR})}{m - 1}} \leq 0.02.$$

By dividing by two and squaring, we get

$$z_{1-\alpha/2}^2 \frac{\widehat{P}_{WR}(1 - \widehat{P}_{WR})}{m - 1} \leq 0.0001,$$

which gives

$$m - 1 \geq z_{1-\alpha/2}^2 \frac{\widehat{P}_{WR}(1 - \widehat{P}_{WR})}{0.0001}.$$

For a 95% confidence interval, and with an estimator of P of 0.3 coming from a source external to the survey, we have $z_{1-\alpha/2} = 1.96$, and

$$m = 1 + 1.96^2 \times \frac{0.3 \times 0.7}{0.0001} = 8068.36.$$

The sample size (m=8069) is therefore larger than the population size, which is possible (but not prudent) since the sampling is with replacement.

b) Design without replacement.

If the design is of size n, the length of the (estimated) confidence interval at a level $1 - \alpha$ for a mean is given by

$$\text{CI}(1 - \alpha) = \left[\widehat{\overline{Y}} - z_{1-\alpha/2}\sqrt{\frac{N - n}{N}\frac{s_y^2}{n}}, \widehat{\overline{Y}} + z_{1-\alpha/2}\sqrt{\frac{N - n}{N}\frac{s_y^2}{n}} \right].$$

For a proportion P and denoting $\widehat{P}_{WOR}$ as the estimator of the proportion for the design without replacement, we therefore have

$$\text{CI}(1 - \alpha) = \left[\widehat{P}_{WOR} - z_{1-\alpha/2}\sqrt{\frac{N - n}{N}\frac{\widehat{P}_{WOR}(1 - \widehat{P}_{WOR})}{n - 1}}, \right.$$

$$\left. \widehat{P}_{WOR} + z_{1-\alpha/2}\sqrt{\frac{N - n}{N}\frac{\widehat{P}_{WOR}(1 - \widehat{P}_{WOR})}{n - 1}} \right].$$

So the total length of the confidence interval does not surpass 0.02, it is necessary and sufficient that

$$2z_{1-\alpha/2}\sqrt{\frac{N-n}{N}\frac{\widehat{P}_{WOR}(1-\widehat{P}_{WOR})}{n-1}} \leq 0.02.$$

By dividing by two and by squaring, we get

$$z_{1-\alpha/2}^2\frac{N-n}{N}\frac{\widehat{P}_{WOR}(1-\widehat{P}_{WOR})}{n-1} \leq 0.0001,$$

which gives

$$(n-1)\times 0.0001 - z_{1-\alpha/2}^2\frac{N-n}{N}\widehat{P}_{WOR}(1-\widehat{P}_{WOR}) \geq 0,$$

or again

$$n\left\{0.0001 + z_{1-\alpha/2}^2\frac{1}{N}\widehat{P}_{WOR}(1-\widehat{P}_{WOR})\right\}$$

$$\geq 0.0001 + z_{1-\alpha/2}^2\widehat{P}_{WOR}(1-\widehat{P}_{WOR}),$$

or

$$n \geq \frac{0.0001 + z_{1-\alpha/2}^2\widehat{P}_{WOR}(1-\widehat{P}_{WOR})}{\left\{0.0001 + z_{1-\alpha/2}^2\frac{1}{N}\widehat{P}_{WOR}(1-\widehat{P}_{WOR})\right\}}.$$

For a 95% confidence interval, and with an *a priori* estimator of P of 0.3 coming from a source external to the survey, we have

$$n \geq \frac{0.0001 + 1.96^2 \times 0.30 \times 0.70}{\left\{0.0001 + 1.96^2 \times \frac{1}{1500} \times 0.30 \times 0.70\right\}} = 1264.98.$$

Here, a sample size of 1265 is sufficient. The obtained approximation justifies the hypothesis of a normal distribution for $\widehat{P}_{WOR}$. The impact of the finite population correction $(1-n/N)$ can therefore be decisive when the population size is small and the desired accuracy is relatively high.

2. If the proportion of affected workers is not estimated *a priori*, we are placed in the most unfavourable situation, that is, one where the variance is greatest: this leads to a likely excessive size n, but ensures that the length of the confidence interval is not longer than the fixed threshold of 0.02. For the design without replacement, this returns to taking a proportion of 50%. In this case, by adapting the calculations from 1-(b), we find $n \geq 1298$. We thus note that a significant variation in the proportion (from 30% to 50%) involves only a minimal variation in the sample size (from 1265 to 1298).

Exercise 2.3 *Probability of inclusion and design with replacement*

In a simple random design with replacement of fixed size m in a population of size N,

1. Calculate the probability that an individual k is selected at least once in a sample.
2. Show that
$$\Pr(k \in S) = \frac{m}{N} + O\left(\frac{m^2}{N^2}\right),$$

   when m/N is small. Recall that a function $f(n)$ of n is of order of magnitude $g(n)$ (noted $f(n) = O(g(n))$) if and only if $f(n)/g(n)$ is limited, that is to say there exists a quantity M such that, for any $n \in \mathbb{N}$, $|f(n)|/g(n) \leq M$.
3. What are the conclusions ?

Solution

1. We obtain this probability from the complementary event:
$$\Pr\left(k \in S\right) = 1 - \Pr\left(k \notin S\right) = 1 - \left(1 - \frac{1}{N}\right)^m.$$

2. Then, we derive

$$\Pr\left(k \in S\right) = 1 - \left(1 - \frac{1}{N}\right)^m = 1 - \sum_{j=0}^{m} \binom{m}{j}\left(-\frac{1}{N}\right)^{m-j}$$

$$= 1 - \left\{\sum_{j=0}^{m-2} \binom{m}{j}\left(-\frac{1}{N}\right)^{m-j} - \frac{m}{N} + 1\right\} = \frac{m}{N} - \sum_{j=0}^{m-2} \binom{m}{j}\left(-\frac{1}{N}\right)^{m-j}$$

$$= \frac{m}{N} + O\left(\frac{m^2}{N^2}\right).$$

3. We conclude that if the sampling rate m/N is small, $(m/N)^2$ is negligible in relation to m/N. We then again find the probability of inclusion of a sample without replacement, because the two modes of sampling become indistinguishable.

Exercise 2.4 *Sample size*

What sample size is needed if we choose a simple random sample to find, within two percentage points (at least) and with 95 chances out of 100, the proportion of Parisians that wear glasses ?

Solution

There are two reasonable positions from which to deal with these issues:

- The size of Paris is very large: the sampling rate is therefore negligible.
- Obviously not having any *a priori* information on the population sought after, we are placed in a situation which leads to a maximum sample size (strong 'precautionary' stance), having $P = 50\ \%$. If the reality is different (which is almost certain), we have *in fine* a lesser uncertainty than was fixed at the start (2 percentage points).

We set n in a way so that

$$1.96 \times \sqrt{\frac{P(1-P)}{n}} = 0.02, \text{ with } P = 0.5,$$

hence $n = 2\ 401$ people.

Exercise 2.5 *Number of clerics*

We want to estimate the number of clerics in the French population. For that, we choose to select n individuals using a simple random sample. If the true proportion (unknown) of clerics in the population is $0.1\ \%$, how many people must be selected to obtain a coefficient of variation CV of $5\ \%$?

Solution

By definition:

$$\mathrm{CV} = \frac{\sigma(Np)}{NP} = \frac{\sigma(p)}{P},$$

where P is the true proportion to estimate ($0.1\ \%$ here) and p its unbiased estimator, which is the proportion of clerics in the selected sample. A CV of $5\ \%$ corresponds to a reasonably 'average' accuracy. In fact,

$$\mathrm{var}(p) \approx \frac{P(1-P)}{n} \quad (f \text{ a priori negligible compared to 1}).$$

Therefore,

$$\mathrm{CV} = \sqrt{\frac{(1-P)}{nP}} \approx \frac{1}{\sqrt{nP}} = 0.05,$$

which gives

$$n = \frac{1}{0.001} \times \frac{1}{0.05^2} = 400\ 000.$$

This large size, impossible in practice to obtain, is a direct result of the scarcity of the sub-population studied.

Exercise 2.6 *Size for proportions*

In a population of 4 000 people, we are interested in two proportions:

P_1 = proportion of individuals owning a dishwasher,
P_2 = proportion of individuals owning a laptop computer.

According to 'reliable' information, we know *a priori* that:

$$45\ \% \le P_1 \le 65\ \%, \quad \text{and} \quad 5\ \% \le P_2 \le 10\ \%.$$

What does the sample size n have to be within the framework of a simple random sample if we want to know *at the same time* P_1 near $\pm\ 2\ \%$ and P_2 near $\pm\ 1\ \%$, with a confidence level of 95 % ?

Solution

We estimate without bias $P_i, (i = 1, 2)$ by the proportion p_i calculated in the sample:

$$\text{var}(p_i) = \left(1 - \frac{n}{N}\right) \frac{1}{n} \frac{N}{N-1}\ P_i(1 - P_i).$$

We want

$$1.96 \times \sqrt{\text{var}(p_1)} \le 0.02, \quad \text{and} \quad 1.96 \times \sqrt{\text{var}(p_2)} \le 0.01.$$

In fact ,

$$\max_{45\,\% \le P_1 \le 65\,\%} P_1(1 - P_1) = 0.5(1 - 0.5) = 0.25,$$

and

$$\max_{5\,\% \le P_2 \le 10\,\%} P_2(1 - P_2) = 0.1(1 - 0.1) = 0.09.$$

The maximum value of $P_i(1 - P_i)$ is 0.25 (see Figure 2.1) and leads to a maximum n (as a security to reach at least the desired accuracy). It is *jointly* necessary that

Fig. 2.1. Variance according to the proportion: Exercise 2.6

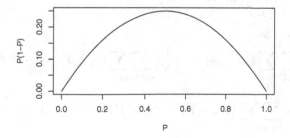

$$\begin{cases} \left(1 - \dfrac{n}{N}\right) \dfrac{1}{n} \dfrac{N}{N-1} \times 0.25 \leq \left(\dfrac{0.02}{1.96}\right)^2 \\[3mm] \left(1 - \dfrac{n}{N}\right) \dfrac{1}{n} \dfrac{N}{N-1} \times 0.09 \leq \left(\dfrac{0.01}{1.96}\right)^2, \end{cases}$$

which implies that

$$\begin{cases} n \geq 1\ 500.62 \\ n \geq 1\ 854.74. \end{cases}$$

The condition on the accuracy of p_2 being the most demanding, we conclude in choosing: $n = 1\ 855$.

Exercise 2.7 *Estimation of the population variance*

Show that

$$\sigma_y^2 = \frac{1}{N} \sum_{k \in U} (y_k - \overline{Y})^2 = \frac{1}{2N^2} \sum_{k \in U} \sum_{\substack{\ell \in U \\ \ell \neq k}} (y_k - y_\ell)^2. \tag{2.1}$$

Use this equality to (easily) find an unbiased estimator of the population variance S_y^2 in the case of simple random sampling where $S_y^2 = N\sigma_y^2/(N-1)$.

Solution

A first manner of showing this equality is the following:

$$\frac{1}{2N^2} \sum_{k \in U} \sum_{\substack{\ell \in U \\ \ell \neq k}} (y_k - y_\ell)^2 = \frac{1}{2N^2} \sum_{k \in U} \sum_{\ell \in U} (y_k - y_\ell)^2$$

$$= \frac{1}{2N^2} \left(\sum_{k \in U} \sum_{\ell \in U} y_k^2 + \sum_{k \in U} \sum_{\ell \in U} y_\ell^2 - 2 \sum_{k \in U} \sum_{\ell \in U} y_k y_\ell \right)$$

$$= \frac{1}{N} \sum_{k \in U} y_k^2 - \frac{1}{N^2} \sum_{k \in U} \sum_{\ell \in U} y_k y_\ell = \frac{1}{N} \sum_{k \in U} y_k^2 - \overline{Y}^2$$

$$= \frac{1}{N} \sum_{k \in U} (y_k - \overline{Y})^2 = \sigma_y^2.$$

A second manner is:

$$\frac{1}{2N^2} \sum_{k \in U} \sum_{\substack{\ell \in U \\ \ell \neq k}} (y_k - y_\ell)^2 = \frac{1}{2N^2} \sum_{k \in U} \sum_{\ell \in U} (y_k - \overline{Y} - y_\ell + \overline{Y})^2$$

$$= \frac{1}{2N^2} \sum_{k \in U} \sum_{\ell \in U} \left\{ (y_k - \overline{Y})^2 + (y_\ell - \overline{Y})^2 - 2(y_k - \overline{Y})(y_\ell - \overline{Y}) \right\}$$

$$= \frac{1}{2N} \sum_{k \in U} (y_k - \overline{Y})^2 + \frac{1}{2N} \sum_{\ell \in U} (y_\ell - \overline{Y})^2 + 0 = \sigma_y^2.$$

The unbiased estimator of σ_y^2 is

$$\widehat{\sigma}_y^2 = \frac{1}{2N^2} \sum_{k \in S} \sum_{\substack{\ell \in S \\ \ell \neq k}} \frac{(y_k - y_\ell)^2}{\pi_{k\ell}},$$

where $\pi_{k\ell}$ is the second-order inclusion probability. With a simple design without replacement of fixed sample size,

$$\pi_{k\ell} = \frac{n(n-1)}{N(N-1)},$$

thus

$$\widehat{\sigma}_y^2 = \frac{N(N-1)}{n(n-1)} \frac{1}{2N^2} \sum_{k \in S} \sum_{\substack{\ell \in S \\ \ell \neq k}} (y_k - y_\ell)^2.$$

By adapting (2.1) with the sample S (in place of U), we get:

$$\frac{1}{2n^2} \sum_{k \in S} \sum_{\substack{\ell \in S \\ \ell \neq k}} (y_k - y_\ell)^2 = \frac{1}{n} \sum_{k \in S} (y_k - \widehat{\overline{Y}})^2,$$

where

$$\widehat{\overline{Y}} = \frac{1}{n} \sum_{k \in S} y_k.$$

Therefore

$$\widehat{\sigma}_y^2 = \frac{(N-1)}{N} \frac{1}{n-1} \sum_{k \in S} \left(y_k - \widehat{\overline{Y}} \right)^2 = \frac{N-1}{N} s_y^2.$$

We get

$$\widehat{\sigma}_y^2 = \frac{N-1}{N} s_y^2, \quad \text{and} \quad \widehat{S}_y^2 = \frac{N}{N-1} \widehat{\sigma}_y^2 = s_y^2.$$

This result is well-known and takes longer to show if we do not use the equality (2.1).

Exercise 2.8 *Repeated survey*

We consider a population of 10 service-stations and are interested in the price of a litre of high-grade petrol at each station. The prices during two consecutive months, May and June, appears in Table 2.2.

1. We want to estimate the evolution of the average price per litre between May and June. We choose as a parameter the difference in average prices.
   Method 1: we sample n stations ($n < 10$) in May and n stations in June, the two samples being completely independent ;
   Method 2: we sample n stations in May and we again question these stations in June (*panel* technique).
   Compare the efficiency of the two concurrent methods.

Table 2.2. Price per litre of high-grade petrol: Exercise 2.8

Station	1	2	3	4	5	6	7	8	9	10
May	5.82	5.33	5.76	5.98	6.20	5.89	5.68	5.55	5.69	5.81
June	5.89	5.34	5.92	6.05	6.20	6.00	5.79	5.63	5.78	5.84

2. The same question, if we this time want to estimate an average price during the combined May-June period.
3. If we are interested in the average price in Question 2, would it not be better to select instead of 10 records twice with Method 1 (10 per month), directly 20 records without worrying about the months (Method 3) ? No calculation is necessary.

N.B.: Question 3 is related to *stratification*.

Solution

1. We denote $\bar{p}_m$ as the simple average of the recorded prices among the n stations for month m (m = May or June).
   We have:

$$\text{var}(\bar{p}_m) = \frac{1-f}{n} S_m^2,$$

where S_m^2 is the variance of the 10 prices relative to month m.

- Method 1. We estimate without bias the evolution of prices by $\bar{p}_{\text{June}} - \bar{p}_{\text{May}}$ (the two estimators are calculated on two different *a priori* samples) and

$$\text{var}_1(\bar{p}_{\text{June}} - \bar{p}_{\text{May}}) = \frac{1-f}{n}\left(S_{\text{May}}^2 + S_{\text{June}}^2\right).$$

Indeed, the covariance is null because the two samples (and therefore the two estimators $\bar{p}_{\text{May}}$ and $\bar{p}_{\text{June}}$) are independent.

- Method 2. We have only one sample (the panel). Still, we estimate the evolution of prices without bias by $\bar{p}_{\text{June}} - \bar{p}_{\text{May}}$, and

$$\text{var}_2(\bar{p}_{\text{June}} - \bar{p}_{\text{May}}) = \frac{1-f}{n}\left(S_{\text{May}}^2 + S_{\text{June}}^2 - 2S_{\text{May, June}}\right).$$

This time, there is a covariance term, with:

$$\text{cov}\left(\bar{p}_{\text{May}}, \bar{p}_{\text{June}}\right) = \frac{1-f}{n} S_{\text{May, June}},$$

where $S_{\text{May, June}}$ represents the true empirical covariance between the 10 records in May and the 10 records in June. We therefore have:

$$\frac{\text{var}_1(\bar{p}_{\text{June}} - \bar{p}_{\text{May}})}{\text{var}_2(\bar{p}_{\text{June}} - \bar{p}_{\text{May}})} = \frac{S_{\text{May}}^2 + S_{\text{June}}^2}{S_{\text{May}}^2 + S_{\text{June}}^2 - 2S_{\text{May, June}}}.$$

After calculating, we find:

$$\left. \begin{aligned} S_{\text{May}}^2 &= 0.05601 \\ S_{\text{June}}^2 &= 0.0564711 \\ S_{\text{May, June}} &= 0.0550289 \end{aligned} \right\} \Rightarrow \frac{\text{var}_1(\overline{p}_{\text{June}} - \overline{p}_{\text{May}})}{\text{var}_2(\overline{p}_{\text{June}} - \overline{p}_{\text{May}})} \approx (6.81)^2.$$

The use of a panel allows for the division of the standard error by 6.81. This enormous gain is due to the very strong correlation between the prices of May and June ($\rho \approx 0.98$): a station where high-grade petrol is expensive in May remains expensive in June compared to other stations (and vice versa). We easily verify this by calculating the true average prices in May (5.77) and June (5.84): if we compare the monthly average prices, only Station 3 changes position between May and June.

2. The average price for the two-month period is estimated without bias, with the two methods, by:

$$\overline{p} = \frac{\overline{p}_{\text{May}} + \overline{p}_{\text{June}}}{2}.$$

- Method 1:

$$\text{var}_1(\overline{p}) = \frac{1}{4} \times \frac{1-f}{n}[S_{\text{May}}^2 + S_{\text{June}}^2].$$

- Method 2:

$$\text{var}_2(\overline{p}) = \frac{1}{4} \times \frac{1-f}{n}[S_{\text{May}}^2 + S_{\text{June}}^2 + 2S_{\text{May, June}}].$$

This time, the covariance is added (due to the '+' sign appearing in $\overline{p}$).

In conclusion, we have

$$\frac{\text{var}_1(\overline{p})}{\text{var}_2(\overline{p})} = \frac{S_{\text{May}}^2 + S_{\text{June}}^2}{S_{\text{May}}^2 + S_{\text{June}}^2 + 2S_{\text{May, June}}} = (0.71)^2 = 0.50.$$

The use of a panel proves to be ineffective: with equal sample sizes, we lose 29 % of accuracy.

As the variances vary in $1/n$, if we consider that the total cost of a survey is proportional to the sample size, this result amounts to saying that for a given variance, Method 1 allows a saving of 50 % of the budget in comparison to Method 2: this is obviously strongly significant.

3. Method 1 remains the best. Indeed, Method 3 amounts to selecting a simple random sample of size $2n$ in a population of size $2N$, whereas Method 1 amounts to having two strata each of size N and selecting n individuals in each stratum: the latter instead gives a proportional allocation.

In fact, we know that for a fixed total sample ($2n$ here), to estimate a combined average, stratification with proportional allocation is always preferable to simple random sampling.

Exercise 2.9 *Candidates in an election*

In an election, there are two candidates. The day before the election, an opinion poll (simple random sample) is taken among n voters, with n equal to at least 100 voters (the voter population is very large compared to the sample size). The question is to find out the necessary difference in percentage points between the two candidates so that the poll produces the name of the winner (known by census the next day) 95 times out of 100. Perform the numeric application for some values of n.

Hints: Consider that the loser of the election is A and that the percentage of votes he receives on the day of the election is P_A ; the day of the sample, we denote $\widehat{P}_A$ as the percentage of votes obtained by this candidate A.

We will convince ourselves of the fact that the problem above posed in 'common terms' can be clearly expressed using a statistical point of view: find the critical region so that the probability of declaring A as the winner on the day of the sample (while P_A is in reality less than 50 %) is less than 5 %.

Solution

In adopting the terminology of test theory, we want a 'critical region' of the form $]c, +\infty[$, the problem being to find c, with:

$$\Pr[\widehat{P}_A > c | P_A < 50\,\%] \leq 5\,\%$$

(the event $P_A < 50\,\%$ is by definition certain; it is presented for reference). Indeed, the rule that will decide on the date of the sample who would win the following day can only be of type '$\widehat{P}$ greater than a certain level'. We make the hypothesis that $\widehat{P}_A \sim \mathcal{N}(P_A,\, \sigma_A^2)$, with:

$$\sigma_A^2 = \frac{P_A(1 - P_A)}{n}.$$

This approximation is justified because n is 'sufficiently large' ($n \geq 100$). We try to find c such that:

$$\Pr\left[\frac{\widehat{P}_A - P_A}{\sigma_A} > \frac{c - P_A}{\sigma_A} \middle| P_A < 50\,\%\right] \leq 5\,\%.$$

However, P_A remains unknown. In reality, it is the maximum of these probabilities that must be considered among all P_A possible, meaning all $P_A < 0.5$. Therefore, we try to find c such that:

$$\max_{\{P_A\}} \Pr\left[\mathcal{N}(0.1) > \frac{c - P_A}{\sigma_A} \middle| P_A < 0.5\right] \leq 0.05.$$

Now, the quantity

$$\frac{c - P_A}{\sqrt{\frac{P_A(1 - P_A)}{n}}}$$

is clearly a decreasing function of P_A (for $P_A < 0.5$). We see that the maximum of the probability is attained for the minimum $(c - P_A)/\sigma_A$, or in other words the maximum P_A (subject to $P_A < 0.5$). Therefore, we have $P_A = 50\%$. We try to find c satisfying:

$$\Pr\left[\mathcal{N}(0,1) > \frac{c - 0.5}{\sqrt{\frac{0.25}{n}}}\right] \leq 0.05.$$

Consulting a quantile table of the normal distribution shows that it is necessary for:

$$\frac{c - 0.5}{\sqrt{\frac{0.25}{n}}} = 1.65.$$

Conclusion: The critical region is

$$\left\{\widehat{P}_A > \frac{1}{2} + 1.65\sqrt{\frac{0.25}{n}}\right\}, \quad \text{that is} \quad \left\{\widehat{P}_A > \frac{1}{2} + \frac{1.65}{2\sqrt{n}}\right\}.$$

The difference in percentage points therefore must be at least the following:

$$\widehat{P}_A - \widehat{P}_B = 2\widehat{P}_A - 1 \geq \frac{1.65}{\sqrt{n}}.$$

If the difference in percentage points is *at least* equal to $1.65/\sqrt{n}$, then we have less than a 5 % chance of declaring A the winner on the day of the opinion poll while in reality he will lose on the day of the elections, that is, we have at least a 95 % chance of making the right prediction. Table 2.3 contains several numeric applications. The case $n = 900$ corresponds to the opinion poll sample size traditionally used for elections.

Table 2.3. Numeric applications: Exercise 2.9

n	100	400	900	2000	5000	10000
$1.65/\sqrt{n}$	16.5	8.3	5.5	3.7	2.3	1.7

Exercise 2.10 *Select-reject method*

Select a sample of size 4 in a population of size 10 using a simple random design without replacement with the select-reject method. This method is due to Fan et al. (1962) and is described in detail in Tillé (2001, p. 74). The procedure consists of sequentially reading the frame. At each stage, we decide whether or not to select a unit of observation with the following probability:

$$\frac{\text{number of units remaining to select in the sample}}{\text{number of units remaining to examine in the population}}.$$

Use the following observations of a uniform random variable over $[0, 1]$:

| 0.375489 0.624004 0.517951 0.0454450 0.632912 |
| 0.246090 0.927398 0.32595 0.645951 0.178048 |

Solution

Noting k as the observation number and j as the number of units already selected at the start of stage k, the algorithm is described in Table 2.4. The sample is composed of units $\{1, 4, 6, 8\}$.

Table 2.4. Select-reject method: Exercise 2.10

k	u_k	j	$\dfrac{n-j}{N-(k-1)}$	I_k
1	0.375489	0	$4/10 = 0.4000$	1
2	0.624004	1	$3/9 = 0.3333$	0
3	0.517951	1	$3/8 = 0.3750$	0
4	0.045450	1	$3/7 = 0.4286$	1
5	0.632912	2	$2/6 = 0.3333$	0
6	0.246090	2	$2/5 = 0.4000$	1
7	0.927398	3	$1/4 = 0.2500$	0
8	0.325950	3	$1/3 = 0.3333$	1
9	0.645951	4	$0/2 = 0.0000$	0
10	0.178048	4	$0/1 = 0.0000$	0

Exercise 2.11 *Sample update method*

In selecting a sample according to a simple design without replacement, there exist several algorithms. One method proposed by McLeod and Bellhouse (1983), works in the following manner:

- We select the first n units of the list.
- We then examine the case of record $(n + 1)$. We select unit $n + 1$ with a probability $n/(n + 1)$. If unit $n + 1$ is selected, we remove one unit from the sample that we selected at random and with equal probabilities.
- For the units k, where $n + 1 < k \leq N$, we maintain this rule. Unit k is selected with probability n/k. If unit k is selected, we remove one unit from the sample that we selected at random and with equal probabilities.

1. We denote $\pi_\ell^{(k)}$ as the probability that individual ℓ is in the sample at stage k, where $(\ell \leq k)$, meaning after we have examined the case of record k $(k \geq n)$. Show that $\pi_\ell^{(k)} = n/k$. (It can be interesting to proceed in a recursive manner.)
2. Verify that the final probability of inclusion is indeed that which we obtain for a design with equal probabilities of fixed size.
3. What is interesting about this method?

Solution

1. • If $k = n$, then $\pi_\ell^{(k)} = 1 = n/n$, for all $\ell \leq n$.
   • If $k = n + 1$, then we have directly $\pi_{n+1}^{(n+1)} = n/(n+1)$. Furthermore, for $\ell < k$,

$$\pi_\ell^{(n+1)} = \Pr[\text{unit } \ell \text{ being in the sample at stage } (n+1)]$$
$$= \Pr[\text{unit } (n+1) \text{ not being selected at stage } n]$$
$$+ \Pr[\text{unit } (n+1) \text{ being selected at stage } n]$$
$$\times \Pr[\text{unit } \ell \text{ not being removed at stage } n]$$
$$= 1 - \frac{n}{n+1} + \frac{n}{n+1} \times \frac{n-1}{n} = \frac{n}{n+1}.$$

   • If $k > n+1$, we use a recursive proof. We suppose that, for all $\ell \leq k-1$,

$$\pi_\ell^{(k-1)} = \frac{n}{k-1}, \tag{2.2}$$

and we are going to show that if (2.2) is true then, for all $\ell \leq k$,

$$\pi_\ell^{(k)} = \frac{n}{k}. \tag{2.3}$$

The initial conditions are confirmed since we have proven (2.3) for $k = n$ and $k = n + 1$. If $\ell = k$, then the algorithm directly gives

$$\pi_k^{(k)} = \frac{n}{k}.$$

   • If $\ell < k$, then we calculate in the sample, using Bayes' theorem,

$$\pi_\ell^k = \Pr[\text{unit } \ell \text{ being in the sample at stage } k]$$
$$= \Pr[\text{unit } k \text{ not being selected at stage } k]$$
$$\times \Pr[\text{unit } \ell \text{ being in the sample at stage } k-1]$$
$$+ \Pr[\text{unit } k \text{ being selected at stage } k]$$
$$\times \Pr[\text{unit } \ell \text{ being in the sample at stage } k-1]$$
$$\times \Pr[\text{unit } \ell \text{ not being removed at stage } k]$$
$$= \left(1 - \frac{n}{k}\right) \times \pi_\ell^{(k-1)} + \frac{n}{k} \times \pi_\ell^{(k-1)} \times \frac{n-1}{n}$$
$$= \pi_\ell^{(k-1)} \frac{k-1}{k} = \frac{n}{k}.$$

2. At the end of the algorithm $k = N$ and therefore $\pi_\ell^{(N)} = n/N$, for all $\ell \in U$.

3. What is interesting about this algorithm is that it permits the selection of a sample of fixed size n with equal probabilities without replacement and without having to know *a priori* the size of the population N. For example, we can sample a list that is being filled 'on the fly' without needing to wait for everything to be complete before starting the selection procedure. We remark that systematic sampling can be put into place without the population being complete but, in this case, the sample is not necessarily of fixed size.

Exercise 2.12 *Domain estimation*

In a population of size N, we sample n individuals by simple random sampling. We consider a subpopulation D (meaning a 'domain') of size N_D, and we denote n_D as the (random) sample size for D. With the selected sample S being decomposable into two parts S_D and $S_{\overline{D}}$, where S_D is the intersection of S and the domain, find the conditional distribution of S_D given n_D (n_D is therefore the cardinality of S_D). What is the practical conclusion?

Solution

$$p(s_D \mid n_D) = \frac{\Pr(\text{selecting } s_D \text{ and obtaining a size } n_D)}{\Pr(\text{obtaining a size } n_D)}$$

If s_D is indeed of size n_D, the numerator is quite simply $\Pr(\text{selecting } s_D)$. If s_D is not of size n_D, the numerator is null. We are now placed in the first case. In fact:

$$p(s_D) = \sum_{s \supset s_D} p(s) = \frac{\text{Number of } s \text{ containing } s_D}{\binom{N}{n}}.$$

The number of s containing s_D is $\binom{N-N_D}{n-n_D}$ because, in order to go from s_D to s, it is necessary and sufficient to choose $(n - n_D)$ individuals to select outside of the domain D, that is, in a group of size $N - N_D$. Furthermore:

$$\Pr(\text{obtaining a size } n_D) = \sum_{\text{card}(s \cap D) = n_D} p(s) = \frac{\#\{s \mid \text{card}(s \cap D) = n_D\}}{\binom{N}{n}}.$$

Counting the s such that $\text{card}(s \cap D) = n_D$ brings us back to selecting n_D individuals in D, (there are $\binom{N_D}{n_D}$ possible cases) and $(n - n_D)$ individuals outside of D (there are $\binom{N-N_D}{n-n_D}$ possible cases). Therefore

$$\Pr(\text{obtaining a size } n_D) = \frac{\binom{N_D}{n_D} \binom{N-N_D}{n-n_D}}{\binom{N}{n}},$$

which is a hypergeometric distribution.

Finally, we get:

$$p[s_D \mid n_D] = \frac{1}{\binom{N_D}{n_D}}.$$

Practical conclusion:

We indeed see that it is the distribution of a simple random sampling of size n_D in a population of size N_D. Thus, all the calculations of bias and variance, if they are conditional on n_D, follow directly from the standard results of simple random sampling, meaning that it is sufficient to continue with the classic formulas in considering that all magnitudes involved are relative to D (we replace n by n_D, N by N_D, S_y^2 by S_{yD}^2, etc.).

Exercise 2.13 *Variance of a domain estimator*

Having carried out a simple random sample in a finite population, we are interested in estimating a total Y_0 in a given domain U_0 of the population. We introduce the variable y^* which is

$$y_k^* = \begin{cases} y_k & \text{if } k \in U_0 \\ 0 & \text{otherwise.} \end{cases}$$

1. Throughout this question, the domain size N_0 is unknown and the individuals in the domain are not identifiable *a priori*. The sample size is denoted as n.

   a) Give the expressions of the unbiased estimator $\widehat{Y}_0$ of the total and its variance.

   b) Show that

   $$(N-1)S_y^{*2} = (N_0 - 1)S_{y0}^2 + N_0\overline{Y}_0^2\left(1 - \frac{N_0}{N}\right),$$

   where S_{y0}^2 is the population variance of y_k^* (or of y_k) in the domain U_0 and S_y^{*2} is the population variance of y_k^* in the entire population.

   c) Deduce that, when N_0 is very large,

   $$\text{var}\left(\widehat{Y}_0\right) \approx \frac{N^2}{n}\left(1 - \frac{n}{N}\right)\left(P_0 S_{y0}^2 + P_0 Q_0 \overline{Y}_0^2\right),$$

   where $P_0 = N_0/N$ and $Q_0 = 1 - P_0$.

2. Throughout this question, the domain size N_0 is known, as we henceforth assume that the individuals in the domain are identifiable *a priori* in the survey frame. Recall that the sampling is simple random in the population.

   a) Give the expressions of the classic unbiased estimator $\widehat{\widehat{Y}}_0$ of the total and its conditional variance given n_0. We denote n_0 as the (random) sample size of individuals in the domain U_0, and we consider that n is sufficiently large so that the probability of obtaining a null n_0 is negligible.

b) We want to compare the performances of $\widehat{\widehat{Y}}_0$ and $\widehat{Y}_0$. For that, we set $n_0 = nP_0$, and we use this value in the expression $\mathrm{var}\left(\widehat{\widehat{Y}}_0|n_0\right)$. Justify this manner of proceeding. Deduce that

$$\mathrm{var}\left(\widehat{\widehat{Y}}_0|n_0\right) \approx \mathrm{var}\left(\widehat{\widehat{Y}}_0\right) \approx \frac{N^2}{n}\left(1 - \frac{n}{N}\right)P_0 S_{y0}^2.$$

c) Show that these approximations lead to

$$\frac{\mathrm{var}\left(\widehat{\widehat{Y}}_0\right)}{\mathrm{var}\left(\widehat{Y}_0\right)} \approx \frac{C_0^2}{C_0^2 + Q_0},$$

where $C_0 = S_{y0}/\overline{Y}_0$ is the coefficient of variation of y_k in the domain U_0. What do you conclude?

3. In a population of given individuals, we wish to estimate the total number of men in the socio-professional category 'employees'. We never have at our disposal any information relating to gender except, obviously, in the sample.

   a) Suppose that we do not know the total number of employees in the population. In what way is this question related to the previous problem (in particular, specify the variable y^* that was used) ?

   b) What is the relative gain in accuracy obtained when we suddenly have at our disposal the information 'total number of employees in the population' ?

   c) How can we estimate this gain? What problem(s) do we face?

Solution

1. a) The estimator is given by

$$\widehat{Y}_0 = N\widehat{\overline{Y}}^* \quad \text{where} \quad \widehat{\overline{Y}}^* = \frac{1}{n}\sum_{k \in S} y_k^*.$$

We get

$$\mathrm{E}(\widehat{Y}_0) = N\overline{Y}^* = \sum_{k \in U} y_k^* = \sum_{k \in U_0} y_k = Y_0$$

(the estimator $\widehat{Y}_0$ is therefore unbiased), and

$$\mathrm{var}[\widehat{Y}_0] = \mathrm{var}[N\widehat{\overline{Y}}^*] = N^2 \frac{N - n}{Nn} S_y^{*2},$$

where S_y^{*2} is the population variance (unknown) of y_k^*.

b) We have

$$(N-1)S_y^{*2}$$

$$= \sum_{k \in U} (y_k^* - \overline{Y}^*)^2$$

$$= \sum_{k \in U} (y_k^* - \overline{Y}_0 + \overline{Y}_0 - \overline{Y}^*)^2$$

$$= \sum_{k \in U} (y_k^* - \overline{Y}_0)^2 + N(\overline{Y}_0 - \overline{Y}^*)^2 + 2(\overline{Y}_0 - \overline{Y}^*) \sum_{k \in U} (y_k^* - \overline{Y}_0)$$

$$= \sum_{k \in U_0} (y_k^* - \overline{Y}_0)^2 + \sum_{k \in U \setminus U_0} (y_k^* - \overline{Y}_0)^2 + N(\overline{Y}_0 - \overline{Y}^*)^2$$

$$+ 2(\overline{Y}_0 - \overline{Y}^*)N(\overline{Y}^* - \overline{Y}_0)$$

$$= (N_0 - 1)S_{y0}^2 + (N - N_0)\overline{Y}_0^2 - N(\overline{Y}_0 - \overline{Y}^*)^2.$$

In fact

$$N(\overline{Y}_0 - \overline{Y}^*)^2 = N\left(\overline{Y}_0 - \overline{Y}_0\frac{N_0}{N}\right)^2 = N\overline{Y}_0^2\left(1 - \frac{N_0}{N}\right)^2,$$

which gives

$$(N-1)S_y^{*2} = (N_0 - 1)S_{y0}^2 + N_0\left(1 - \frac{N_0}{N}\right)\overline{Y}_0^2.$$

c) If N_0 is very large, then $N_0 \approx (N_0 - 1)$ and $N \approx (N - 1)$:

$$\text{var}(\widehat{Y}_0) \approx N\frac{N-n}{Nn}\left[(N_0 - 1)S_{y0}^2 + N_0\left(1 - \frac{N_0}{N}\right)\overline{Y}_0^2\right]$$

$$\approx N^2\frac{N-n}{Nn}\left(P_0 S_{y0}^2 + P_0 Q_0 \overline{Y}_0^2\right).$$

2. a) We have

$$\widehat{\overline{Y}}_0 = N_0 \widehat{\overline{Y}}_0,$$

where

$$\widehat{\overline{Y}}_0 = \frac{1}{n_0} \sum_{k \in U_0 \cap S} y_k,$$

$$n_0 = \#(U_0 \cap S),$$

and

$$\text{var}\left(\widehat{\overline{Y}}_0 | n_0\right) = N_0^2 \frac{N_0 - n_0}{N_0 n_0} S_{y0}^2.$$

Indeed, in this conditional approach, everything happens as if we had completed a simple random survey of n_0 individuals in U_0 (see Exercise 2.12).

b) Since n_0 follows a hypergeometric distribution, we have $E(n_0) = nP_0$. The value n_0 does not appear in var$(\widehat{Y}_0)$: to compare similar expressions, it is thus legitimate to substitute $E(n_0)$ with n_0, which is random. We thus assimilate var $\left(\widehat{\widehat{Y}}_0 | n_0\right)$ to var $\left(\widehat{\widehat{Y}}_0\right)$. Since $N_0 = NP_0$, we get

$$\text{var}\left(\widehat{\widehat{Y}}_0\right) \approx P_0^2 N^2 \frac{NP_0 - nP_0}{NP_0 nP_0} S_{y0}^2 = P_0 N^2 \frac{N - n}{Nn} S_{y0}^2.$$

Note that we would reach the same expression by starting from the unconditional variance var $\left(\widehat{\widehat{Y}}_0\right)$ and by replacing, in the first approximation, the term $E(1/n_0)$ with $1/E(n_0)$.

c) The relationship between the two variances is:

$$\frac{\text{var}\left(\widehat{\widehat{Y}}_0\right)}{\text{var}\left(\widehat{Y}_0\right)} \approx \frac{P_0 S_{y0}^2}{P_0 S_{y0}^2 + P_0 Q_0 \overline{Y}_0^2} = \frac{C_0^2}{C_0^2 + Q_0} < 1.$$

We conclude that the knowledge of N_0 permits having a more efficient estimator. The 'gain' is all the more important when C_0 is small, meaning that the domain groups similar individuals (according to y_k), and/or that Q_0 is large, or in other words that the domain is of small size.

3. a) We initially define for the entire population the variable

$$y_k = \begin{cases} 1 & \text{if } k \text{ is male} \\ 0 & \text{otherwise.} \end{cases}$$

Being interested in the domain U_0 of the employees, we will define y_k^* as previously, which comes back to writing:

$$y_k^* = \begin{cases} y_k & \text{if } k \text{ is an employee} \\ 0 & \text{otherwise,} \end{cases}$$

that is to say:

$$y_k^* = \begin{cases} 1 & \text{if } k \text{ is male and an employee} \\ 0 & \text{if } k \text{ is not an employee or not male.} \end{cases}$$

Then, $E(\widehat{Y}_0) = N_{h0}$ is the number of male employees in the population.

b) N_0 is the total number of employees (male + female) henceforth known. The domain U_0 is then defined by the group of employees (male and female). The variable y_k being defined as above, the relative gain from one method to another is

$$\frac{\text{var}\left(\widehat{\overline{Y}}_0\right)}{\text{var}\left(\widehat{Y}_0\right)} = \frac{C_0^2}{C_0^2 + Q_0},$$

with

$$C_0 = \frac{S_{y0}}{\overline{Y}_0},$$

and

$$\overline{Y}_0 = \frac{N_{h0}}{N_0} = P_0^h,$$

which is the proportion of men among the employees. As

$$S_{y0}^2 \approx P_0^h(1 - P_0^h),$$

we have

$$C_0^2 = \frac{1 - P_0^h}{P_0^h},$$

and $Q_0 = 1 - P_0$, the proportion of non-employees in the total population (and not only in the domain).

c) We can estimate without bias (or nearly, because n_0 can be null with a negligible probability) P_0^h by n_{h0}/n_0 and P_0 by n_0/n. However, the gain is a non-linear function of P_0^h and P_0. The estimator of the gain is therefore biased and the estimation of the associated variance has to rely on a linearisation technique if n is large.

Exercise 2.14 *Complementary sampling*

Let U be a population of size N. We define the following sampling distribution: we first select a sample S_1 according to a simple design without replacement of fixed size n_1.

1. We then select a sample S_2 in U outside of S_1 according to a simple random design without replacement of fixed size n_2. The final sample S consists of S_1 and S_2. Give the sampling distribution of S. What is interesting about this result?

2. We then select a sample S_3 from S_1, according to a simple random design without replacement of fixed size n_3 where $(n_3 < n_1)$. Give the sampling distribution of S_3 (in relation to U). What is interesting about this result?

3. Using again the framework from Question 1, we define the estimator of $\overline{Y}$ by:

$$\widehat{\overline{Y}}_\theta = \theta\widehat{\overline{Y}}_1 + (1 - \theta)\widehat{\overline{Y}}_2,$$

with $0 < \theta < 1$,

$$\widehat{\overline{Y}}_1 = \frac{1}{n_1} \sum_{k \in S_1} y_k \text{ and } \widehat{\overline{Y}}_2 = \frac{1}{n_2} \sum_{k \in S_2} y_k.$$

Show that, for any θ, $\widehat{\overline{Y}}_\theta$ estimates $\overline{Y}$ without bias.

4. Give the optimal estimator (as θ) in the class of estimators of the form $\widehat{\overline{Y}}_\theta$.

Solution

1. We have of course $S_1 \subset S, S_2 \subset S$, and $S_1 \cap S_2 = \emptyset$. Therefore, for s of size $n = n_1 + n_2$, we have ($\#S$ indicates the size of the sample S)

$$\Pr(S = s) = \sum_{s_1 \subset s | \#s_1 = n_1} \Pr(S_1 = s_1)\Pr(S_2 = s\backslash s_1 | S_1 = s_1)$$

$$= \binom{n_1 + n_2}{n_1}\binom{N}{n_1}^{-1}\binom{N - n_1}{n_2}^{-1}$$

$$= \frac{(n_1 + n_2)!}{n_1! n_2!} \times \frac{n_1!(N - n_1)!}{N!} \times \frac{n_2!(N - n_1 - n_2)!}{(N - n_1)!}$$

$$= \frac{(n_1 + n_2)!(N - n_1 - n_2)!}{N!} = \binom{N}{n}^{-1}.$$

The sampling of $S_1 \cup S_2$ is therefore carried out according to a simple random design of fixed size $n = n_1 + n_2$. If we want to increase the sample size already selected using a simple design (for example, to increase the accuracy of an estimator, or because we notice a lower response rate than expected), it is sufficient to reselect a sample according to a simple design among the units that were not selected at the time of the first sampling.

2. The probability of selecting s_3 is calculated as follows using the conditional probabilities.

$$\Pr(S_3 = s_3) = \sum_{s_1 | s_3 \subset s_1} \Pr(S_1 = s_1)\Pr(S_3 = s_3 | S_1 = s_1)$$

$$= \binom{N - n_3}{n_1 - n_3}\binom{N}{n_1}^{-1}\binom{n_1}{n_3}^{-1}$$

$$= \frac{(N - n_3)!}{(n_1 - n_3)!(N - n_1)!} \times \frac{n_1!(N - n_1)!}{N!} \times \frac{n_3!(n_1 - n_3)!}{n_1!}$$

$$= \binom{N}{n_3}^{-1}.$$

Here once again, we find the distribution characterising the simple random sampling of size n_3 in a population of size N. In practice, to 'calibrate' a sample, this property can be used to compete with that shown in 1. We use *a priori* the sample s_3, but if its size proves to be insufficient, we call upon s_1 in its group. If we iterate the process, we can set up a group of nested samples, all coming from simple random sampling and using first of all the smallest and then eventually the others as reserve samples, and in relation to the needs as dictated by the field.

3. *Method 1:*

$$E(\widehat{\overline{Y}}_\theta) = \theta E(\widehat{\overline{Y}}_1) + (1 - \theta)E(\widehat{\overline{Y}}_2).$$

The conditional expectation $E(\widehat{\overline{Y}}_2|S_1)$ is the expectation of a mean in a simple random sample without replacement of fixed size from the population $U\backslash S_1$, which is therefore the true mean of this population, being:

$$E(\widehat{\overline{Y}}_2|S_1) = \frac{1}{N - n_1} \sum_{U\backslash S_1} y_k = \frac{N\overline{Y} - n_1\widehat{\overline{Y}}_1}{N - n_1},$$

and therefore

$$E(\widehat{\overline{Y}}_2) = EE(\widehat{\overline{Y}}_2|S_1) = \frac{N\overline{Y} - n_1 E[\widehat{\overline{Y}}_1]}{N - n_1} = \frac{N\overline{Y} - n_1\overline{Y}}{N - n_1} = \overline{Y}.$$

Thus

$$E(\widehat{\overline{Y}}_\theta) = \theta E(\widehat{\overline{Y}}_1) + (1 - \theta)E(\widehat{\overline{Y}}_2) = \theta\overline{Y} + (1 - \theta)\overline{Y} = \overline{Y}.$$

Method 2:

We can also use the results from 1., which avoids conditional expectations. Indeed, we can express the simple mean on S of the form

$$\widehat{\overline{Y}} = \frac{n_1\widehat{\overline{Y}}_1 + n_2\widehat{\overline{Y}}_2}{n},$$

thus

$$\widehat{\overline{Y}}_2 = \frac{n\widehat{\overline{Y}} - n_1\widehat{\overline{Y}}_1}{n_2}.$$

We therefore get

$$\widehat{\overline{Y}}_\theta = \theta\widehat{\overline{Y}}_1 + (1 - \theta)\frac{n\widehat{\overline{Y}} - n_1\widehat{\overline{Y}}_1}{n_2} = \left[\theta - \frac{n_1}{n_2}(1 - \theta)\right]\widehat{\overline{Y}}_1 + (1 - \theta)\frac{n\widehat{\overline{Y}}}{n_2}.$$

Since $E(\widehat{\overline{Y}}_1) = E(\widehat{\overline{Y}}) = \overline{Y}$,

$$E(\widehat{\overline{Y}}_\theta) = \left[\theta - \frac{n_1}{n_2}(1 - \theta)\right]E[\widehat{\overline{Y}}_1] + (1 - \theta)\frac{nE[\widehat{\overline{Y}}]}{n_2}$$

$$= \left[\theta - \frac{n_1}{n_2}(1 - \theta)\right]\overline{Y} + (1 - \theta)\frac{n\overline{Y}}{n_2}$$

$$= \overline{Y}.$$

4. Since $\widehat{\overline{Y}}_\theta$ is unbiased, we find θ that minimises the variance of $\widehat{\overline{Y}}_\theta$.
   Method 1:

$$\operatorname{var}\left(\widehat{\overline{Y}}_\theta\right) = \theta^2 \operatorname{var}(\widehat{\overline{Y}}_1) + (1-\theta)^2 \operatorname{var}(\widehat{\overline{Y}}_2) + 2\theta(1-\theta)\operatorname{cov}(\widehat{\overline{Y}}_1, \widehat{\overline{Y}}_2),$$

$$\operatorname{var}(\widehat{\overline{Y}}_2) = \operatorname{var}E(\widehat{\overline{Y}}_2|S_1) + E\operatorname{var}(\widehat{\overline{Y}}_2|S_1),$$

now

$$\operatorname{var}(\widehat{\overline{Y}}_2|S_1) = \left(1 - \frac{n_2}{N-n_1}\right)\frac{S_y'^2}{n_2},$$

where $S_y'^2$ is the population variance of y_k in $U\backslash S_1$. Since S_1 is derived from a simple random sample without replacement, it is clear that $U\backslash S_1$ is as well, and $E(S_y'^2) = S_y^2$. Therefore,

$$\operatorname{var}(\widehat{\overline{Y}}_2) = \operatorname{var}\left(\frac{N\overline{Y} - n_1\widehat{\overline{Y}}_1}{N-n_1}\right) + E\left[\left(1 - \frac{n_2}{N-n_1}\right)\frac{S_y'^2}{n_2}\right]$$

$$= \left(\frac{n_1}{N-n_1}\right)^2 \operatorname{var}\left(\widehat{\overline{Y}}_1\right) + \left(1 - \frac{n_2}{N-n_1}\right)\frac{E\left\lfloor S_y'^2\right\rfloor}{n_2}$$

$$= \left(\frac{n_1}{N-n_1}\right)^2 \frac{N-n_1}{Nn_1}S_y^2 + \left(1 - \frac{n_2}{N-n_1}\right)\frac{S_y^2}{n_2}$$

$$= \frac{N-n_2}{Nn_2}S_y^2.$$

We notice that it is the variance of a simple random sample of size n_2 in the complete population. Therefore

$$\operatorname{var}(\widehat{\overline{Y}}_1) = \frac{N-n_1}{Nn_1}S_y^2 \quad \text{and} \quad \operatorname{var}(\widehat{\overline{Y}}_2) = \frac{N-n_2}{Nn_2}S_y^2.$$

Furthermore,

$$\operatorname{cov}(\widehat{\overline{Y}}_1, \widehat{\overline{Y}}_2) = E\left[\operatorname{cov}(\widehat{\overline{Y}}_1, \widehat{\overline{Y}}_2|S_1)\right] + \operatorname{cov}[E(\widehat{\overline{Y}}_1|S_1), E(\widehat{\overline{Y}}_2|S_1)],$$

where now $\widehat{\overline{Y}}_1$ is constant conditionally to S_1, thus

$$\operatorname{cov}(\widehat{\overline{Y}}_1, \widehat{\overline{Y}}_2|S_1) = 0, \text{ and } E(\widehat{\overline{Y}}_1|S_1) = \widehat{\overline{Y}}_1,$$

and

$$\operatorname{cov}(\widehat{\overline{Y}}_1, \widehat{\overline{Y}}_2) = 0 + \operatorname{cov}\left(\widehat{\overline{Y}}_1, \frac{N\overline{Y} - n_1\widehat{\overline{Y}}_1}{N-n_1}\right) = -\frac{n_1}{N-n_1}\frac{N-n_1}{Nn_1}S_y^2$$

$$= -\frac{1}{N}S_y^2.$$

Therefore,

$$\text{var}\left(\widehat{\overline{Y}}_\theta\right) = \frac{S_y^2}{N}\left[\theta^2\frac{N-n_1}{n_1} + (1-\theta)^2\frac{N-n_2}{n_2} - 2\theta(1-\theta)\right].$$

The optimal value of θ is obtained by differentiating $\text{var}\left(\widehat{\overline{Y}}_\theta\right)$ with respect to θ and setting the derivative equal to zero, which gives

$$2\theta^*\frac{N-n_1}{n_1} - 2(1-\theta^*)\frac{N-n_2}{n_2} - 2(1-2\theta^*) = 0,$$

and we get

$$\theta^* = \frac{n_1}{n}.$$

Method 2:
We use the expression $\widehat{\overline{Y}}_\theta$ as a function of $\widehat{\overline{Y}}_1$ and $\widehat{\overline{Y}}$, which avoids the tedious calculation of the variance of $\widehat{\overline{Y}}_2$. We very easily verify that

$$\widehat{\overline{Y}}_\theta = \delta\widehat{\overline{Y}}_1 + (1-\delta)\widehat{\overline{Y}},$$

with

$$\delta = \frac{n}{n_2}\theta - \frac{n_1}{n_2}.$$

$$\text{var}\left(\widehat{\overline{Y}}_\theta\right) = \delta^2\text{var}\left(\widehat{\overline{Y}}_1\right) + (1-\delta)^2\text{var}\left(\widehat{\overline{Y}}\right) + 2\delta(1-\delta)\text{cov}\left(\widehat{\overline{Y}}_1,\widehat{\overline{Y}}\right).$$

Now,

$$\text{cov}\left(\widehat{\overline{Y}}_1,\widehat{\overline{Y}}\right) = \text{Ecov}\left(\widehat{\overline{Y}}_1,\widehat{\overline{Y}}|S_1\right) + \text{cov}\left[\text{E}(\widehat{\overline{Y}}_1|S_1), \text{E}(\widehat{\overline{Y}}|S_1)\right]$$

$$= \text{E}(0) + \text{cov}\left[\widehat{\overline{Y}}_1, \frac{n_1}{n}\widehat{\overline{Y}}_1 + \frac{n_2}{n}\text{E}(\widehat{\overline{Y}}_2|S_1)\right]$$

$$= \text{cov}\left(\widehat{\overline{Y}}_1, \frac{n_1}{n}\widehat{\overline{Y}}_1 + \frac{n_2}{n}\frac{N\overline{Y} - n_1\widehat{\overline{Y}}_1}{N-n_1}\right)$$

$$= \frac{n_1}{n}\frac{N-n}{N-n_1}\text{var}\left(\widehat{\overline{Y}}_1\right) = \frac{n_1}{n}\frac{N-n}{N-n_1}\left(1-\frac{n_1}{N}\right)\frac{S_y^2}{n_1}.$$

Finally,

$$\text{var}\left(\widehat{\overline{Y}}_\theta\right)$$

$$= \delta^2\left(1-\frac{n_1}{N}\right)\frac{S_y^2}{n_1} + (1-\delta)^2\left(1-\frac{n}{N}\right)\frac{S_y^2}{n}$$

$$+ 2\delta(1-\delta)\frac{n_1}{n}\frac{N-n}{N-n_1}\left(1-\frac{n_1}{N}\right)\frac{S_y^2}{n_1}$$

$$= \frac{S_y^2}{N}\left[\delta^2\left(\frac{N-n_1}{n_1}\right) + (1-\delta)^2\left(\frac{N-n}{n}\right) + 2\delta(1-\delta)\left(\frac{N-n}{n}\right)\right]$$

$$= \frac{S_y^2}{N}\left[\left(\frac{N-n}{n}\right) + \frac{N(n-n_1)}{nn_1}\delta^2\right].$$

As $(n - n_1 > 0)$, var $\left[\widehat{\overline{Y}}_\theta\right]$ is manifestly minimal for $\delta^2 = 0$, being

$$\frac{n}{n_2}\theta^* - \frac{n_1}{n_2} = 0,$$

therefore

$$\theta^* = \frac{n_1}{n}.$$

We indeed find again the same θ^* as with Method 1, in a little more 'elegant' fashion. No matter the method, the optimal estimator must be the simple mean of the sample S, being $\widehat{\overline{Y}}$. Therefore, when we select samples repeatedly (by simple random sampling each time), the best estimator is still the most simple, meaning that one which we naturally get by combining all the samples *in fine*.

Exercise 2.15 *Capture-recapture*

In surveys, it sometimes happens that the population size is ignored by the survey taker. One method to remediate this is the following: we identify, among the total population of size N (unknown), M individuals. We then allow these individuals to 'mix' with the total population, and we select n individuals by simple random sampling in the total population after mixing. We then pick out from this sample m individuals belonging to the first 'marked' population.

1. What is the distribution of m ; what is its expected value and variance?
2. What is the probability that m is equal to zero? We suppose n is small with respect to M and with respect to $N - M$.
3. Considering the expectation of m, give a natural estimator $\widehat{N}$ of N in the case where m is not equal to zero. We verify that in practice this occurs if n and M are 'sufficiently large'.
4. Calculate $\mathcal{M} = \mathrm{E}(m \mid m > 0)$ and $\mathcal{V} = \mathrm{var}\,(m \mid m > 0)$. In using a Taylor expansion of m around $\mathcal{M}$, approach $\mathrm{E}(\widehat{N} \mid m > 0)$ by considering n 'large' (and, consequently, N 'particularly large').
5. Conclude about the eventual bias of $\widehat{N}$.

This method, called 'capture-recapture' (see Thompson, 1992), can be used, for example, to estimate the number of wild animals of a certain type in a large forest (we control M, the number of marked animals, and obviously n).

Solution

1. The random variable m follows a hypergeometric distribution with parameters N, M, n:

$$\Pr(m = x) = \frac{\binom{M}{x} \times \binom{N-M}{n-x}}{\binom{N}{n}},$$

for all $x = \max(0, M-N+n)$, 1, 2, $\ldots$, $\min(n, M)$. We can obviously calculate the moments directly by using the previous expression. We can also notice that m/n is the classical unbiased estimator of the true proportion of M/N 'marked' individuals. Hence:

$$E\left(\frac{m}{n}\right) = \frac{M}{N}, \quad \text{and therefore} \quad E(m) = n\frac{M}{N}.$$

The variance is

$$\operatorname{var}\left(\frac{m}{n}\right) = \left(1 - \frac{n}{N}\right) \frac{1}{n} \frac{N}{N-1} \left(\frac{M}{N}\right) \left(1 - \frac{M}{N}\right).$$

If N is large, we then have:

$$\operatorname{var}(m) \approx \left(1 - \frac{n}{N}\right) \left(n\frac{M}{N}\right) \left(1 - \frac{M}{N}\right).$$

2. The probability that m is null is:

$$\Pr(m = 0) = \frac{\binom{M}{0}\binom{N-M}{n}}{\binom{N}{n}} = \frac{\binom{N-M}{n}}{\binom{N}{n}} = \frac{(N-M)^{[n]}}{N^{[n]}} \approx \left(\frac{N-M}{N}\right)^n,$$

where $N^{[n]} = N \times (N-1) \times \cdots \times (N-n+1)$. This probability is negligible when M and n are sufficiently large.

3. Since

$$\frac{M}{N} = E\left(\frac{m}{n}\right),$$

we can use:

$$\widehat{N} = M\frac{n}{m},$$

but only if $m > 0$. In practice, this is almost certainly confirmed if M and n are sufficiently large according to Question 2. If $m = 0$, we do not use any estimation (in concrete terms, we continue with the process from the beginning, until $m > 0$).

4. As

$$E(m) = E(m \mid m = 0)\Pr(m = 0) + E(m \mid m > 0)\Pr(m > 0),$$

we have

$$\mathcal{M} = E(m \mid m > 0) = \frac{E(m)}{\Pr(m > 0)} = n\frac{M}{N}\frac{1}{1 - \Pr(m = 0)},$$

and

$$\mathcal{V} = \operatorname{var}(m \mid m > 0) = E(m^2 \mid m > 0) - [E(m \mid m > 0)]^2$$

$$= \frac{E(m^2)}{\Pr(m > 0)} - \mathcal{M}^2$$

$$= \frac{\operatorname{var}(m) + [E(m)]^2}{\Pr(m > 0)} - \mathcal{M}^2$$

$$= \frac{1}{\Pr(m > 0)} \left[\operatorname{var}(m) - \frac{\Pr(m = 0)}{\Pr(m > 0)} \left(n\,\frac{M}{N} \right)^2 \right].$$

Furthermore, we have:

$$\frac{1}{m} = \frac{1}{\mathcal{M}\left(1 + \frac{m - \mathcal{M}}{\mathcal{M}}\right)}, \qquad \text{for all } m > 0 \ (\mathcal{M} > 0).$$

Now, the term

$$\Delta = \frac{m - \mathcal{M}}{\mathcal{M}},$$

conditional on $m > 0$, is of null expectation, by construction. Furthermore:

$$\operatorname{var}(\Delta \mid m > 0) = \frac{1}{\mathcal{M}^2} \operatorname{var}(m \mid m > 0) = \frac{\operatorname{var}(m)}{\Pr(m > 0)\mathcal{M}^2} - \Pr(m = 0).$$

If n and M are large, then $\Pr(m = 0)$ is negligible with respect to the first term of this difference. Since n is large, we can write

$$\left(1 + \frac{m - \mathcal{M}}{\mathcal{M}}\right)^{-1} \approx 1 - \frac{m - \mathcal{M}}{\mathcal{M}} + \left(\frac{m - \mathcal{M}}{\mathcal{M}}\right)^2 + \ldots$$

With n large, we neglect the terms of order 3 and above, of order of magnitude by $(nM/N)^{-3/2}$. Thus, for $m > 0$,

$$\widehat{N} = \frac{Mn}{m} \approx \frac{Mn}{\mathcal{M}} (1 - \Delta + \Delta^2),$$

and

$$E(\widehat{N} \mid m > 0) \approx \frac{Mn}{\mathcal{M}} (1 + E(\Delta^2 \mid m > 0)) = N\Pr(m > 0) \left(1 + \frac{\mathcal{V}}{\mathcal{M}^2}\right).$$

5. The estimator is then biased. The bias results from the conjunction of two elements: on the one hand, we are restricted at $m > 0$, and on the other hand the random variable m is in the denominator of the estimator. If n is large, the bias is small because

$$\Pr(m > 0) = 1 - \Pr(m = 0)$$

approaches 1 and that

$$\frac{\mathcal{V}}{\mathcal{M}^2} \text{ varies by } 1/n$$

and therefore approaches zero. The estimator $\widehat{N}$ thus appears as an interesting estimator of N. It would remain to calculate its variance.

Exercise 2.16 *Subsample and covariance*

We consider a simple random sample without replacement of size n in a population U of size N (sample denoted as S). We also consider two individuals k and ℓ *distinct*.

1. Show that:
$$\Pr[k \in S \text{ and } \ell \notin S] = \frac{n(N-n)}{N(N-1)}.$$

2. In the previous sample S, we select by simple random sampling n_1 individuals. We denote S_1 as the sample obtained and S_2 as the complementary sample of S_1 in S. Let k and ℓ be any two distinct individuals belonging to the sample S (we thus work 'conditionally on S'). What is $\Pr(k \in S_1$ and $\ell \in S_2 \mid S)$? (Hint: use Question 1.)

3. If k and ℓ are any two elements (but *distinct*) in the population, show that
$$\Pr[k \in S_1 \text{ and } \ell \in S_2] = \frac{n_1(n-n_1)}{N(N-1)}.$$

4. Show, in the conditions of Question 2, that we can consider S_1 as a simple random sample of size n_1 selected from a population of size N. (Hint: calculate $\Pr(S_1 = s_1)$.)

5. By using the results from Question 1, calculate, first of all for k different from ℓ and then for k equal to ℓ, the following:
$$\text{cov}\left(I\{k \in S_1\}, I\{\ell \in S_2\}\right),$$
where $I\{A\}$ represents the indicator for event A.

6. Deduce that:
$$\text{cov}\left(\widehat{\overline{Y}}_1, \widehat{\overline{Y}}_2\right) = -\frac{S_y^2}{N},$$
where $\widehat{\overline{Y}}_\ell$ is the simple mean of a real variable y_k calculated in the sample S_ℓ ($\ell = 1, 2$), and S_y^2 is the population variance of y_k.

7. Calculate $\text{cov}\left(\widehat{\overline{Y}}, \widehat{\overline{Y}}_1\right)$ where $\widehat{\overline{Y}}$ is the simple mean of y_k calculated in S.

Solution

1. Since

$$\Pr(k \in S \text{ and } \ell \notin S) + \Pr(k \in S \text{ and } \ell \in S)$$
$$= \Pr(k \in S \text{ and } (\ell \in S \text{ or } \ell \notin S)) = \Pr(k \in S),$$

we have

$$\Pr(k \in S \text{ and } \ell \notin S) = \pi_k - \pi_{k\ell} = \frac{n}{N} - \frac{n(n-1)}{N(N-1)} = \frac{n(N-n)}{N(N-1)}.$$

A second method consists of writing:

$$\Pr(k \in S \text{ and } \ell \notin S) = \sum_{\substack{s \ni k \\ s \not\ni \ell}} p(s) = \frac{\#\{s | k \in s \text{ and } \ell \notin s\}}{\binom{N}{n}}.$$

Now the number of samples s containing k but not ℓ is $\binom{N-2}{n-1}$. Indeed, k being in s, there remain $(n-1)$ individuals to select in the population outside of k and of ℓ.

2. The sample S is fixed: we can use the previous result by considering that the population here is the sample S and that the sample is S_1:

$$\Pr(k \in S_1 \text{ and } \ell \notin S_1 \mid S) = \frac{n_1(n-n_1)}{n(n-1)}, \text{ with } k \text{ and } \ell \in S.$$

It remains to state that $(\ell \notin S_1 \mid S)$ is equivalent to $(\ell \in S_2 \mid S)$, seeing that S_1 and S_2 form a partition of S.

3. As for all S:

$$\Pr[k \in S_1 \text{ and } \ell \in S_2 \mid S] = \frac{n_1(n-n_1)}{n(n-1)}, \text{ if } k \text{ and } \ell \in S.$$

We have

$$\Pr[k \in S_1 \text{ and } \ell \in S_2]$$
$$= \sum_{s | k \in s \text{ and } \ell \in s} \Pr[k \in S_1 \text{ and } \ell \in S_2 \mid S = s] \Pr[S = s]$$
$$= \frac{n_1(n-n_1)}{n(n-1)} \binom{N}{n}^{-1} \binom{N-2}{n-2}$$
$$= \frac{n_1(n-n_1)}{n(n-1)} \frac{n(n-1)}{N(N-1)} = \frac{n_1(n-n_1)}{N(N-1)}.$$

A faster approach, but less natural, consists of stating the result from Question 2 obtained by depending on the lone fact that k and ℓ are in S (the integral composition of S does not provide anything). Also,

$$\Pr(k \in S_1 \text{ and } \ell \in S_2 | k \in S \text{ and } \ell \in S) = \frac{n_1(n - n_1)}{n(n - 1)},$$

which leads to

$$
\begin{aligned}
\Pr&(k \in S_1 \text{ and } \ell \in S_2) \\
&= \Pr(k \in S_1 \text{ and } \ell \in S_2 | k \in S \text{ and } \ell \in S)\Pr(k \in S \text{ and } \ell \in S) \\
&= \frac{n_1(n - n_1)}{n(n - 1)} \frac{n(n - 1)}{N(N - 1)} = \frac{n_1(n - n_1)}{N(N - 1)}.
\end{aligned}
$$

4. The probability of selecting s_1 is:

$$\Pr(S_1 = s_1) = \sum_{s \supset s_1} \Pr(S_1 = s_1 \mid S = s) \; \Pr(S = s) = \frac{\binom{N - n_1}{n - n_1}}{\binom{n}{n_1}\binom{N}{n}} = \frac{1}{\binom{N}{n_1}}.$$

This result is characteristic of a simple random sampling of size n_1 in a population of size N.

5. If $k \neq \ell$, then

$$
\begin{aligned}
\operatorname{cov}&(I\{k \in S_1\}, I\{\ell \in S_2\}) \\
&= \operatorname{E}(I\{k \in S_1\}I\{\ell \in S_2\}) - (\operatorname{E}I\{k \in S_1\})\,(\operatorname{E}I\{\ell \in S_2\}) \\
&= \Pr[k \in S_1 \text{ and } \ell \in S_2] - \Pr[k \in S_1]\Pr[\ell \in S_2] \\
&= \frac{n_1(n - n_1)}{N(N - 1)} - \frac{n_1}{N}\frac{n - n_1}{N} = \frac{n_1(n - n_1)}{N^2(N - 1)}.
\end{aligned}
$$

If $k = \ell$,

$$\operatorname{cov}(I\{k \in S_1\}, I\{k \in S_2\}) = -\Pr(k \in S_1)\,\Pr(k \in S_2) = -\frac{n_1(n - n_1)}{N^2}.$$

6. We let $n_2 = n - n_1$:

$$
\begin{aligned}
\operatorname{cov}&\left(\widehat{\overline{Y}}_1, \widehat{\overline{Y}}_2\right) \\
&= \operatorname{cov}\left(\sum_{k \in U} \frac{y_k I\{k \in S_1\}}{n_1}, \sum_{\ell \in U} \frac{y_\ell I\{\ell \in S_2\}}{n_2}\right) \\
&= \frac{1}{n_1(n - n_1)} \sum_{k \in U} \sum_{\ell \in U} \operatorname{cov}\left(I\{k \in S_1\}, I\{\ell \in S_2\}\right) y_k y_\ell \\
&= \frac{1}{n_1(n - n_1)}\left[-\frac{n_1(n - n_1)}{N^2}\sum_{k \in U} y_k^2 + \frac{n_1(n - n_1)}{N^2(N - 1)}\sum_{k \in U}\sum_{\substack{\ell \in U \\ \ell \neq k}} y_k y_\ell\right] \\
&= -\frac{1}{N^2}\left(\sum_{k \in U} y_k^2 - \frac{1}{N - 1}\sum_{k \in U}\sum_{\substack{\ell \in U \\ \ell \neq k}} y_k y_\ell\right) = -\frac{S_y^2}{N}.
\end{aligned}
$$

This is a second method that depends only on the result from Question 4, meaning that S_1 is a simple random sample of size n_1 in a population of size N and that, by analogy, S_2 is a simple random sample of size n_2 in a population of size $N - n_1$.

$$\mathrm{cov}(\widehat{\overline{Y}}_1, \widehat{\overline{Y}}_2) = \mathrm{E}_{S_1}\mathrm{cov}(\widehat{\overline{Y}}_1, \widehat{\overline{Y}}_2 | S_1) + \mathrm{cov}_{S_1}\left[\mathrm{E}(\widehat{\overline{Y}}_1 | S_1), \mathrm{E}(\widehat{\overline{Y}}_2 | S_1)\right].$$

Now, conditionally on S_1, $\widehat{\overline{Y}}_1$ is constant:

$$\mathrm{cov}(\widehat{\overline{Y}}_1, \widehat{\overline{Y}}_2) = \mathrm{cov}_{S_1}\left[\widehat{\overline{Y}}_1, \mathrm{E}(\widehat{\overline{Y}}_2 | S_1)\right].$$

We have

$$\mathrm{E}(\widehat{\overline{Y}}_2 | S_1) = \frac{1}{N - n_1} \sum_{k \in U \setminus S_1} y_k = \frac{N\overline{Y} - n_1 \widehat{\overline{Y}}_1}{N - n_1}.$$

Ultimately,

$$\mathrm{cov}(\widehat{\overline{Y}}_1, \widehat{\overline{Y}}_2) = \mathrm{cov}_{S_1}\left(\widehat{\overline{Y}}_1, -\frac{n_1 \widehat{\overline{Y}}_1}{N - n_1}\right) = -\frac{n_1}{N - n_1}\mathrm{var}_{S_1}\left(\widehat{\overline{Y}}_1\right)$$

$$= -\frac{n_1}{N - n_1}\left(1 - \frac{n_1}{N}\right)\frac{S_y^2}{n_1} = -\frac{S_y^2}{N}.$$

7. Finally, the covariance is

$$\mathrm{cov}\left(\widehat{\overline{Y}}, \widehat{\overline{Y}}_1\right) = \mathrm{cov}\left(\frac{n_1}{n}\widehat{\overline{Y}}_1 + \frac{n_2}{n}\widehat{\overline{Y}}_2, \widehat{\overline{Y}}_1\right) \quad \text{(with } n_2 = n - n_1\text{)}$$

$$= \frac{n_1}{n}\mathrm{var}(\widehat{\overline{Y}}_1) + \frac{n_2}{n}\mathrm{cov}(\widehat{\overline{Y}}_1, \widehat{\overline{Y}}_2)$$

$$= \frac{n_1}{n}\left(1 - \frac{n_1}{N}\right)\frac{S_y^2}{n_1} - \frac{n_2}{n}\frac{S_y^2}{N}$$

$$= \left(1 - \frac{n}{N}\right)\frac{S_y^2}{n} = \mathrm{var}\left(\widehat{\overline{Y}}\right) = \mathrm{cov}(\widehat{\overline{Y}}, \widehat{\overline{Y}}).$$

From this fact, $\widehat{\overline{Y}}$ and $(\widehat{\overline{Y}} - \widehat{\overline{Y}}_1)$ appear to be uncorrelated, which is quite surprising.

Exercise 2.17 *Recapture with replacement*

The objective is to estimate the number of rats present on an island. We set up a trap which is installed at a location selected at random on the island. When a rat is trapped, it is marked and then released. If, for 50 captured rats, we count 42 distinctly marked rats, estimate using the maximum likelihood method the number of rats living on the island, assuming that the 50 rats were captured at random and with replacement.

Note: the maximum likelihood solution can be obtained through searching using, for example, a spreadsheet.

Solution

In this approach, N is the parameter to estimate and r is the random variable for which it is necessary to express the density and to later maximize. We denote $f_N(r)$ as the probability of obtaining r distinct rats in m trials with replacement (m is a controlled size, known and non-random) in a population of size N. This model is reasonable under the conditions of the process. We note that there are $\binom{N}{r} = N!/r!(N-r)!$ ways of choosing the list of r rats involved. Thus,

$$f_N(r) = \frac{N!}{r!(N-r)!}g_N(r),$$

where $g_N(r)$ is the probability of obtaining r distinct and properly identified rats in m trials with replacement (valid expression because all rats have, for each trial, the same probability of being selected). This list of rats being fixed, the universe Ω of possibilities is formed by the group of mappings of $\{1,\ldots,m\}$ to $\{1,\ldots,N\}$ (we assume that the r rats listed are identified by the first r integers). We have $m \geq r$, and in fact

$$g_N(r) = \sum_{\omega \in \mathrm{FAV}} p(\omega),$$

where $p(\omega)$ is the probability of obtaining a given mapping ω and FAV is the group of favourable mappings. We have $p(\omega) = N^{-m}$, for all ω. It remains to calculate the total number of favourable cases. It is exactly a question of the number of surjective mappings of $\{1,\ldots,m\}$ in $\{1,\ldots,r\}$, which is equal to $r!$ multiplied by the Stirling number of second kind $\mathfrak{s}_m^{(r)}$, which is:

$$\mathfrak{s}_m^{(r)} = \frac{1}{r!}\sum_{i=1}^{r}\binom{r}{i}i^m(-1)^{r-i}.$$

The Stirling number of second kind is equal to the number of ways of finding a group of m elements in r non-empty parts (see Stanley, 1997). However, the calculation of $\mathfrak{s}_m^{(r)}$ does not interest us here. Indeed, $\mathfrak{s}_m^{(r)}$ does not depend on N but only on m and r. Eventually we obtain

$$f_N(r) = \frac{N!}{(N-r)!N^m}\mathfrak{s}_m^{(r)}, r = 1,\ldots,\min(m,N). \tag{2.4}$$

We are going to maximize the function $f_N(r)$ for N. Now, maximizing $f_N(r)$ for N comes back to maximizing

$$\frac{N!}{(N-r)!N^m} = \frac{\prod_{i=0}^{r-1}(N-i)}{N^m},$$

as $\mathfrak{s}_m^{(r)}$ does not depend on N. When $m = 50$ and $r = 42$, we find the solution through a search (see Table 2.5). The solution of the maximum likelihood is

Table 2.5. Search for the solution of maximum likelihood: Exercise 2.17

N	$\frac{N!}{(N-r)!N^m} \times 10^{21}$	N	$\frac{N!}{(N-r)!N^m} \times 10^{21}$	N	$\frac{N!}{(N-r)!N^m} \times 10^{21}$
100	3.97038	120	6.49114	140	7.05245
101	4.13269	121	6.56558	141	7.03671
102	4.29281	122	6.63468	142	7.01773
103	4.45037	123	6.69847	143	6.99563
104	4.60503	124	6.75702	144	6.97054
105	4.75645	125	6.81037	145	6.94259
106	4.90434	126	6.85860	146	6.91191
107	5.04842	127	6.90178	147	6.87864
108	5.18844	128	6.94001	148	6.84288
109	5.32416	129	6.97339	149	6.80478
110	5.45539	130	7.00200	150	6.76444
111	5.58193	131	7.02597	151	6.72199
112	5.70364	132	7.04541	152	6.67755
113	5.82038	133	7.06042	153	6.63122
114	5.93203	134	7.07115	154	6.58311
115	6.03850	135	7.07770	155	6.53335
116	6.13971	**136**	**7.08021**	156	6.48202
117	6.23561	137	7.07881	157	6.42924
118	6.32616	138	7.07363	158	6.37510
119	6.41134	139	7.06479	159	6.31970

therefore $N = 136$. Another manner of tackling the problem consists of setting the first derivative of the logarithm of the likelihood function equal to zero

$$\frac{d\left(\frac{\log[\prod_{i=0}^{r-1}(N-i)]}{N^m}\right)}{dN} = \frac{d}{dN}\left[\sum_{i=0}^{r-1}\log(N-i) - m\log N\right] = \sum_{i=0}^{r-1}\frac{1}{N-i} - \frac{m}{N} = 0,$$

which gives

$$\sum_{i=0}^{r-1}\frac{N}{N-i} = m.$$

We obtain a non-linear equation that we can also solve by trial and error. Obviously, we obtain the same result.

Exercise 2.18 *Collection*

Your child would like to collect pictures of football players sold in sealed packages. The complete collection consists of 350 distinct pictures. Each package contains one picture 'at random' in a totally independent manner from one package to another. Purchasing X packages is similar to taking X samples with replacement and with equal probability in the population of size $N = 350$. To simplify, your child does not trade any pictures.

1. What is the probability distribution of the number of pictures to purchase in order to obtain exactly r different players?
2. How many photos must be purchased on average in order to obtain the complete collection?

Solution

1. In Exercise 2.17, we saw that if n_S represents the number of distinct units obtained by selecting m units with replacement in a population of size N, then

$$p_m(r) = \Pr(n_S = r) = \frac{N!}{(N-r)!N^m} \mathfrak{s}_m^{(r)},$$

where $r = 1, \ldots, \min(m, N)$ and $\mathfrak{s}_m^{(r)}$ is a Stirling number of second kind,

$$\mathfrak{s}_m^{(r)} = \frac{1}{r!} \sum_{i=1}^{r} \binom{r}{i} i^m (-1)^{r-i}.$$

If we let X be the random variable representing the number of drawings necessary to obtain r distinct individuals, then

$\Pr[X = m]$
$= \Pr[\text{selecting } r - 1 \text{ distinct units in } m - 1 \text{ samples with replacement}]$
$\quad \times \Pr[\text{selecting in the } m\text{th sample a unit not yet selected}$
$\quad \text{knowing that } r - 1 \text{ distinct units have already been selected}]$
$= p_{m-1}(r-1) \times \dfrac{N-r+1}{N}$
$= \dfrac{N!}{(N-r)!N^m} \mathfrak{s}_{m-1}^{(r-1)},$

for $m = r, r+1, \ldots$.

2. We know the probability distribution of the random variable X. We now wish to calculate its expected value in the case $r = N$, which corresponds to the complete collection. In the case of any r, we have

$$E(X) = \sum_{m=r}^{\infty} m \frac{N!}{(N-r)!N^m} \mathfrak{s}_{m-1}^{(r-1)}.$$

Since

$$\sum_{m=r}^{\infty} \Pr[X = m] = 1,$$

we have

$$\sum_{m=r}^{\infty} \frac{\mathfrak{s}_{m-1}^{(r-1)}}{N^m} = \frac{(N-r)!}{N!} = \prod_{i=0}^{r-1} \frac{1}{(N-i)}. \tag{2.5}$$

By differentiating Identity (2.5) with respect to N (for the right-hand side, use the logarithmic derivative), we easily obtain

$$\sum_{m=r}^{\infty} m \frac{\mathfrak{s}_{m-1}^{(r-1)}}{N^{m+1}} = \left(\sum_{j=0}^{r-1} \frac{1}{N-j} \right) \frac{(N-r)!}{N!}.$$

We then get

$$\mathrm{E}(X) = \frac{N!N}{(N-r)!} \sum_{m=r}^{\infty} m \frac{\mathfrak{s}_{m-1}^{(r-1)}}{N^{m+1}} = \sum_{j=0}^{r-1} \frac{N}{N-j}.$$

For the complete collection:

$$\mathrm{E}(X) = \sum_{j=0}^{r-1} \frac{N}{N-j} = N \sum_{j=1}^{N} \frac{1}{j} = 350 \times \sum_{j=1}^{350} \frac{1}{j} \approx 350 \times (\log 350 + \gamma),$$

where γ is Euler's constant, approximately 0.5772. We get $\mathrm{E}(X) \approx 2252$ pictures.

Exercise 2.19 *Proportion of students*

A sample of 100 students is chosen using a simple random design without replacement from a population of 1000 students. We are then interested in the results obtained by these students in an exam. There are two possible results: success or failure. The outcome is presented in Table 2.6.

Table 2.6. Sample of 100 students: Exercise 2.19

	Men	Women	Total
Success	$n_{11} = 35$	$n_{12} = 25$	$n_{1.} = 60$
Failure	$n_{21} = 20$	$n_{22} = 20$	$n_{2.} = 40$
Total	$n_{.1} = 55$	$n_{.2} = 45$	$n = 100$

1. Estimate the success rate for men and for women.
2. Calculate the approximate bias of the estimated success rates.
3. Estimate the mean square error of these success rates.
4. Give the 95% confidence intervals for the success rate for men R_M and for women R_W. What can we say about their respective positions?

5. What confidence intervals must be considered in order for the true values R_M and R_W to be inside the disjoint confidence intervals? Comment on this.
6. Using the estimation results by domain, find a more simple result for Questions 2 and 3.

Solution

The notation for different proportions in the population U is presented in Table 2.7.

Table 2.7. Notation for different proportions: Exercise 2.19

	Men	Women	Total
Success	P_{11}	P_{12}	$P_{1.}$
Failure	P_{21}	P_{22}	$P_{2.}$
Total	$P_{.1}$	$P_{.2}$	1

1. The success rate for men is naturally estimated by:

$$r_M = \frac{\widehat{P}_{11}}{\widehat{P}_{.1}} = \frac{n_{11}}{n_{.1}} = \frac{35}{55} \approx 63.6\%.$$

The success rate for women is estimated by:

$$r_W = \frac{\widehat{P}_{12}}{\widehat{P}_{.2}} = \frac{n_{12}}{n_{.2}} = \frac{25}{45} \approx 55.6\%.$$

These two estimators are ratios. Indeed, the denominators of these estimators are random.

2. Since the sample size n is 100, we can consider without hesitation that n is large. The bias of a ratio $r = \widehat{Y}/\widehat{X}$ is given by

$$B(r) = E(r) - R \approx R\left(\frac{S_x^2}{\overline{X}^2} - \frac{S_{xy}}{\overline{X}\,\overline{Y}}\right)\frac{1-f}{n},$$

where

$$x_k = \begin{cases} 1 \text{ if the individual is a man (resp. a woman)} \\ 0 \text{ otherwise,} \end{cases}$$

and

$$y_k = \begin{cases} 1 \text{ if the individual is a man (resp. a woman) who succeeded} \\ 0 \text{ otherwise,} \end{cases}$$

for all $k \in U$. For example, we have for the men:

$$S_x^2 = \frac{1}{N-1} \left(\sum_{k \in U} x_k^2 - N\overline{X}^2 \right) = \frac{1}{N-1} \left(NP_{.1} - NP_{.1}^2 \right)$$

$$= \frac{N}{N-1} P_{.1} \left(1 - P_{.1} \right),$$

and

$$S_{xy} = \frac{1}{N-1} \left(\sum_{k \in U} x_k y_k - N\overline{X}\ \overline{Y} \right)$$

$$= \frac{1}{N-1} \left(NP_{11} - NP_{.1}P_{11} \right)$$

$$= \frac{N}{N-1} P_{11} \left(1 - P_{.1} \right).$$

We therefore have

$$\frac{S_x^2}{\overline{X}^2} - \frac{S_{xy}}{\overline{X}\ \overline{Y}} = \frac{1}{P_{.1}^2} \frac{N}{N-1} P_{.1} \left(1 - P_{.1} \right) - \frac{1}{P_{.1}P_{11}} \frac{N}{N-1} P_{11} \left(1 - P_{.1} \right) = 0.$$

The bias is thus approximately null: $\mathrm{B}(r) \approx 0$.

3. Since n is large, the mean square error, similar to the variance, is given by the approximation

$$\mathrm{MSE}(r) \approx \frac{1-f}{n\overline{X}^2} \left(S_y^2 - 2RS_{xy} + R^2 S_x^2 \right),$$

where

$$R = \frac{\overline{Y}}{\overline{X}}.$$

For the men, we get

$$\mathrm{MSE}(r_M) \approx \mathrm{var}(r_M)$$

$$\approx \frac{1-f}{nP_{.1}^2} \frac{N}{N-1} \left\{ P_{11}(1 - P_{11}) - 2\frac{P_{11}}{P_{.1}} P_{11}(1 - P_{.1}) + \frac{P_{11}^2}{P_{.1}^2} P_{.1}(1 - P_{.1}) \right\}$$

$$= \frac{1-f}{nP_{.1}^2} \frac{N}{N-1} \left\{ P_{11} - 2\frac{P_{11}^2}{P_{.1}} + \frac{P_{11}^2}{P_{.1}} \right\}$$

$$= \frac{1-f}{nP_{.1}} \frac{N}{N-1} \frac{P_{11}}{P_{.1}} \left\{ 1 - \frac{P_{11}}{P_{.1}} \right\}.$$

The estimator (slightly biased) directly becomes

$$\widehat{\mathrm{MSE}}(r_M) = \frac{1-f}{n\widehat{P}_{.1}} \frac{N}{N-1} \frac{\widehat{P}_{11}}{\widehat{P}_{.1}} \left\{ 1 - \frac{\widehat{P}_{11}}{\widehat{P}_{.1}} \right\}.$$

We get

- For the men:

$$\widehat{\mathrm{MSE}}(r_M) = \frac{1 - \frac{1}{10}}{100\frac{55}{100}} \frac{1000}{999} \frac{35}{55} \left\{ 1 - \frac{35}{55} \right\} = 0.00379041.$$

- For the women:

$$\widehat{\mathrm{MSE}}(r_W) = \frac{1 - \frac{1}{10}}{100\frac{45}{100}} \frac{1000}{999} \frac{25}{45} \left\{ 1 - \frac{25}{45} \right\} = 0.0049432148.$$

4. With 95 chances out of 100 (roughly), we have the estimated intervals

$$\widehat{\mathrm{CI}}(R_M; 0.95) = \left[r_M - 1.96\sqrt{\widehat{\mathrm{MSE}}(r_M)}, r_M + 1.96\sqrt{\widehat{\mathrm{MSE}}(r_M)} \right]$$
$$= [0.636 - 0.121; 0.636 + 0.121] = [0.515; 0.757],$$

$$\widehat{\mathrm{CI}}(R_W; 0.95) = \left[r_W - 1.96\sqrt{\widehat{\mathrm{MSE}}(r_W)}, r_W + 1.96\sqrt{\widehat{\mathrm{MSE}}(r_W)} \right]$$
$$= [0.556 - 0.138; 0.556 + 0.138] = [0.418; 0.694].$$

The sample size is not very large, but we can consider it to be *a priori* sufficient to approach the distribution of ratios by the normal distribution. Therefore, the two intervals overlap very considerably: we cannot say that the ratios R_M and R_W are significantly different, considering the selected sample size (that is, we do not find two disjoint intervals).

5. With 40 chances out of 100 (roughly), we have the estimated intervals

$$\widehat{\mathrm{CI}}(R_M; 0.40) = \left[r_M - 0.52\sqrt{\widehat{\mathrm{MSE}}(r_M)}, r_M + 0.52\sqrt{\widehat{\mathrm{MSE}}(r_M)} \right]$$
$$= [0.636 - 0.032; 0.636 + 0.032] = [0.604; 0.668],$$

$$\widehat{\mathrm{CI}}(R_W; 0.40) = \left[r_W - 0.52\sqrt{\widehat{\mathrm{MSE}}(r_W)}, r_W + 0.52\sqrt{\widehat{\mathrm{MSE}}(r_W)} \right]$$
$$= [0.556 - 0.037; 0.556 + 0.037] = [0.519; 0.593].$$

The establishment of such intervals is an exercise of style which does not represent much in practice. Except for an 'absolute miracle', we indeed have $R_M \neq R_W$ (why would we think otherwise?). The question is to find out if the confidence interval actually confirms this evidence or not. If the two intervals do not overlap, the produced statistic could be used as evidence in confirming that $R_M \neq R_W$. If they overlap such as in Question 4, we find the statistic has no usefulness. We can only say that the sample was not large enough to reject the equality hypothesis of the ratios. Obviously, the 40% intervals allow to significantly separate R_M from R_W, but the probability of covering the true values is so poor that we cannot seriously refer to it.

6. The approach (bias and variance) of the previous questions relied upon direct calculations carried out starting from the ratio. However, we can note that we are precisely in the situation of mean estimation in a domain: for example, if we go back to the notation from Question 2, R_M is the mean of y_k for the domain of men ($x_k = 1$). We know that the estimated mean for the domain (here r_M) has, as expected value, the true value R_M as soon as we use a conditional expectation to the sample size matching up with the domain (here $n_{.1}$). A problem occurs when $n_{.1} = 0$, in which case we cannot calculate r_M, but this situation can only occur with a negligible probability (here $n = 100$). Thus, for all $n_{.1} > 0$, we have

$$E\left[r_M \mid n_{.1}\right] = R_M,$$

and therefore

$$E\left[r_M\right] = E_{n_{.1}} E\left[r_M \mid n_{.1}\right] = E_{n_{.1}}\left[R_M\right] = R_M,$$

where $E_{n_{.1}}[.]$ is the expectation in relation to the hypergeometric distribution of the random variable $n_{.1}$ in a population of size $N_{.1}$ (excluding the case where $n_{.1} = 0$). The bias is approximately null. For the conditional variance, we use the characteristic expression for a simple random sample of size $n_{.1}$:

$$\mathrm{var}\left[r_M \mid n_{.1}\right] = \left(1 - \frac{n_{.1}}{N_{.1}}\right) \frac{S_1}{n_{.1}},$$

where

$$S_1 = \frac{N_{11}}{N_{.1}}\left(1 - \frac{N_{11}}{N_{.1}}\right) = \frac{P_{11}}{P_{.1}}\left(1 - \frac{P_{11}}{P_{.1}}\right),$$

seeing that it is a question of a proportion. The unconditional variance is obtained by

$$\mathrm{var}\left[r_M\right] = E_{n_{.1}}\mathrm{var}\left[r_M \mid n_{.1}\right] + \mathrm{var}_{n_{.1}} E\left[r_M \mid n_{.1}\right]$$

$$= E_{n_{.1}}\mathrm{var}\left[r_M \mid n_{.1}\right]$$

$$= E_{n_{.1}}\left[\left(1 - \frac{n_{.1}}{N_{.1}}\right)\frac{1}{n_{.1}}\right]\frac{P_{11}}{P_{.1}}\left(1 - \frac{P_{11}}{P_{.1}}\right)$$

$$= \left(E\left[\frac{1}{n_{.1}}\right] - \frac{1}{N_{.1}}\right)\frac{P_{11}}{P_{.1}}\left(1 - \frac{P_{11}}{P_{.1}}\right).$$

In the first approximation, as n is large:

$$E\left[\frac{1}{n_{.1}}\right] \approx \frac{1}{E[n_{.1}]} = \frac{1}{nP_{.1}}.$$

Since $N_{.1} = NP_{.1}$, we finally get

$$\mathrm{var}\left[r_M\right] = \frac{1 - f}{nP_{.1}}\frac{P_{11}}{P_{.1}}\left(1 - \frac{P_{11}}{P_{.1}}\right),$$

and we indeed find the variance found in Question 3 apart from a factor $N/(N - 1)$ (this factor is obviously close to 1).

Exercise 2.20 *Sampling with replacement and estimator improvement*

Consider a population of size N. We perform simple random sampling with replacement of size $m = 3$. We denote $\tilde{S}$ as the random sample selection (with repetitions). For example, with $N = 5$, $\tilde{S}$ can have as values

$$(1, 2, 5), (1, 3, 4), (2, 4, 4), (2, 2, 3), (2, 3, 3), (3, 3, 3).$$

(we consider two samples containing the same units in a different order to be distinct). Consider the reduction function $r(.)$, which suppresses from the sample the information concerning any multiplicity of units. For example:

$$r((2, 2, 3)) = \{2, 3\}, \quad r((2, 3, 3)) = \{2, 3\}, \quad r((3, 3, 3)) = \{3\}.$$

We denote S as the random sample without replacement obtained by suppressing the information concerning the multiplicity of units (in S, the order of individuals does not matter).

1. Calculate the probability R_i that sample $\tilde{S}$ contains exactly i distinct individuals ($i = 1, 2$, or 3).
2. Show that the design of S conditional on its size $\#S$ is a simple random design without replacement of fixed size.
3. Give the sampling design for S, that is, the list of all possible values of S and the probabilities associated with those values.
4. Consider the following two estimators:
   The mean with repetition

$$\widetilde{Y} = \frac{1}{3} \sum_{k \in \tilde{S}} y_k,$$

   the mean calculated on distinct values

$$\widehat{Y} = \frac{1}{\#S} \sum_{k \in S} y_k.$$

   Calculate the expected values and the variances for these estimators. Make a conclusion.

Solution

1. The probability of having three distinct individuals is

$$R_3 = \frac{N-1}{N} \times \frac{N-2}{N} = \frac{(N-1)(N-2)}{N^2}.$$

   In fact, with the first individual (any) being selected, there is a probability $1/N$ that the second individual is identical to the first. Furthermore, with the first two individuals (distinct) having been selected, there is a probability $2/N$ that the third individual is also one of the first two. Another method consists of counting the number of distinct trios of elements (there are $N(N-1)(N-2)$ combinations) and multiplying this number by the probability of obtaining any given trio, which is $1/N^3$. The probability of getting the same unit three times is

$$R_1 = \frac{1}{N} \times \frac{1}{N} = \frac{1}{N^2}.$$

The probability of obtaining two distinct individuals is obtained by the difference:

$$R_2 = 1 - R_1 - R_3 = \frac{N^2 - (N-1)(N-2) - 1}{N^2} = \frac{3(N-1)}{N^2}.$$

2. For reasons of symmetry between the units, the design of S conditional on $\#S$ had to be simple. However, we are going to calculate this conditional design rigorously. The design of S is obtained from the design of $\tilde{S}$. Conditional on the size j of S, we have, for $j = 1, 2, 3$:

$$\Pr(S = s | \#S = j) = \begin{cases} \dfrac{\Pr(S = s)}{R_j} = \dfrac{\sum_{\tilde{s}|r(\tilde{s})=s} \tilde{p}(\tilde{s})}{R_j} & \text{if } \#s = j \\ 0 & \text{otherwise,} \end{cases}$$

where $\tilde{p}(\tilde{s})$ is the probability of obtaining an ordered sample with repetition $\tilde{s}$. Since the sampling is done with replacement, we have $\tilde{p}(\tilde{s}) = 1/N^3$, for all $\tilde{s}$, which is:

$$\Pr(S = s | \#S = j) = \begin{cases} \dfrac{1}{R_j} \times \#\{\tilde{s}|r(\tilde{s}) = s\} \times \dfrac{1}{N^3} & \text{if } \#s = j \\ 0 & \text{otherwise.} \end{cases}$$

- If $j = 1$, and $\#s = 1$, then $\#\{\tilde{s}|r(\tilde{s}) = s\} = 1$.
- If $j = 2$, and $\#s = 2$, then $\#\{\tilde{s}|r(\tilde{s}) = s\} = 6$. In fact, if $s = \{a, b\}$, we can have, for $\tilde{S}$:

$$(a, a, b) \text{ or } (a, b, a) \text{ or } (b, a, a) \text{ or } (a, b, b) \text{ or } (b, a, b) \text{ or } (b, b, a).$$

- If $j = 3$, and $\#s = 3$, then $\#\{\tilde{s}|r(\tilde{s}) = s\} = 3! = 6$.

We can then calculate the probability $p(s)$ of selecting s, conditional on $\#S$:
- If $j = 1$, and $\#s = 1$, then

$$\Pr(S = s | \#S = 1) = \frac{1/N^3}{1/N^2} = \binom{N}{1}^{-1}.$$

- If $j = 2$, and $\#s = 2$, then

$$\Pr(S = s | \#S = 2) = \frac{\frac{6}{N^3}}{\frac{3(N-1)}{N^2}} = 2\frac{1}{N(N-1)} = \binom{N}{2}^{-1}.$$

- If $j = 3$, and $\#s = 3$, then

$$\Pr(S = s | \#S = 3) = \frac{\frac{6}{N^3}}{\frac{(N-1)(N-2)}{N^2}} = \frac{6}{N(N-1)(N-2)} = \binom{N}{3}^{-1}.$$

The design conditional on $\#S$ is simple without replacement of fixed size equal to $\#S$.

3. Being conditional on $\#S$, the sample is simple without replacement of fixed size and we have:

$$\begin{aligned} p(s) &= \Pr(S = s) \\ &= \Pr(S = s | \#S = \#s)\Pr(\#S = \#s) \\ &= \begin{cases} R_1 \times \binom{N}{1}^{-1} = \frac{1}{N^3} & \text{if } \#s = 1 \\ R_2 \times \binom{N}{2}^{-1} = \frac{6}{N^3} & \text{if } \#s = 2 \\ R_3 \times \binom{N}{3}^{-1} = \frac{6}{N^3} & \text{if } \#s = 3. \end{cases} \end{aligned}$$

4. The estimator $\widetilde{Y}$ is the classical estimator obtained by sampling with replacement of size 3. It is unbiased and

$$\text{var}(\widetilde{Y}) = \frac{\sigma_y^2}{3} = \frac{N-1}{3N}S_y^2,$$

where

$$\sigma_y^2 = \frac{1}{N}\sum_{k\in U}(y_k - \overline{Y})^2,$$

$$\overline{Y} = \frac{1}{N}\sum_{k\in U}y_k,$$

and

$$S_y^2 = \frac{N}{N-1}\sigma_y^2.$$

The estimator $\widehat{\overline{Y}}$ is more particular to treat, but we have

$$\text{E}(\widehat{\overline{Y}}) = \text{E}(\widehat{\overline{Y}}|\#S = 1)R_1 + \text{E}(\widehat{\overline{Y}}|\#S = 2)R_2 + \text{E}(\widehat{\overline{Y}}|\#S = 3)R_3.$$

Being conditional on the size of S, the design is simple without replacement with fixed size, $\text{E}(\widehat{\overline{Y}}|\#S = \alpha) = \overline{Y}$, for $\alpha = 1, 2, 3$, and therefore $\text{E}(\widehat{\overline{Y}}) = \overline{Y}$. Moreover,

$$\text{var}(\widehat{Y}) = \text{E}\{\text{var}(\widehat{Y}|\#S)\} + \text{var}\underbrace{\left\{\text{E}(\widehat{Y}|\#S)\right\}}_{=\widetilde{Y}}$$

$$= \text{E}\{\text{var}(\widehat{Y}|\#S)\}$$

$$= \text{var}(\widehat{Y}|\#S = 1)R_1 + \text{var}(\widehat{Y}|\#S = 2)R_2 + \text{var}(\widehat{Y}|\#S = 3)R_3$$

$$= \frac{N-1}{N}\frac{S_y^2}{1}R_1 + \frac{N-2}{N}\frac{S_y^2}{2}R_2 + \frac{N-3}{N}\frac{S_y^2}{3}R_3$$

$$= \frac{S_y^2}{N}\left[(N-1)R_1 + \frac{N-2}{2}R_2 + \frac{N-3}{3}R_3\right]$$

$$= \frac{S_y^2(2N-1)(N-1)}{6N^2}$$

$$= \left(1 - \frac{1}{2N}\right)\text{var}(\widetilde{Y}).$$

Thus, $\widehat{Y}$ appears to be systematically more efficient than $\widetilde{Y}$.

Exercise 2.21 *Variance of the variance*

In a simple random design *without* replacement, give the first- through fourth-order inclusion probabilities. Next, give the variance for the estimator of the sampling variance. Simplify the expression for the case where N is very large, then suppose that y is distributed according to a normal distribution in U. What can we say about the estimator of the variance if n is 'large'?

Solution

If we denote I_i as the indicator variable for the presence of unit i in sample S, we have

$$I_i = \begin{cases} 1 \text{ if } i \in S \\ 0 \text{ if } i \notin S. \end{cases}$$

The first- through fourth-order inclusion probabilities are:

$$\pi_1 = \text{E}(I_i) = \frac{n}{N}, i = 1, \ldots, N,$$

$$\pi_2 = \text{E}(I_iI_j) = \frac{n(n-1)}{N(N-1)}, j \neq i,$$

$$\pi_3 = \text{E}(I_iI_jI_k) = \frac{n(n-1)(n-2)}{N(N-1)(N-2)}, j \neq i, k \neq i, k \neq j,$$

and

$$\pi_4 = \text{E}(I_iI_jI_kI_\ell) = \frac{n(n-1)(n-2)(n-3)}{N(N-1)(N-2)(N-3)},$$

$$j \neq i, k \neq i, \ell \neq i, k \neq j, \ell \neq j, \ell \neq k.$$

The corrected variance in the sample is:

$$s_y^2 = \frac{1}{n-1} \sum_{i \in S} \left(y_i - \widehat{\overline{Y}} \right)^2,$$

where

$$\widehat{\overline{Y}} = \frac{1}{n} \sum_{i \in S} y_i.$$

This estimator is unbiased for the corrected variance in the population

$$E\left(s_y^2 \right) = S_y^2, \tag{2.6}$$

where

$$S_y^2 = \frac{1}{N-1} \sum_{i \in U} (y_i - \overline{Y})^2 = \frac{N}{N-1} \sigma_y^2,$$

and

$$\overline{Y} = \frac{1}{N} \sum_{k \in U} y_k.$$

In fact, since s_y^2 can also be written (see Exercise 2.7),

$$s_y^2 = \frac{1}{2n(n-1)} \sum_{i \in S} \sum_{j \in S} (y_i - y_j)^2,$$

we get

$$E(s_y^2) = \frac{1}{2n(n-1)} \sum_{i \in U} \sum_{j \in U} (y_i - y_j)^2 E(I_i I_j)$$

$$= \frac{1}{2N(N-1)} \sum_{i \in U} \sum_{j \in U} (y_i - y_j)^2 = S_y^2.$$

To calculate the variance of s_y^2 following the sampling, we suppose that the population mean $\overline{Y}$ is null, without sacrificing the general nature of the solution (we can still set $Y_i = Z_i + \overline{Y}$, with $\overline{Z} = 0$). We also denote

$$\mu_4 = \frac{1}{N} \sum_{i \in U} (y_i - \overline{Y})^4.$$

Preliminary calculations
We will subsequently use the following four results:

1. If $\overline{Y} = 0$, then

$$\frac{1}{N^2} \sum_{i \in U} \sum_{\substack{j \in U \\ j \neq i}} y_i^2 y_j^2 = \sigma_y^4 - \frac{\mu_4}{N}. \tag{2.7}$$

In fact,

$$\frac{1}{N^2} \sum_{i \in U} \sum_{\substack{j \in U \\ j \neq i}} y_i^2 y_j^2 = \frac{1}{N^2} \sum_{i \in U} \sum_{j \in U} y_i^2 y_j^2 - \frac{1}{N^2} \sum_{i \in U} y_i^4 = \sigma_y^4 - \frac{\mu_4}{N}.$$

2. If $\overline{Y} = 0$, then

$$\frac{1}{N} \sum_{i \in U} \sum_{\substack{j \in U \\ j \neq i}} y_i^3 y_j = -\mu_4. \tag{2.8}$$

In fact, seeing as $\sum_{j \in U} y_j = 0$,

$$\frac{1}{N} \sum_{i \in U} \sum_{\substack{j \in U \\ j \neq i}} y_i^3 y_j = \frac{1}{N} \sum_{i \in U} \sum_{j \in U} y_i^3 y_j - \frac{1}{N} \sum_{i \in U} y_i^4 = -\mu_4.$$

3. If $\overline{Y} = 0$, then

$$\frac{1}{N^2} \sum_{i \in U} \sum_{\substack{j \in U \\ j \neq i}} \sum_{\substack{k \in U \\ k \neq i \\ k \neq j}} y_i^2 y_j y_k = \frac{2\mu_4}{N} - \sigma_y^4. \tag{2.9}$$

Indeed, as

$$\frac{1}{N^2} \sum_{i \in U} \sum_{j \in U} \sum_{k \in U} y_i^2 y_j y_k = 0 = \frac{1}{N^2} \sum_{i \in U} y_i^4 + \frac{1}{N^2} \sum_{i \in U} \sum_{\substack{j \in U \\ j \neq i}} y_i^2 y_j^2$$

$$+ \frac{2}{N^2} \sum_{i \in U} \sum_{\substack{j \in U \\ j \neq i}} y_i^3 y_j + \frac{1}{N^2} \sum_{i \in U} \sum_{\substack{j \in U \\ j \neq i}} \sum_{\substack{k \in U \\ k \neq i \\ k \neq j}} y_i^2 y_j y_k,$$

with the results from (2.7) and (2.9) we have:

$$0 = \frac{\mu_4}{N} + \sigma_y^4 - \frac{\mu_4}{N} - 2\frac{\mu_4}{N} + \frac{1}{N^2} \sum_{i \in U} \sum_{\substack{j \in U \\ j \neq i}} \sum_{\substack{k \in U \\ k \neq i \\ k \neq j}} y_i^2 y_j y_k.$$

Therefore,

$$\frac{1}{N^2} \sum_{i \in U} \sum_{\substack{j \in U \\ j \neq i}} \sum_{\substack{k \in U \\ k \neq i \\ k \neq j}} y_i^2 y_j y_k = \frac{2\mu_4}{N} - \sigma_y^4.$$

4. If $\overline{Y} = 0$, then

$$\frac{1}{N^2} \sum_{i \in U} \sum_{\substack{j \in U \\ j \neq i}} \sum_{\substack{k \in U \\ k \neq i \\ k \neq j}} \sum_{\substack{\ell \in U \\ \ell \neq i \\ \ell \neq j \\ \ell \neq k}} y_i y_j y_k y_\ell = -3 \left(\frac{2\mu_4}{N} - \sigma_y^4 \right).$$

In fact, since

$$\frac{1}{N^2} \sum_{i \in U} \sum_{j \in U} \sum_{k \in U} \sum_{\ell \in U} y_i y_j y_k y_\ell = 0$$

$$= \frac{1}{N^2} \sum_{i \in U} y_i^4 + \frac{3}{N^2} \sum_{i \in U} \sum_{\substack{j \in U \\ j \neq i}} y_i^2 y_j^2 + \frac{4}{N^2} \sum_{i \in U} \sum_{\substack{j \in U \\ j \neq i}} y_i^3 y_j$$

$$+ \frac{6}{N^2} \sum_{i \in U} \sum_{\substack{j \in U \\ j \neq i}} \sum_{\substack{k \in U \\ k \neq i \\ k \neq j}} y_i^2 y_j y_k + \frac{1}{N^2} \sum_{i \in U} \sum_{\substack{j \in U \\ j \neq i}} \sum_{\substack{k \in U \\ k \neq i \\ k \neq j}} \sum_{\substack{\ell \in U \\ \ell \neq i \\ \ell \neq j \\ \ell \neq k}} y_i y_j y_k y_\ell,$$

by the results from (2.7), (2.8), and (2.9), we have

$$0 = \frac{\mu_4}{N} + 3 \left(\sigma_y^4 - \frac{\mu_4}{N} \right) - \frac{4\mu_4}{N} + 6 \left(\frac{2\mu_4}{N} - \sigma_y^4 \right)$$

$$+ \frac{1}{N^2} \sum_{i \in U} \sum_{\substack{j \in U \\ j \neq i}} \sum_{\substack{k \in U \\ k \neq i \\ k \neq j}} \sum_{\substack{\ell \in U \\ \ell \neq i \\ \ell \neq j \\ \ell \neq k}} y_i y_j y_k y_\ell.$$

Thus

$$\frac{1}{N^2} \sum_{i \in U} \sum_{\substack{j \in U \\ j \neq i}} \sum_{\substack{k \in U \\ k \neq i \\ k \neq j}} \sum_{\substack{\ell \in U \\ \ell \neq i \\ \ell \neq j \\ \ell \neq k}} y_i y_j y_k y_\ell = -3 \left(\frac{2\mu_4}{N} - \sigma_y^4 \right).$$

These preliminary calculations will be used to calculate the variance which can be divided into two parts according to:

$$\mathrm{var} \left(s_y^2 \right) = \mathrm{E} \left(s_y^4 \right) - \left\{ \mathrm{E} \left(s_y^2 \right) \right\}^2.$$

Since $\mathrm{E} \left(s_y^2 \right)$ is given by (2.6), we must calculate

$$\mathrm{E} \left(s_y^4 \right) = \mathrm{E} \left(\frac{1}{n-1} \sum_{i \in S} y_i^2 - \frac{n}{n-1} \widehat{\overline{Y}}^2 \right)^2$$

$$= \frac{n^2}{(n-1)^2} \mathrm{E} \left(\frac{1}{n} \sum_{i \in S} y_i^2 - \widehat{\overline{Y}}^2 \right)^2$$

$$= \frac{n^2}{(n-1)^2} \left(A - 2B + C \right),$$

where

$$A = \mathrm{E}\left(\frac{1}{n}\sum_{i \in S} y_i^2\right)^2, \quad B = \mathrm{E}\left(\frac{1}{n}\sum_{i \in S} y_i^2\right)\widehat{\overline{Y}}^2, \quad \text{and} \quad C = \mathrm{E}\left(\widehat{\overline{Y}}^4\right).$$

Calculation of the 3 terms A, B and C

1. Calculation of A

$$A = \mathrm{E}\left(\frac{1}{n}\sum_{i \in S} y_i^2\right)^2 = \mathrm{E}\left(\frac{1}{n^2}\sum_{i \in S} y_i^4 + \frac{1}{n^2}\sum_{i \in S}\sum_{\substack{j \in S \\ j \neq i}} y_i^2 y_j^2\right)$$

$$= \frac{1}{n^2}\sum_{i \in U} y_i^4 \pi_1 + \frac{1}{n^2}\sum_{i \in U}\sum_{\substack{j \in U \\ j \neq i}} y_i^2 y_j^2 \pi_2.$$

By Result (2.7),

$$A = \frac{N\pi_1\mu_4}{n^2} + \frac{N^2\pi_2}{n^2}\left(\sigma_y^4 - \frac{\mu_4}{N}\right) = \frac{N(\pi_1 - \pi_2)}{n^2}\mu_4 + \frac{N^2\pi_2}{n^2}\sigma_y^4. \qquad (2.10)$$

2. Calculation of B

$$B = \mathrm{E}\left(\frac{1}{n}\sum_{i \in S} y_i^2\right)\widehat{\overline{Y}}^2 = \mathrm{E}\left(\frac{1}{n^3}\sum_{i \in S}\sum_{j \in S}\sum_{k \in S} y_i^2 y_j y_k\right)$$

$$= \mathrm{E}\left(\frac{1}{n^3}\sum_{i \in S} y_i^4\right) + \mathrm{E}\left(\frac{1}{n^3}\sum_{i \in S}\sum_{\substack{j \in S \\ j \neq i}} y_i^2 y_j^2\right)$$

$$+ \mathrm{E}\left(\frac{1}{n^3}\sum_{i \in S}\sum_{\substack{j \in S \\ j \neq i}}\sum_{\substack{k \in S \\ k \neq i \\ k \neq j}} y_i^2 y_j y_k\right) + \mathrm{E}\left(\frac{2}{n^3}\sum_{i \in S}\sum_{\substack{j \in S \\ j \neq i}} y_i^3 y_j\right)$$

$$= \frac{1}{n^3}\sum_{i \in U} y_i^4 \pi_1 + \frac{1}{n^3}\sum_{i \in U}\sum_{\substack{j \in U \\ j \neq i}} y_i^2 y_j^2 \pi_2$$

$$+ \frac{1}{n^3}\sum_{i \in U}\sum_{\substack{j \in U \\ j \neq i}}\sum_{\substack{k \in U \\ k \neq i \\ k \neq j}} y_i^2 y_j y_k \pi_3 + \frac{2}{n^3}\sum_{i \in U}\sum_{\substack{j \in U \\ j \neq i}} y_i^3 y_j \pi_2.$$

Through Results (2.7), (2.8), and (2.9), we get:

$$B = \frac{A}{n} + \frac{N^2 \pi_3}{n^3}\left(\frac{2\mu_4}{N} - \sigma_y^4\right) - \frac{2N\pi_2}{n^3}\mu_4$$

$$= \frac{A}{n} + \frac{2N(\pi_3 - \pi_2)}{n^3}\mu_4 - \frac{N^2\pi_3}{n^3}\sigma_y^4$$

$$= \frac{N(\pi_1 - 3\pi_2 + 2\pi_3)}{n^3}\mu_4 + \frac{N^2(\pi_2 - \pi_3)}{n^3}\sigma_y^4. \qquad (2.11)$$

3. Calculation of C

$$C = E\left(\widehat{\overline{Y}}^4\right)$$

$$= E\left(\frac{1}{n^4}\sum_{i \in S}\sum_{j \in S}\sum_{k \in S}\sum_{\ell \in S} y_i y_j y_k y_\ell\right)$$

$$= E\left(\frac{1}{n^4}\sum_{i \in S} y_i^4\right) + E\left(\frac{3}{n^4}\sum_{i \in S}\sum_{\substack{j \in S \\ j \neq i}} y_i^2 y_j^2\right) + E\left(\frac{4}{n^4}\sum_{i \in S}\sum_{\substack{j \in S \\ j \neq i}} y_i^3 y_j\right)$$

$$+ E\left(\frac{6}{n^4}\sum_{i \in S}\sum_{\substack{j \in S \\ j \neq i}}\sum_{\substack{k \in S \\ k \neq i \\ k \neq j}} y_i^2 y_j y_k\right) + E\left(\frac{1}{n^4}\sum_{i \in S}\sum_{\substack{j \in S \\ j \neq i}}\sum_{\substack{k \in S \\ k \neq i \\ k \neq j}}\sum_{\substack{\ell \in S \\ \ell \neq i \\ \ell \neq j \\ \ell \neq k}} y_i y_j y_k y_\ell\right)$$

By calculating the expectations, we have

$$C = \frac{1}{n^4}\sum_{i \in U} y_i^4 \pi_1 + \frac{3}{n^4}\sum_{i \in U}\sum_{\substack{j \in U \\ j \neq i}} y_i^2 y_j^2 \pi_2 + \frac{4}{n^4}\sum_{i \in U}\sum_{\substack{j \in U \\ j \neq i}} y_i^3 y_j \pi_2$$

$$+ \frac{6}{n^4}\sum_{i \in U}\sum_{\substack{j \in U \\ j \neq i}}\sum_{\substack{k \in U \\ k \neq i \\ k \neq j}} y_i^2 y_j y_k \pi_3 + \frac{1}{n^4}\sum_{i \in U}\sum_{\substack{j \in U \\ j \neq i}}\sum_{\substack{k \in U \\ k \neq i \\ k \neq j}}\sum_{\substack{\ell \in U \\ \ell \neq i \\ \ell \neq j \\ \ell \neq k}} y_i y_j y_k y_\ell \pi_4.$$

Finally, by Results (2.7), (2.8), and (2.9), we get:

$$C = \frac{N\pi_1}{n^4}\mu_4 + \frac{3N^2\pi_2}{n^4}\left(\sigma_y^4 - \frac{\mu_4}{N}\right) - \frac{4N\pi_2}{n^4}\mu_4$$

$$+ \frac{6N^2\pi_3}{n^4}\left(\frac{2\mu_4}{N} - \sigma_y^4\right) - \frac{3N^2\pi_4}{n^4}\left(\frac{2\mu_4}{N} - \sigma_y^4\right)$$

$$= \frac{N(\pi_1 - 7\pi_2 + 12\pi_3 - 6\pi_4)}{n^4}\mu_4 + \frac{3N^2(\pi_2 - 2\pi_3 + \pi_4)}{n^4}\sigma_y^4. \qquad (2.12)$$

From Expressions (2.10), (2.11), (2.12) and (2.6), we finally have the variance of the estimator of the population variance.

$$
\begin{aligned}
&\text{var}(s_y^2) \\
&= \frac{n^2}{(n-1)^2}\left(A - 2B + C\right) - S_y^4 \\
&= \frac{n^2}{(n-1)^2}\left\{ \frac{N(\pi_1 - \pi_2)}{n^2}\mu_4 + \frac{N^2\pi_2}{n^2}\sigma_y^4 \right. \\
&\quad - 2\left[\frac{N(\pi_1 - 3\pi_2 + 2\pi_3)}{n^3}\mu_4 + \frac{N^2(\pi_2 - \pi_3)}{n^3}\sigma_y^4\right] \\
&\quad + \left. \frac{N(\pi_1 - 7\pi_2 + 12\pi_3 - 6\pi_4)}{n^4}\mu_4 + \frac{3N^2(\pi_2 - 2\pi_3 + \pi_4)}{n^4}\sigma_y^4 \right\} - S_y^4 \\
&= \frac{N(N-n)}{n(n-1)(N-1)^2(N-2)(N-3)} \\
&\quad \times \left\{ \mu_4(N-1)\left[N(n-1) - (n+1)\right] - \sigma_y^4\left[N^2(n-3) + 6N - 3(n+1)\right] \right\}.
\end{aligned}
\tag{2.13}
$$

With simple random sampling, we estimate the sampling variance by:

$$
\widehat{\text{var}}(\widehat{\overline{Y}}) = \left(1 - \frac{n}{N}\right)\frac{s_y^2}{n},
$$

an estimator that has the sampling variance:

$$
\text{var}(\widehat{\text{var}}(\widehat{\overline{Y}})) = \left(1 - \frac{n}{N}\right)^2 \frac{1}{n^2}\text{var}(s_y^2),
$$

where $\text{var}(s_y^2)$ is defined in (2.13). So, this expression is surprisingly complex for a problem that *a priori* had appeared to be simple. If N approaches toward infinity (in practice N is 'very large'), we get the valuable expression for a design with replacement:

$$
\text{var}(s_y^2) \approx \frac{1}{n}\left\{\mu_4 - \frac{n-3}{n-1}\sigma_y^4\right\}.
\tag{2.14}
$$

If the variable y has a normal distribution in population U, then we know furthermore that $\mu_4 = 3\sigma_y^4$, and we get

$$
\text{var}(s_y^2) \approx \frac{2\sigma_y^4}{n-1}.
$$

Finally, in the case:

$$
\text{var}(\widehat{\text{var}}(\widehat{\overline{Y}})) \approx \left(1 - \frac{n}{N}\right)^2 \frac{1}{n^2}\frac{2\sigma_y^4}{n-1}.
$$

The standard deviation of $\widehat{\text{var}}(\overline{Y})$ varies by $1/n^{3/2}$. If n is large, this standard deviation is $\sqrt{n}$ times smaller than $\widehat{\text{var}}(\overline{Y})$: this is the reason for which in practice we content ourselves with the calculation of $\widehat{\text{var}}(\overline{Y})$, which we judge to be sufficiently accurate.

3

Sampling with Unequal Probabilities

3.1 Calculation of inclusion probabilities

If we have an auxiliary variable $x_k > 0$, $k \in U$, 'sufficiently' proportional to the variable y_k, it is often interesting to select the units with unequal probabilities proportional to x_k. To do this, we first calculate the inclusion probabilities according to

$$\pi_k = n \frac{x_k}{\sum_{\ell \in U} x_\ell}. \tag{3.1}$$

If Expression (3.1) gives $\pi_k > 1$, the corresponding units are selected in the sample (with an inclusion probability equal to 1), and we then recalculate the π_k according to (3.1) on the remaining units.

3.2 Estimation and variance

The Horvitz-Thompson estimator of the total is

$$\widehat{Y}_\pi = \sum_{k \in S} \frac{y_k}{\pi_k},$$

and its variance is:

$$\text{var}(\widehat{Y}_\pi) = \sum_{k \in U} \sum_{\ell \in U} \frac{y_k}{\pi_k} \frac{y_\ell}{\pi_\ell} \Delta_{k\ell},$$

where $\Delta_{k\ell} = \pi_{k\ell} - \pi_k \pi_\ell$, and $\pi_{k\ell}$ is the second-order inclusion probability. If $k = \ell$, then $\pi_{kk} = \pi_k$. To obtain a positive estimate of the variance (see page 4), a sufficient constraint is to have $\Delta_{k\ell} \leq 0$ for all $k \neq \ell$ in U. This constraint is called the Sen-Yates-Grundy constraint.

There exist several algorithms that allow for the selection of units with unequal probabilities. Two books give a brief overview of such methods: Brewer and Hanif (1983) and Gabler (1990). The most well-known methods are systematic sampling (Madow, 1948), sampling with replacement (Hansen and Hurwitz, 1943), the method of Sunter (1977) and Sunter (1986). As well, the method of Brewer (1975) presents an interesting approach. The representation through a splitting method (see on this topic Deville and Tillé, 1998) allows for the rewriting of methods in a standardised manner and the creation of new algorithms.

EXERCISES

Exercise 3.1 *Design and inclusion probabilities*

Let there exist a population $U = \{1, 2, 3\}$ with the following design:

$$p(\{1, 2\}) = \frac{1}{2}, p(\{1, 3\}) = \frac{1}{4}, p(\{2, 3\}) = \frac{1}{4}.$$

Give the first-order inclusion probabilities. Give the variance-covariance matrix $\boldsymbol{\Delta}$ of indicator variables for inclusion in the sample. Give the variance matrix of the unbiased estimator for the total.

Solution

Clearly, we have:

$$\pi_1 = \frac{3}{4}, \pi_2 = \frac{3}{4}, \pi_3 = \frac{1}{2}.$$

Notice that $\pi_1 + \pi_2 + \pi_3 = 2$. In fact, the design is of fixed size and $n = 2$. Finally, we directly obtain the

$$\Delta_{k\ell} = \operatorname{cov}(I_k, I_\ell) = \begin{cases} \pi_{k\ell} - \pi_k \pi_\ell & \text{if } k \neq \ell \\ \pi_k(1 - \pi_k) & \text{if } k = \ell \end{cases}$$

$$\Delta_{11} = \frac{3}{4}\left(1 - \frac{3}{4}\right) = \frac{3}{16}, \quad \Delta_{12} = \frac{1}{2} - \frac{3}{4} \times \frac{3}{4} = \frac{-1}{16},$$

$$\Delta_{13} = \frac{1}{4} - \frac{3}{4} \times \frac{1}{2} = \frac{-1}{8}, \quad \Delta_{22} = \frac{3}{4}\left(1 - \frac{3}{4}\right) = \frac{3}{16},$$

$$\Delta_{23} = \frac{1}{4} - \frac{3}{4} \times \frac{1}{2} = \frac{-1}{8}, \quad \Delta_{33} = \frac{1}{2}\left(1 - \frac{1}{2}\right) = \frac{1}{4},$$

which gives the positive symmetric matrix:

$$\boldsymbol{\Delta} = \begin{pmatrix} 3/16 & -1/16 & -1/8 \\ -1/16 & 3/16 & -1/8 \\ -1/8 & -1/8 & 1/4 \end{pmatrix}.$$

If we denote $\mathbf{u}$ as the column vector of $y_k/\pi_k, k = 1, \ldots, N$, and $\mathbf{1}$ as the column vector of $I_k, k = 1, \ldots, N$, we have

$$\text{var}\left(\sum_{k \in S} \frac{y_k}{\pi_k}\right) = \text{var}(\mathbf{u}'\mathbf{1}) = \mathbf{u}'\text{var}(\mathbf{1})\mathbf{u} = \mathbf{u}'\mathbf{\Delta}\mathbf{u}.$$

Exercise 3.2 *Variance of indicators and design of fixed size*

Given a sampling design for a population U, we denote I_k as the random indicator variable for the presence of unit k in the sample, and

$$\Delta_{k\ell} = \begin{cases} \text{var}(I_k) & \text{if } \ell = k \\ \text{cov}(I_k, I_\ell) & \text{if } k \neq \ell. \end{cases}$$

Show that if

$$\sum_{k \in U} \sum_{\ell \in U} \Delta_{k\ell} = 0,$$

then the design is of fixed size.

Solution

Denoting n_S as the size, *a priori* random, of the sample S:

$$\sum_{k \in U} \sum_{\ell \in U} \Delta_{k\ell} = \sum_{k \in U} \sum_{\ell \in U} \text{cov}(I_k, I_\ell) = \text{var}\sum_{k \in U} I_k = \text{var}(n_S).$$

$\text{var}(n_S) = 0$ implies that the design is of fixed size.

Exercise 3.3 *Variance of indicators and sampling design*

Consider the variance-covariance matrix $\mathbf{\Delta} = [\Delta_{k\ell}]$ of indicators for the presence of observation units in the sample for a design $p(s)$,

$$\mathbf{\Delta} = \begin{pmatrix} 1 & 1 & 1 & -1 & -1 \\ 1 & 1 & 1 & -1 & -1 \\ 1 & 1 & 1 & -1 & -1 \\ -1 & -1 & -1 & 1 & 1 \\ -1 & -1 & -1 & 1 & 1 \end{pmatrix} \times \frac{6}{25}.$$

1. Is this a design of fixed size?
2. Does this design satisfy the Sen-Yates-Grundy constraints?
3. Calculate the inclusion probabilities of this design knowing that

$$\pi_1 = \pi_2 = \pi_3 > \pi_4 = \pi_5.$$

4. Give the second-order inclusion probability matrix.
5. Give the probabilities associated with all possible samples.

Solution

1. If we denote I_k as the indicator random variable for the presence of unit k in the sample, we have:

$$\Delta_{k\ell} = \text{cov}\left(I_k, I_\ell\right).$$

If the design is of fixed size,

$$\sum_{k \in U} I_k = n,$$

(with n fixed). We then have, for all $\ell \in U$:

$$\sum_{k \in U} \Delta_{k\ell} = \sum_{k \in U} \text{cov}\left(I_k, I_\ell\right) = \text{cov}\left(\sum_{k \in U} I_k, I_\ell\right) = \text{cov}\left(n, I_\ell\right) = 0.$$

In a design of fixed size, the sum of all rows and the sum of all columns in $\Delta_{k\ell}$ are null. We immediately confirm that this is not the case here, and thus the design is not of fixed size.

2. No, because we have some $\Delta_{k\ell} > 0$ for $k \neq \ell$.
3. Since $\text{var}(I_k) = \pi_k(1 - \pi_k) = 6/25$ for all k, we have

$$\pi_k^2 - \pi_k + \frac{6}{25} = 0.$$

Therefore

$$\pi_k = \frac{1 \pm \sqrt{1 - 4 \times \frac{6}{25}}}{2} = \frac{1 \pm \frac{1}{5}}{2},$$

and

$$\pi_1 = \pi_2 = \pi_3 = \frac{3}{5} > \pi_4 = \pi_5 = \frac{2}{5}.$$

4. Since $\pi_{k\ell} = \Delta_{k\ell} + \pi_k \pi_\ell$, for all $k, \ell \in U$, if we let $\boldsymbol{\pi}$ be the column vector of $\pi_k, k \in U$, the second-order inclusion probability matrix is:

$$\boldsymbol{\Pi} = \boldsymbol{\Delta} + \boldsymbol{\pi}\boldsymbol{\pi}'$$

$$= \begin{pmatrix} 1 & 1 & 1 & -1 & -1 \\ 1 & 1 & 1 & -1 & -1 \\ 1 & 1 & 1 & -1 & -1 \\ -1 & -1 & -1 & 1 & 1 \\ -1 & -1 & -1 & 1 & 1 \end{pmatrix} \times \frac{6}{25} + \begin{pmatrix} 9 & 9 & 9 & 6 & 6 \\ 9 & 9 & 9 & 6 & 6 \\ 9 & 9 & 9 & 6 & 6 \\ 6 & 6 & 6 & 4 & 4 \\ 6 & 6 & 6 & 4 & 4 \end{pmatrix} \times \frac{1}{25}$$

$$= \begin{pmatrix} 3 & 3 & 3 & 0 & 0 \\ 3 & 3 & 3 & 0 & 0 \\ 3 & 3 & 3 & 0 & 0 \\ 0 & 0 & 0 & 2 & 2 \\ 0 & 0 & 0 & 2 & 2 \end{pmatrix} \times \frac{1}{5}.$$

5. On the one hand, the second-order inclusion probabilities equal to zero show that certain pairs of units cannot be selected (such as unit 1 with unit 4). On the other hand, certain units are always selected together. Indeed,

$$\Pr(2 \in S | 1 \in S) = \frac{\pi_{12}}{\pi_1} = 1, \quad \text{and} \quad \Pr(3 \in S | 1 \in S) = \frac{\pi_{13}}{\pi_1} = 1.$$

Therefore if unit 1 is selected, units 2 and 3 are selected as well. Likewise if unit 4 is selected, unit 5 is selected as well. By following this reasoning, we see that units 1, 2, and 3 are always selected together, and units 4 and 5 as well. The only two samples having a strictly positive probability are $\{1, 2, 3\}, \{4, 5\}$. The probabilities associated with all the possible samples are given by:

$$p(\{1, 2, 3\}) = \pi_1 = \frac{3}{5}, \quad p(\{4, 5\}) = \pi_4 = \frac{2}{5},$$

and are null for all other samples.

Exercise 3.4 *Estimation of a square root*

Consider a population of 5 individuals. We are interested in a characteristic of interest y which takes the values:

$$y_1 = y_2 = 1, \text{ and } y_3 = y_4 = y_5 = \frac{8}{3}.$$

We define the following design:

$$p(\{1, 2\}) = \frac{1}{2}, p(\{3, 4\}) = p(\{3, 5\}) = p(\{4, 5\}) = \frac{1}{6}.$$

1. Calculate the first- and second-order inclusion probabilities.
2. Give the probability distribution of the π-estimator of the total.
3. Calculate the variance estimator with the Sen-Yates-Grundy expression (we verify that the design is indeed of fixed size). Is this estimator biased? Could we have foreseen this?
4. We propose to estimate the square root of the total (denoted $\sqrt{Y}$), using the square root of the π-estimator $\sqrt{\widehat{Y}_\pi}$. Give the probability distribution of this estimator. Show that it underestimates $\sqrt{Y}$. Could we have foreseen this?
5. Calculate the variance of $\sqrt{\widehat{Y}_\pi}$.

Solution

1. The inclusion probabilities are

$$\pi_1 = \pi_2 = \frac{1}{2}, \quad \pi_3 = \pi_4 = \pi_5 = \frac{1}{3},$$

$$\pi_{12} = \frac{1}{2}, \quad \pi_{34} = \pi_{35} = \pi_{45} = \frac{1}{6}, \pi_{k\ell} = 0, \text{ for all other pairs } (k, \ell).$$

2. The Horvitz-Thompson estimator is

$$\widehat{Y}_\pi = \begin{cases} \dfrac{1}{1/2} + \dfrac{1}{1/2} = 4 & \text{with a probability } \dfrac{1}{2} \\ \dfrac{8/3}{1/3} + \dfrac{8/3}{1/3} = 16 & \text{with a probability } \dfrac{1}{6} + \dfrac{1}{6} + \dfrac{1}{6} = \dfrac{1}{2}. \end{cases}$$

3. The samples all being of size 2, the variance estimator to calculate when we select i and j is

$$\widehat{\text{var}}\left(\widehat{Y}_\pi\right) = \left(\frac{y_i}{\pi_i} - \frac{y_j}{\pi_j}\right)^2 \frac{\pi_i \pi_j - \pi_{ij}}{\pi_{ij}}.$$

Since $\pi_1 = \pi_2 = 1/2$ and $\pi_3 = \pi_4 = \pi_5 = 2 \times 1/6 = 1/3$, and considering the values of y_i, the results are given in Table 3.1. We obtain

Table 3.1. Estimated variances for the samples: Exercise 3.4

s	$p(s)$	$\widehat{\text{var}}(\widehat{Y}_\pi)$
$\{1,2\}$	1/2	0
$\{3,4\}$	1/6	0
$\{3,5\}$	1/6	0
$\{4,5\}$	1/6	0

$\widehat{\text{var}}\left(\widehat{Y}_\pi\right) = 0$, for each possible sample, and thus $\text{E}\,\widehat{\text{var}}\left(\widehat{Y}_\pi\right) = 0$. In fact, it is obvious that $\text{var}\left(\widehat{Y}_\pi\right) > 0$, since $\widehat{Y}_\pi$ varies depending on the selected sample. Therefore $\widehat{\text{var}}\left(\widehat{Y}_\pi\right)$ is biased. The bias follows from the existence of second-order inclusion probabilities equal to zero (see Exercise 3.23).

4. Since

$$\sqrt{\widehat{Y}_\pi} = \begin{cases} 2 & \text{with a probability } 1/2 \\ 4 & \text{with a probability } 1/2, \end{cases}$$

$$\text{E}\left(\sqrt{\widehat{Y}_\pi}\right) = 2 \times \frac{1}{2} + 4 \times \frac{1}{2} = 3 < \sqrt{10} = \sqrt{Y}.$$

There is thus underestimation. This result was foreseeable, as the square root is a concave function, and we know that for each concave function ϕ, we have

$$\mathrm{E}\left[\phi(X)\right] < \phi\left[\mathrm{E}(X)\right].$$

5. The variance is given by

$$\mathrm{var}\left(\sqrt{\widehat{Y}_\pi}\right) = \mathrm{E}\left(\widehat{Y}_\pi\right) - \left[\mathrm{E}\left(\sqrt{\widehat{Y}_\pi}\right)\right]^2 = 10 - 9 = 1.$$

Exercise 3.5 *Variance and concurrent estimates of variance*

Consider a population $U = \{1, 2, 3\}$ and the following design:

$$p(\{1,2\}) = \frac{1}{2}, p(\{1,3\}) = \frac{1}{4}, p(\{2,3\}) = \frac{1}{4}.$$

1. Give the probability distribution of the π-estimator and the Hájek ratio of the mean.
2. Give the probability distributions of the two classical variance estimators of the π-estimator in the case where $y_k = \pi_k, k \in U$.

Solution

1. The first- and second-order inclusion probabilities are respectively

$$\pi_1 = 3/4, \quad \pi_2 = 3/4, \quad \pi_3 = 1/2,$$

and

$$\pi_{12} = 1/2, \quad \pi_{13} = 1/4, \quad \pi_{23} = 1/4.$$

The probability distribution of the π-estimator of the mean is given by

$$\widehat{\overline{Y}}_\pi = \begin{cases} \dfrac{1}{N}\left(\dfrac{y_1}{\pi_1} + \dfrac{y_2}{\pi_2}\right) = \dfrac{1}{3}\left(\dfrac{y_1}{3/4} + \dfrac{y_2}{3/4}\right) & \text{if } S = \{1,2\} \\[2mm] \dfrac{1}{N}\left(\dfrac{y_1}{\pi_1} + \dfrac{y_3}{\pi_3}\right) = \dfrac{1}{3}\left(\dfrac{y_1}{3/4} + \dfrac{y_3}{1/2}\right) & \text{if } S = \{1,3\} \\[2mm] \dfrac{1}{N}\left(\dfrac{y_2}{\pi_2} + \dfrac{y_3}{\pi_3}\right) = \dfrac{1}{3}\left(\dfrac{y_2}{3/4} + \dfrac{y_3}{1/2}\right) & \text{if } S = \{2,3\}, \end{cases}$$

which gives

$$\widehat{\overline{Y}}_\pi = \begin{cases} \dfrac{4}{9}\left(y_1 + y_2\right) & \text{with a probability } \dfrac{1}{2} \\[2mm] \dfrac{1}{9}\left(4y_1 + 6y_3\right) & \text{with a probability } \dfrac{1}{4} \\[2mm] \dfrac{1}{9}\left(4y_2 + 6y_3\right) & \text{with a probability } \dfrac{1}{4}, \end{cases}$$

which comes back to saying that the population size N cannot be estimated with a null variance. The π-estimator of the mean is such that the sum of the weights of the observations is not equal to 1. Nonetheless, this estimator is unbiased. Indeed,

$$\mathrm{E}\left(\widehat{\overline{Y}}_\pi\right) = \frac{1}{2} \times \frac{4}{9}(y_1 + y_2) + \frac{1}{4} \times \frac{1}{9}(4y_1 + 6y_3) + \frac{1}{4} \times \frac{1}{9}(4y_2 + 6y_3)$$
$$= \frac{1}{3}(y_1 + y_2 + y_3) = \overline{Y}.$$

For the Hájek ratio,

- if $S = \{1, 2\}$, then

$$\widehat{\overline{Y}}_H = \left(\frac{1}{\pi_1} + \frac{1}{\pi_2}\right)^{-1}\left(\frac{y_1}{\pi_1} + \frac{y_2}{\pi_2}\right)$$
$$= \left(\frac{1}{3/4} + \frac{1}{3/4}\right)^{-1}\left(\frac{y_1}{3/4} + \frac{y_2}{3/4}\right)$$
$$= \frac{1}{2}(y_1 + y_2),$$

- if $S = \{1, 3\}$, then

$$\widehat{\overline{Y}}_H = \left(\frac{1}{\pi_1} + \frac{1}{\pi_3}\right)^{-1}\left(\frac{y_1}{\pi_1} + \frac{y_3}{\pi_3}\right)$$
$$= \left(\frac{1}{3/4} + \frac{1}{1/2}\right)^{-1}\left(\frac{y_1}{3/4} + \frac{y_3}{1/2}\right)$$
$$= \frac{1}{5}(2y_1 + 3y_3),$$

- if $S = \{2, 3\}$, then

$$\widehat{\overline{Y}}_H = \left(\frac{1}{\pi_2} + \frac{1}{\pi_3}\right)^{-1}\left(\frac{y_2}{\pi_2} + \frac{y_3}{\pi_3}\right)$$
$$= \left(\frac{1}{3/4} + \frac{1}{1/2}\right)^{-1}\left(\frac{y_2}{3/4} + \frac{y_3}{1/2}\right)$$
$$= \frac{1}{5}(2y_2 + 3y_3).$$

The probability distribution of the Hájek ratio is thus given by:

$$\widehat{\overline{Y}}_H = \begin{cases} \dfrac{1}{2}(y_1 + y_2) & \text{if } S = \{1, 2\}, \text{ with a probability } \dfrac{1}{2} \\[2mm] \dfrac{1}{5}(2y_1 + 3y_3) & \text{if } S = \{1, 3\}, \text{ with a probability } \dfrac{1}{4} \\[2mm] \dfrac{1}{5}(2y_2 + 3y_3) & \text{if } S = \{2, 3\}, \text{ with a probability } \dfrac{1}{4}. \end{cases}$$

Here, the sum of the affected weights of the observations is 1 by construction, in order that the population size N is perfectly estimated (that is, with a null variance). However, the estimator is biased. Indeed,

$$
E\left(\widehat{\overline{Y}}_H\right) = \frac{1}{2} \times \frac{1}{2}(y_1 + y_2) + \frac{1}{4} \times \frac{1}{5}(2y_1 + 3y_3) + \frac{1}{4} \times \frac{1}{5}(2y_2 + 3y_3)
$$

$$
= \frac{1}{20}(7y_1 + 7y_2 + 6y_3) = \overline{Y} + \frac{y_1}{60} + \frac{y_2}{60} - \frac{2y_3}{60}.
$$

2. If $y_k = \pi_k, k \in U$, then $y_k/\pi_k = 1, k \in U$, and

$$
\widehat{\overline{Y}}_\pi = \frac{1}{N}\sum_{k \in S}\frac{y_k}{\pi_k} = \frac{1}{N}\sum_{k \in S}1 = \frac{n}{N},
$$

whatever the selected sample is. We therefore have $\mathrm{var}\left(\widehat{\overline{Y}}_\pi\right) = 0$.
We now calculate the two classical variance estimators. The Sen-Yates-Grundy estimator, which estimates without bias the variance in the case of sampling with fixed size (here $n = 2$), is given by:

$$
\widehat{\mathrm{var}}_2\left(\widehat{\overline{Y}}_\pi\right) = \frac{1}{2N^2}\sum_{k \in S}\sum_{\substack{\ell \in S \\ \ell \neq k}}\left(\frac{y_k}{\pi_k} - \frac{y_\ell}{\pi_\ell}\right)^2\frac{\pi_k\pi_\ell - \pi_{k\ell}}{\pi_{k\ell}}.
$$

Since $y_k/\pi_k = 1, k \in U$, we have

$$
\left(\frac{y_k}{\pi_k} - \frac{y_\ell}{\pi_\ell}\right) = 0,
$$

for all k, ℓ and therefore $\widehat{\mathrm{var}}_2\left(\widehat{\overline{Y}}_\pi\right) = 0$.
The (unbiased) Horvitz-Thompson variance estimator is given by:

$$
\widehat{\mathrm{var}}_1\left(\widehat{\overline{Y}}_\pi\right) = \frac{1}{N^2}\sum_{k \in S}\frac{y_k^2}{\pi_k^2}(1 - \pi_k) + \frac{1}{N^2}\sum_{k \in S}\sum_{\substack{\ell \in S \\ \ell \neq k}}\frac{y_k y_\ell}{\pi_k\pi_\ell}\frac{\pi_{k\ell} - \pi_k\pi_\ell}{\pi_{k\ell}}.
$$

If $S = \{1, 2\}$, and knowing that $y_k = \pi_k, k \in U$, we get

$$
\widehat{\mathrm{var}}_1\left(\widehat{\overline{Y}}_\pi\right) = \frac{1}{N^2}\left((1 - \pi_1) + (1 - \pi_2) + 2 \times \left(1 - \frac{\pi_1\pi_2}{\pi_{12}}\right)\right)
$$

$$
= \frac{1}{N^2}\left(4 - \pi_1 - \pi_2 - 2 \times \frac{\pi_1\pi_2}{\pi_{12}}\right)
$$

$$
= \frac{1}{9}\left(4 - \frac{3}{4} - \frac{3}{4} - 2 \times \frac{3/4 \times 3/4}{1/2}\right) = \frac{1}{36}.
$$

If $S = \{1, 3\}$, inspired by the previous result, we have

$$\widehat{\text{var}}_1\left(\widehat{\overline{Y}}_\pi\right) = \frac{1}{N^2}\left(4 - \pi_1 - \pi_3 - 2 \times \frac{\pi_1\pi_3}{\pi_{13}}\right)$$
$$= \frac{1}{9}\left(4 - \frac{3}{4} - \frac{1}{2} - 2 \times \frac{3/4 \times 1/2}{1/4}\right) = -\frac{1}{36}.$$

Finally, if $S = \{2, 3\}$, we get

$$\widehat{\text{var}}_1\left(\widehat{\overline{Y}}_\pi\right) = \frac{1}{N^2}\left(4 - \pi_2 - \pi_3 - 2 \times \frac{\pi_2\pi_3}{\pi_{23}}\right)$$
$$= \frac{1}{9}\left(4 - \frac{3}{4} - \frac{1}{2} - 2 \times \frac{3/4 \times 1/2}{1/4}\right) = -\frac{1}{36}.$$

The probability distribution of $\widehat{\text{var}}_1\left(\widehat{\overline{Y}}_\pi\right)$ is therefore:

$$\widehat{\text{var}}_1\left(\widehat{\overline{Y}}_\pi\right) = \begin{cases} 1/36 \text{ with a probability of } 1/2 \text{ (if } S = \{1,2\}) \\ -1/36 \text{ with a probability of } 1/2 \text{ (if } S = \{1,3\} \text{ or } \{2,3\}). \end{cases}$$

It is obviously preferable, in the present case, to use $\widehat{\text{var}}_2\left(\widehat{\overline{Y}}_\pi\right)$ which precisely estimates here the variance of the mean estimator. In fact, $\widehat{\text{var}}_1\left(\widehat{\overline{Y}}_\pi\right)$ is unbiased, but on the one hand it has a strictly positive variance, and on the other hand, it can take a negative value, which is unacceptable in practice. However, even if it is manifestly preferable to use $\widehat{\text{var}}_2\left(\widehat{\overline{Y}}_\pi\right)$ in this example, there does not exist a theoretical result that shows that, in general, $\widehat{\text{var}}_2\left(\widehat{\overline{Y}}_\pi\right)$ has a smaller variance than $\widehat{\text{var}}_1\left(\widehat{\overline{Y}}_\pi\right)$.

Exercise 3.6 *Unbiased estimation*

Consider a random design without replacement that is applied to a population U of size N. We denote $\pi_k, k \in U$, and $\pi_{k\ell}, k, \ell \in U, k \neq \ell$, respectively, as the first- and second-order inclusion probabilities, strictly positive, and S as the random sample. Consider the following estimator:

$$\widehat{\theta} = \frac{1}{N^2}\sum_{k \in S}\frac{y_k}{\pi_k} + \frac{1}{N^2}\sum_{k \in S}\sum_{\substack{\ell \in S \\ \ell \neq k}}\frac{y_\ell}{\pi_{k\ell}}.$$

For what function of interest is this estimator unbiased?

Solution
Let I_k be 1 if k is in S and 0 otherwise.

$$
\begin{aligned}
\mathrm{E}\left(\hat{\theta}\right) &= \mathrm{E}\left(\frac{1}{N^2}\sum_{k\in U}\frac{y_k}{\pi_k}I_k + \frac{1}{N^2}\sum_{k\in U}\sum_{\substack{\ell\in U\\\ell\neq k}}\frac{y_\ell}{\pi_{k\ell}}I_kI_\ell\right)\\
&= \frac{1}{N^2}\sum_{k\in U}\frac{y_k}{\pi_k}\mathrm{E}\left(I_k\right) + \frac{1}{N^2}\sum_{k\in U}\sum_{\substack{\ell\in U\\\ell\neq k}}\frac{y_\ell}{\pi_{k\ell}}\mathrm{E}(I_kI_\ell)\\
&= \frac{1}{N^2}\sum_{k\in U}y_k + \frac{1}{N^2}\sum_{k\in U}\sum_{\substack{\ell\in U\\\ell\neq k}}y_\ell = \frac{1}{N^2}\sum_{k\in U}\sum_{\ell\in U}y_\ell = \frac{1}{N}\sum_{k\in U}\overline{Y} = \overline{Y}.
\end{aligned}
$$

Exercise 3.7 *Concurrent estimation of the population variance*

For a design without replacement with strictly positive inclusion probabilities, construct at least two unbiased estimators for σ_y^2. We can use the expression obtained in Exercise 2.7. In this context, why would we try to estimate σ_y^2 without bias?

Solution
Since we can write

$$
\sigma_y^2 = \frac{1}{2N^2}\sum_{k\in U}\sum_{\substack{\ell\in U\\\ell\neq k}}\left(y_k - y_\ell\right)^2,
$$

we directly have an unbiased estimator for σ_y^2 by

$$
\widehat{\sigma}_{y1}^2 = \frac{1}{2N^2}\sum_{k\in S}\sum_{\substack{\ell\in S\\\ell\neq k}}\frac{(y_k - y_\ell)^2}{\pi_{k\ell}},
$$

as the second-order inclusion probabilities $\pi_{k\ell}$ are all strictly positive. However, this estimator is not the only unbiased estimator for σ_y^2. Indeed, we can also write

$$
\begin{aligned}
\sigma_y^2 &= \frac{1}{N}\sum_{k\in U}y_k^2 - \overline{Y}^2 = \frac{1}{N}\sum_{k\in U}y_k^2 - \frac{1}{N^2}\sum_{k\in U}\sum_{\ell\in U}y_ky_\ell\\
&= \frac{1}{N}\sum_{k\in U}y_k^2 - \frac{1}{N^2}\sum_{k\in U}\sum_{\substack{\ell\in U\\\ell\neq k}}y_ky_\ell - \frac{1}{N^2}\sum_{k\in U}y_k^2\\
&= \frac{N-1}{N^2}\sum_{k\in U}y_k^2 - \frac{1}{N^2}\sum_{k\in U}\sum_{\substack{\ell\in U\\\ell\neq k}}y_ky_\ell.
\end{aligned}
$$

This last expression allows us to construct the following estimator:

$$\widehat{\sigma}_{y2}^2 = \frac{N-1}{N^2} \sum_{k \in S} \frac{y_k^2}{\pi_k} - \frac{1}{N^2} \sum_{k \in S} \sum_{\substack{\ell \in S \\ \ell \neq k}} \frac{y_k y_\ell}{\pi_{k\ell}}.$$

These two estimators are generally different. We prefer *a priori* the estimator $\widehat{\sigma}_{y1}^2$ which has the advantage of always being positive and which takes the value zero if y_k is constant on U, but a complete comparative study would call for the calculation of variances for these two estimators.

Another solution consists of writing, from the Horvitz-Thompson mean estimator $\widehat{\overline{Y}}_\pi$:

$$\text{var}\left(\widehat{\overline{Y}}_\pi\right) = \text{E}\left(\widehat{\overline{Y}}_\pi^2\right) - \overline{Y}^2,$$

thus

$$\overline{Y}^2 = \text{E}\left(\widehat{\overline{Y}}_\pi^2\right) - \text{var}\left(\widehat{\overline{Y}}_\pi\right).$$

We can therefore construct an unbiased estimator $\widehat{(\overline{Y}^2)}$ for $\overline{Y}^2$:

$$\widehat{(\overline{Y}^2)} = \widehat{\overline{Y}}_\pi^2 - \widehat{\text{var}}\left(\widehat{\overline{Y}}_\pi\right),$$

where $\widehat{\text{var}}\left(\widehat{\overline{Y}}_\pi\right)$ is an unbiased estimator for $\text{var}\left(\widehat{\overline{Y}}_\pi\right)$. A family of unbiased estimators for σ_y^2 is therefore

$$\widehat{\sigma}_{y3}^2 = \frac{1}{N} \sum_{k \in S} \frac{y_k^2}{\pi_k} - \widehat{\overline{Y}}_\pi^2 + \widehat{\text{var}}\left(\widehat{\overline{Y}}_\pi\right).$$

In the case for designs of fixed size, we know two concurrent expressions for $\widehat{\text{var}}\left(\widehat{\overline{Y}}_\pi\right)$: the Horvitz-Thompson estimator and the Sen-Yates-Grundy estimator.

The testing of an unbiased estimator for σ_y^2 allows us to compare the performance of a complex design of size n with that of a simple random design without replacement of the same size, which serves as a reference design. The variance of a simple design is

$$\text{var}\left(\widehat{\overline{Y}}_{SRS}\right) = \frac{N-n}{Nn} \frac{N}{N-1} \sigma_y^2.$$

After a survey is carried out using a complex design, we are able to estimate $\text{var}\left(\widehat{\overline{Y}}_{SRS}\right)$ if and only if we have an unbiased estimator for σ_y^2. The ratio

$$\text{DEFF} = \frac{\text{var}\left(\widehat{\overline{Y}}_\pi\right)}{\text{var}\left(\widehat{\overline{Y}}_{SRS}\right)},$$

is called the *design effect,* and acts as a performance indicator of a sampling design for a given variable of interest y. The design effect can be estimated as a simple ratio; we get the estimator for $\mathrm{var}\left(\widehat{\overline{Y}}_{SRS}\right)$ by replacing σ_y^2 with one of its unbiased estimators.

Exercise 3.8 *Systematic sampling*

A population is comprised of 6 households with respective sizes 2, 4, 3, 9, 1 and 2 (the size x_k of household k is the number of people included). We select 3 households without replacement, with a probability proportional to its size.

1. Give, in fractional form, the inclusion probabilities of the 6 households in the sampling frame (note: we may recalculate certain probabilities).
2. Carry out the sampling using a systematic method.
3. Using the sample obtained in 2., give an estimation for the mean size $\overline{X}$ of households; was the result predictable?

Solution

1. For all k:
$$\pi_k = 3\frac{x_k}{X}, \text{ with } X = 21.$$
   Therefore
$$\pi_k = \frac{x_k}{7}, k \in U.$$
   A problem arises for unit 4 because $\pi_4 > 1$. We assign the value 1 to π_4 and for the other units we recalculate the $\pi_k, k \neq 4$, according to:
$$\pi_k = 2\frac{x_k}{X - 9} = 2\frac{x_k}{12} = \frac{x_k}{6}.$$
   Finally, the inclusion probabilities are presented in Table 3.2. We can verify that
$$\sum_{k=1}^{6} \pi_k = 3.$$

Table 3.2. Inclusion probabilities: Exercise 3.8

k	1	2	3	4	5	6
π_k	1/3	2/3	1/2	1	1/6	1/3

2. We select a random number between 0 and 1, and we are interested in the cumulative probabilities presented in Table 3.3. We advance in this list using a sampling interval of 1. In each case, we obtain *in fine* three distinct individuals (including household 4).

Table 3.3. Cumulative inclusion probabilities: Exercise 3.8

k	1	2	3	4	5	6
$\sum_{j \leq k} \pi_j$	1/3	1	3/2	5/2	8/3	3

3. We have:

$$\widehat{\overline{X}} = \frac{1}{6} \sum_{k \in S} \frac{x_k}{\pi_k} = \frac{1}{6} \left[\frac{x_{k_1}}{1} + \frac{x_{k_2}}{x_{k_2}/6} + \frac{x_{k_3}}{x_{k_3}/6} \right],$$

with $k_1 = 4$ (household 4 is definitely chosen) and k_2 and k_3 being the other two selected households

$$\widehat{\overline{X}} = \frac{1}{6} [9 + 6 + 6] = 3.5 = \overline{X}.$$

This result was obvious, as x_k and π_k are perfectly proportional, by construct (we have a null variance, thus a 'perfect' estimator for the estimation of the mean size $\overline{X}$).

Exercise 3.9 *Systematic sampling of businesses*

In a small municipality, we listed six businesses for which total sales (variable x_k) are respectively 40, 10, 8, 1, 0.5 and 0.5 million Euros. With the aim of estimating total paid employment, select three businesses at random and without replacement, with unequal probabilities according to total sales, using systematic sampling (by justifying your process). To do this, we use the following result for a uniform random variable between [0, 1]: 0.83021. What happens if we modify the order of the list?

Solution

The sampling by unequal probabilities, proportional to total sales (auxiliary variable) is *a priori* justified by the (reasonable) hypothesis that there is a somewhat proportional relationship between total sales and paid employment. The choice of systematic sampling is justified by the simplicity of the method. Since

$$\sum_{k \in U} x_k = 60,$$

and

$$\pi_1 = \frac{n x_1}{\sum_{\ell \in U} x_\ell} = 3 \times \frac{40}{60} = 2 > 1,$$

unit 1 is selected with certainty and removed from the population. Since

$$\sum_{k \in U \setminus \{1\}} x_k = 20,$$

and

$$\pi_2 = \frac{(n-1)x_2}{\sum_{\ell \in U \setminus \{1\}} x_\ell} = 2 \times \frac{10}{20} = 1,$$

unit 2 is selected with certainty and removed from the population. It remains to select one unit among units 3, 4, 5, 6.

$$\sum_{k \in U \setminus \{1,2\}} x_k = 10,$$

$$\pi_3 = \frac{(n-2)x_3}{\sum_{\ell \in U \setminus \{1,2\}} x_\ell} = \frac{8}{10} = 0.8, \quad \pi_4 = \frac{(n-2)x_4}{\sum_{\ell \in U \setminus \{1,2\}} x_\ell} = \frac{1}{10} = 0.1,$$

$$\pi_5 = \frac{(n-2)x_5}{\sum_{\ell \in U \setminus \{1,2\}} x_\ell} = \frac{0,5}{10} = 0.05, \quad \pi_6 = \frac{(n-2)x_6}{\sum_{\ell \in U \setminus \{1,2\}} x_\ell} = \frac{0,5}{10} = 0.05.$$

The cumulative inclusion probabilities are (denoting $V_k = \sum_{i=3}^{k} \pi_k$)

$$V_3 = 0.8, V_4 = 0.9, V_5 = 0.95, V_6 = 1.$$

Since $V_3 = 0.8 \leq 0.83021 \leq V_4 = 0.9$, we select unit 4. The final sample selected is $\{1, 2, 4\}$. If we modify the order of the list, the two largest units $(x = 40)$ and $(x = 10)$ are always kept with certainty, whatever the initial order. With the number selected at random between 0 and 1, everything depends upon the position of the unit for which $x = 8$ when we consider the four remaining units $(x = 0.5; 0.5; 1; 8)$. If this unit is in position 2, 3, or 4, then it is always selected (easy to verify). If it is in position 1, then anything is possible: we could select any of the three other individuals, depending on their appropriate positions (more precisely, we always select the individual found in the second position). The order of the list therefore influences upon the selected sample.

Exercise 3.10 *Systematic sampling and variance*

Consider a population U comprised of six units. We know the values of an auxiliary characteristic x for all the units in the population:

$$x_1 = 200, x_2 = 80, x_3 = 50, x_4 = 50, x_5 = 10, x_6 = 10.$$

1. Calculate the first-order inclusion probabilities proportional to x_k for a sample size $n = 4$. Consider 0.48444 to be a value chosen from a uniform random variable on the interval $[0, 1]$. Select a sample with unequal probabilities and without replacement of size 4 by means of systematic sampling, keeping the initial order of the list.
2. Give the second-order inclusion probability matrix (initial order of the list fixed).

3. We assume that a variable of interest y takes the following values:

$$y_1 = 80, y_2 = 50, y_3 = 30, y_4 = 25, y_5 = 10, y_6 = 5.$$

Construct a table having, by row each sample s possible and by column the sampling probabilities $p(s)$, the respective estimators for the total $\widehat{Y}(s)$ and the variance $\widehat{\mathrm{var}}(\widehat{Y})(s)$ (in the Sen-Yates-Grundy form). Calculate, based on this table, the expected values $\mathrm{E}(\widehat{Y})$ and $\mathrm{E}(\widehat{\mathrm{var}}(\widehat{Y}))$. Comment.

Solution

1. Since $X = \sum_{k \in U} x_k = 400$, we calculate $nx_1/X = 4 \times 200/400 = 2 > 1$. We eliminate unit 1 from the population and we must again select 3 units among the 5 remaining. Then, we calculate $\sum_{k \in U \setminus \{1\}} = 200$. As $3 \times 80/200 = 1.2 > 1$, we eliminate unit 2 from the population and once again we must select 2 units among the 4 remaining. Finally, we have $\sum_{k \in U \setminus \{1,2\}} = 120$. $\pi_3 = \pi_4 = 2 \times 50/120 = 5/6$ and $\pi_5 = \pi_6 = 2 \times 10/120 = 1/6$, (denoting $V_k = \sum_{i=3}^{k} \pi_i$) so the cumulative probabilities are $V_3 = 5/6, V_4 = 10/6, V_5 = 11/6, V_6 = 2$. We thus select the sample $\{1, 2, 3, 4\}$, as shown in Figure 3.1.

Fig. 3.1. Systematic sampling of two units: Exercise 3.10

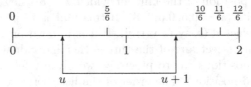

2. The second-order inclusion probability matrix is given by:

$$\begin{pmatrix} - & 1 & 5/6 & 5/6 & 1/6 & 1/6 \\ 1 & - & 5/6 & 5/6 & 1/6 & 1/6 \\ 5/6 & 5/6 & - & 4/6 & 1/6 & 0 \\ 5/6 & 5/6 & 4/6 & - & 0 & 1/6 \\ 1/6 & 1/6 & 1/6 & 0 & - & 0 \\ 1/6 & 1/6 & 0 & 1/6 & 0 & - \end{pmatrix}.$$

The first two lines and the first two columns of this matrix result from the obvious property:

For all k, for all $\ell : \pi_k = 1 \Rightarrow \pi_{k\ell} = \pi_\ell$.

To determine the other values, we are going to consider all the possible contexts: in fixed order, if we denote u as the value chosen at random between 0 and 1, we see that:

- If $0 \le u \le 4/6$, then we fall in the intervals of units 1 and 2.
- If $4/6 < u \le 5/6$, then we fall in the intervals of units 1 and 3.
- If $5/6 < u \le 1$, then we fall in the intervals of units 2 and 4.

Therefore:

$$\pi_{34} = \frac{4}{6} \; ; \; \pi_{35} = \frac{1}{6} \; ; \; \pi_{46} = \frac{1}{6}.$$

The other combinations indeed yield $\pi_{k\ell} = 0$.

3. There are only three possible samples of fixed size (see the matrix). We recall:

$$\widehat{Y}(s) = \sum_{k \in S} \frac{y_k}{\pi_k}$$

$$\widehat{\text{var}}(\widehat{Y})(s) = \frac{1}{2} \sum_{k \in S} \sum_{\substack{\ell \in S \\ \ell \neq k}} \frac{\pi_k \pi_\ell - \pi_{k\ell}}{\pi_{k\ell}} \left(\frac{y_k}{\pi_k} - \frac{y_\ell}{\pi_\ell} \right)^2.$$

The true total Y is 200. We immediately confirm that $\widehat{Y}$ is unbiased for

Table 3.4. Estimated variances according to the samples: Exercise 3.10

s	$\widehat{Y}(s)$	$\widehat{\text{var}}(\widehat{Y})(s)$	$p(s)$
$1,2,3,4$	196	0.75	4/6
$1,2,3,5$	226	−48	1/6
$1,2,4,6$	190	0	1/6

Y, in accordance with the theory:

$$E(\widehat{Y}) = \sum_s p(s)\widehat{Y}(s) = 200.$$

On the other hand, the numerical values of the estimated variances are surprising, in particular for the second sample (we notice that the sum occurring in $\widehat{\text{var}}(\widehat{Y})(s)$ only has to be calculated in fact on the lone pair formed by the two final elements in the samples, with all the other terms being zero). The presence of an estimation $\widehat{\text{var}}(\widehat{Y})$ that is negative is unpleasant, but it is theoretically possible. The fact that $\widehat{\text{var}}(\widehat{Y})$ is equal to zero for the third sample is by 'luck'. Clearly, E $\widehat{\text{var}}(\widehat{Y})$ differs from $\text{var}(\widehat{Y})$, since E $\widehat{\text{var}}(\widehat{Y}) < 0$. That is explained by the presence of three null second-order inclusion probabilities (see Exercise 3.23), which biases the estimator $\widehat{\text{var}}(\widehat{Y})(s)$.

Exercise 3.11 *Systematic sampling and order*

Consider a population of 5 units. We want to select using systematic sampling with unequal probabilities a sample of two units with inclusion probabilities proportional to the following values of X_i

$$1, 1, 6, 6, 6.$$

1. Calculate the first-order inclusion probabilities.
2. Considering the two units where the value of X_i is 1, calculate their second-order inclusion probabilities for every possible permutation of the list. What is the outcome?

Solution

1. With $n = 2$, we have

$$X = \sum_{i=1}^{6} X_i = 20, \quad \pi_i = n\frac{X_i}{X} \quad (i = 1 \text{ to } 6),$$

and $\pi_1 = \pi_2 = 0.1, \pi_3 = \pi_4 = \pi_5 = 0.6$.
2. The 10 possible permutations (there are $\binom{5}{2}$ ways of arranging the '1' values among the 5 cases) are given in Table 3.5. However, we will not carry

Table 3.5. The 10 permutations of the population: Exercise 3.11

1	1 1 6 6 6
2	1 6 1 6 6
3	1 6 6 1 6
4	1 6 6 6 1
5	6 1 1 6 6
6	6 1 6 1 6
7	6 1 6 6 1
8	6 6 1 1 6
9	6 6 1 6 1
10	6 6 6 1 1

out calculations for all possible permutations, as we see rather quickly that the probabilities for lines 1, 4, 5, 8 and 10 are going to be identical. In fact, with systematic sampling, only the relative position of the units matters. For these 5 lines, the two smallest values are consecutive and they appear as five particular breaks of a given circular layout, as shown in Figures 3.2 and 3.3. There only exists one other possible circular order, in which we

Fig. 3.2. Systematic sampling, case 1: the two smallest probabilities are adjacent: Exercise 3.11

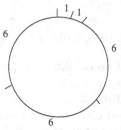

Fig. 3.3. Systematic sampling, case 2: the two smallest probabilities are not adjacent: Exercise 3.11

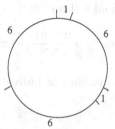

notice that the two smallest units are always separated by only one larger unit. This circular order will allow for the representation of permutations 2, 3, 6, 7 and 9. Once again, for whatever break is determined as the start of the permutation, we obtain the same second-order inclusion probabilities. To cover all situations, we can thus confine our examination to the first two cases, that is the permutations

$$1\ 1\ 6\ 6\ 6$$
$$1\ 6\ 1\ 6\ 6$$

The first-order inclusion probabilities for the two respective permutations are

$$0.1\ 0.1\ 0.6\ 0.6\ 0.6$$
$$0.1\ 0.6\ 0.1\ 0.6\ 0.6$$

In the two cases, each corresponding to a particular permutation from the list at the start, it is impossible to jointly select the two smallest units because the sampling step value is 1. Thus, this is valid for all possible permutations and therefore their second-order inclusion probability is null. The main outcome is that there is no unbiased variance estimator.

Exercise 3.12 *Sunter's method*

We know the values of an auxiliary variable for 10 units in a population. These values are the following:

$$10, 10, 8, 6, 6, 4, 2, 2, 1, 1.$$

Select a sample with unequal probabilities proportional to these values with $n = 4$ units by using the Sunter method. Use the findings of a uniform random variable over $[0, 1]$, given in Table 3.6.

Reminder: The Sunter method consists of scanning an ordered list and, for each record k (k from 1 to N), to proceed as follows:

- Generate a random number u_k between 0 and 1.
- If $k = 1$, retain the individual k if and only if (step 1) $u_1 \leq \pi_1$.
- If $k \geq 2$, retain the individual k if and only if (step k):

$$u_k \leq \frac{n - n_{k-1}}{n - \sum_{i=1}^{k-1} \pi_i} \pi_k,$$

where n_{k-1} represents the number of individuals already selected at the end of step $k - 1$.

After having verified that, in every case, at least one of the first two records is retained, notice that there can be some 'problems' with the 5th record.

Table 3.6. Uniform random numbers: Exercise 3.12

0.375489	0.624004	0.517951	0.045450	0.632912
0.24609	0.927398	0.32595	0.645951	0.178048

Solution

By denoting k as the order of the individual, π_k as the inclusion probability, V_k as the cumulative inclusion probabilities ($V_k = \sum_{i=1}^{k} \pi_i$ and $V_0 = 0$) and n_{k-1} as the number of units selected at the start of step k ($n_0 = 0$), we can describe the steps of the algorithm using Table 3.7. The selected units are 1, 2, 3 and 4. We observe that we selected the units with the largest inclusion probabilities. If the first record is not kept ($u_1 > 0.8$), the calculation for the second record is:

$$\frac{4}{4 - 0.8} \times 0.8 = 1.$$

We are thus certain to retain the second record. In this case, we verify that it is possible for records 3 and 4 not to be retained, in which case, arriving at record 5, we calculate:

$$\frac{3}{4 - 2.72} \times 0.48 = 1.125 > 1.$$

This value is larger than 1: we retain the individual with certainty, but the existence of such a possibility suggests that the algorithm does not respect exactly the π_k fixed at the start (we speak of an 'inexact' algorithm).

Table 3.7. Application of the Sunter method: Exercise 3.12

x_k	π_k	V_k	u_k	n_{k-1}	$\dfrac{n - n_{k-1}}{n - V_{k-1}}\pi_k$	I_k
10	0.8	0.8	0.375489	0	0.8	1
10	0.8	1.6	0.624004	1	0.75	1
8	0.64	2.24	0.517951	2	0.5333	1
6	0.48	2.72	0.045450	3	0.2727	1
6	0.48	3.2	0.632912	4	0	0
4	0.32	3.52	0.246090	4	0	0
2	0.16	3.68	0.927398	4	0	0
2	0.16	3.84	0.325950	4	0	0
1	0.08	3.92	0.645951	4	0	0
1	0.08	4	0.178048	4	0	0
50	4					4

Exercise 3.13 *Sunter's method and second-order probabilities*

In a population of size 6, we know the values of an auxiliary characteristic x for all the units in the population:

$$x_1 = 400, x_2 = x_3 = 15, x_4 = 10, x_5 = x_6 = 5.$$

1. Select from this population a sample of size 3 using the Sunter method (see Exercise 3.12) with unequal inclusion probabilities proportional to the characteristic x. Keep the initial order of the data and use the following findings of a uniform random variable over [0,1]:

$$0.28 \quad 0.37 \quad 0.95 \quad 0.45 \quad 0.83 \quad 0.74.$$

2. Give the following second-order inclusion probabilities: π_{23}, π_{24} (always keeping the initial order of the data).

Solution

1. For all k, a priori

$$\pi_k = 3\frac{x_k}{X} \text{ with } X = 450.$$

Clearly, $3x_1/X > 1$, which leads to $\pi_1 = 1$. We therefore eliminate individual 1 and we start again, which leads to, for $k \geq 2$:

$$\pi_k = 2\frac{x_k}{X}, \text{ with } X = 50.$$

Finally,

$$\pi_1 = 1, \pi_2 = 0.6, \pi_3 = 0.6, \pi_4 = 0.4, \pi_5 = 0.2, \pi_6 = 0.2.$$

If we denote $V_k = \sum_{i=1}^{k} \pi_i$,

$$V_1 = 1, V_2 = 1.6, V_3 = 2.2, V_4 = 2.6, V_5 = 2.8, V_6 = 3.$$

The application of the Sunter algorithm is detailed in Table 3.8.

Table 3.8. Application of the Sunter method: Exercise 3.13

k	π_k	V_k	u_k	n_{k-1}	$\pi_k(n - n_{k-1})/(n - V_{k-1})$	I_k
1	1	1	0.28	0	1	1
2	0.6	1.6	0.37	1	$0.6 \times 2/2 = 3/5$	1
3	0.6	2.2	0.95	2	$0.6 \times 1/(3 - 1.6) = 3/7$	0
4	0.4	2.6	0.45	2	$0.4 \times 1/(3 - 2.2) = 1/2$	1
5	0.2	2.8	0.83	3	0	0
6	0.2	3	0.74	3	0	0

The selected sample is: $\{1,2,4\}$.

2. We know that, (Sunter, 1986) for all $2 \le k < \ell$:

$$\pi_{kl} = \frac{\pi_k \pi_\ell n(n - 1)}{(n - V_{k-1})(n - V_k)} \prod_{i=1}^{k-1} \left(1 - \frac{2\pi_i}{n - V_{i-1}}\right),$$

where $(V_0 = 0)$ as soon as

$$\frac{n - n_k}{n - V_k}\pi_{k+1} \le 1, \tag{3.2}$$

for all $k = 1, 2, \ldots, 5$. For this last inequality, we verify that with the smallest n_k possible for all k (being $n_1 = 1 = n_2 = n_3$, $n_4 = n_5 = 2$), we can only obtain values smaller (or equal) to 1 for the left-hand side. We then have:

$$\pi_{23} = \frac{\pi_2 \pi_3 n(n - 1)}{(n - \pi_1)(n - \pi_1 - \pi_2)} \left(1 - 2\frac{\pi_1}{n}\right) = \frac{9}{35},$$

and

$$\pi_{24} = \frac{\pi_2 \pi_4 n(n - 1)}{(n - \pi_1)(n - \pi_1 - \pi_2)} \left(1 - 2\frac{\pi_1}{n}\right) = \frac{6}{35}.$$

It is quite remarkable to note that in this favourable case where (3.2) is satisfied (which is due to the values of x_k), it is possible to calculate all the second-order inclusion probabilities, starting from a quite simple expression.

Exercise 3.14 *Eliminatory method*

Consider the following sampling design with unequal probabilities: in a population U of size $N \geq 3$, we select one unit with unequal probabilities $\alpha_k, k \in U$ (we have of course $\sum_{k \in U} \alpha_k = 1$). This unit is definitely removed from the population and is not kept in the sample. Among the remaining $N - 1$ units, we select n units according to a simple random sampling design without replacement.

1. Calculate the first- and second-order inclusion probabilities for this sampling design.
2. Do the second-order inclusion probabilities satisfy the Sen-Yates-Grundy conditions?
3. How do we determine the α_k in order to select the units according to the inclusion probabilities $\pi_k, k \in U$, fixed *a priori*?
4. Is this method a) of fixed size, b) without replacement, c) applicable for every vector of inclusion probabilities fixed *a priori*? Explain.
5. We assume, from now on, that we know, for each individual k in the population, an auxiliary information x_k, and we set

$$\alpha_k = \frac{x_k}{X} \text{ for all } k \text{ with } X = \sum_{k \in U} x_k.$$

   a) Having given a sample $s = \{k_1, k_2, \ldots, k_n\}$ (without replacement), show using an appropriate conditioning that its sampling probability is

   $$p(s) = \frac{\sum_{k \in s} \alpha_k}{\binom{N-1}{n-1}}.$$

   b) Express the expected value of the ratio:

   $$\widehat{R}(S) = \frac{\sum_{k \in S} y_k}{\sum_{k \in S} x_k}$$

   as a function of $p(s)$ and $\widehat{R}(s)$.

   c) Calculate the previous expected value and show that the ratio $\widehat{R}$ is an unbiased estimator of $R = \overline{Y}/\overline{X}$.

6. Using the definition of an expected value, prove the two following results:
   a)

   $$A_S = X \frac{\sum_{k \in S} y_k^2}{\sum_{k \in S} x_k} \quad \text{estimates without bias} \quad \sum_{k \in U} y_k^2,$$

   b)

   $$B_S = \frac{X}{\sum_{k \in S} x_k} \frac{N-1}{n-1} \sum_{k \in S} \sum_{\substack{\ell \in S \\ \ell \neq k}} y_k y_\ell,$$

   estimates without bias

$$\sum_{k \in U} \sum_{\substack{\ell \in U \\ \ell \neq k}} y_k y_\ell.$$

7. Deduce an unbiased estimator for $\left(\sum_{k \in U} y_k\right)^2$.

8. Complete by suggesting an unbiased estimator for the *variance* of $\widehat{R}$ (Note: it is not necessary to express the true variance of $\widehat{R}$, which is excessively complicated).

Solution

1. We condition with respect to the result from the first drawing. With a probability α_k, individual k is eliminated, and with a probability $(1 - \alpha_k)$ it is kept for the outcome of a simple random design without replacement of size n among the $(N - 1)$ remaining individuals.

$$\pi_k = (1 - \alpha_k) \frac{n}{N - 1} + \alpha_k \times 0.$$

For the second-order inclusion probability, we condition with respect to the event 'neither k nor ℓ are selected in the first drawing': that is achieved with probability $(1 - \alpha_k - \alpha_\ell)$, with what remains to be considered being the second-order probability in a simple random design without replacement of size n among $(N - 1)$ individuals.

$$\pi_{k\ell} = (1 - \alpha_k - \alpha_\ell) \frac{n(n - 1)}{(N - 1)(N - 2)} + 0.$$

2. Yes, as for all $k \neq \ell$, we have:

$$\pi_k \pi_\ell - \pi_{k\ell}$$

$$= (1 - \alpha_k)(1 - \alpha_\ell) \left(\frac{n}{N - 1}\right)^2 - (1 - \alpha_k - \alpha_\ell) \frac{n(n - 1)}{(N - 1)(N - 2)}$$

$$= (1 - \alpha_k - \alpha_\ell) \left\{ \left(\frac{n}{N - 1}\right)^2 - \frac{n(n - 1)}{(N - 1)(N - 2)} \right\} + \alpha_k \alpha_\ell \left(\frac{n}{N - 1}\right)^2$$

$$= (1 - \alpha_k - \alpha_\ell) \frac{n(N - n - 1)}{(N - 2)(N - 1)^2} + \alpha_k \alpha_\ell \left(\frac{n}{N - 1}\right)^2$$

$$\geq 0.$$

3. We immediately deduce the α_k by using from 1.:

$$\alpha_k = 1 - \pi_k \frac{N - 1}{n}, \quad \text{for all } k = 1, \ldots, N.$$

4. The method is indeed without replacement and of fixed size, but it is only applicable if $0 \leq \alpha_k \leq 1$, for all k, and therefore if:

$$\pi_k \frac{N-1}{n} \leq 1,$$

which is

$$\pi_k \leq \frac{n}{N-1}, \text{ for all } k = 1, \ldots, N.$$

This condition is very restrictive and has little chance of occurring in practice: indeed, with a sample of fixed size n, we have

$$\sum_{k \in U} \pi_k = n,$$

and thus the mean inclusion probability is n/N, which is only slightly less than $n/(N-1)$. Briefly speaking, the only favourable case is that for a sample with 'nearly equal' probabilities: introducing a first sample with unequal probabilities therefore does not generate a sufficient margin of flexibility so that the overall process significantly moves away from a selection with unequal probabilities.

5.   a) The design is

$$p(s) = \sum_{\ell=1}^{n} p(s \mid k_\ell \text{ selected in the first drawing})\alpha_{k_\ell}.$$

If k_ℓ is selected in the first drawing, it remains to select the other $(n-1)$ elements of s by simple random sampling, among $(N-1)$ elements, being:

$$p(s \mid k_\ell) = \frac{1}{\binom{N-1}{n-1}}.$$

Therefore,

$$p(s) = \frac{1}{\binom{N-1}{n-1}} \sum_{k \in s} \alpha_k.$$

b) We have

$$\mathrm{E}[\widehat{R}(S)] = \sum_{s \in \mathcal{S}_n} p(s)\widehat{R}(s),$$

where $\mathcal{S}_n$ is the set of $\binom{N}{n}$ samples without replacement of size n that we can form in a population of size N.

c) The expected value is

$$\mathrm{E}(\widehat{R}) = \sum_{s \in \mathcal{S}_n} \frac{1}{\binom{N-1}{n-1}} \left(\sum_{k \in s} \alpha_k \right) \frac{\left(\sum_{k \in s} y_k \right)}{\left(\sum_{k \in s} x_k \right)}.$$

In fact,

$$\sum_{k \in s} x_k = X \left(\sum_{k \in s} \alpha_k \right) \quad \text{(since } \alpha_k = x_k/X \text{)}.$$

Therefore,

$$E(\widehat{R}) = \frac{1}{X} \frac{1}{\binom{N-1}{n-1}} \sum_{s \in \mathcal{S}_n} \sum_{k \in s} y_k$$

$$= \frac{1}{\binom{N-1}{n-1}} \frac{1}{X} \sum_{s \in \mathcal{S}_n} \sum_{k \in U} y_k I_k, = \frac{1}{\binom{N-1}{n-1}} \frac{1}{X} \sum_{k \in U} \left(\sum_{s \in \mathcal{S}_n} I_k \right) y_k.$$

But $\sum_{s \in \mathcal{S}_n} I_k$ is the number of samples s containing k: all these samples are determined by choosing their $(n-1)$ different elements of k among $(N-1)$ individuals (the restricted population of k), which gives

$$\sum_{s \in \mathcal{S}_n} I_k = \binom{N-1}{n-1}.$$

We finally get

$$E(\widehat{R}) = \frac{\sum_{k \in U} y_k}{X} = R.$$

Thus $\widehat{R}$ estimates without bias R. This property, extremely rare for a ratio, follows directly from the sampling mode.

6. a) This is exactly the same method as in 5.:

$$\sum_{s \in \mathcal{S}_n} p(s) \frac{X}{\left(\sum_{k \in s} x_k \right)} \sum_{k \in s} y_k^2$$

$$= \sum_{s \in \mathcal{S}_n} p(s) \frac{1}{\sum_{k \in s} \alpha_k} \sum_{k \in s} y_k^2 = \sum_{s \in \mathcal{S}_n} \frac{1}{\binom{N-1}{n-1}} \sum_{k \in s} y_k^2 = \sum_{k \in U} y_k^2.$$

b) The expected value is given by

$$\sum_{s \in \mathcal{S}_n} p(s) \frac{X}{\left(\sum_{k \in s} x_k \right)} \frac{N-1}{n-1} \sum_{k \in s} \sum_{\substack{\ell \in s \\ \ell \neq k}} y_k y_\ell$$

$$= \frac{\frac{N-1}{n-1}}{\binom{N-1}{n-1}} \left(\sum_{s \in \mathcal{S}_n} \sum_{k \in s} \sum_{\substack{\ell \in s \\ \ell \neq k}} y_k y_\ell \right).$$

In fact,

$$\sum_{s \in \mathcal{S}_n} \sum_{k \in s} \sum_{\substack{\ell \in s \\ \ell \neq k}} y_k y_\ell = \sum_{k \in U} \sum_{\substack{\ell \in U \\ \ell \neq k}} \left(\sum_{s \in \mathcal{S}_n} I_k I_\ell \right) y_k y_\ell,$$

where U is the total population, and $\sum_{s\in S_n} I_k I_\ell$ counts the samples of S_n such as $(k,\ \ell)\in s$, that are of the number $\binom{N-2}{n-2}$. The calculated expected value is therefore:

$$\frac{\binom{N-2}{n-2}}{\binom{N-1}{n-1}}\frac{N-1}{n-1}\left(\sum_{k\in U}\sum_{\substack{\ell\in U\\ \ell\neq k}} y_k y_\ell\right) = \sum_{k\in U}\sum_{\substack{\ell\in U\\ \ell\neq k}} y_k y_\ell.$$

7. Since

$$\left(\sum_{k\in U} y_k\right)^2 = \sum_{k\in U} y_k^2 + \sum_{k\in U}\sum_{\substack{\ell\in U\\ \ell\neq k}} y_k y_\ell = \mathrm{E}(A_S) + \mathrm{E}(B_S),$$

$A_S + B_S$ is an unbiased estimator of $\left(\sum_{k\in U} y_k\right)^2$.

8.
$$\mathrm{var}(\widehat{R}) = \sum_{s\in S_n} p(s)\,\widehat{R}^2(s) - R^2.$$

Since

$$R^2 = \frac{\left(\sum_{k\in U} y_k\right)^2}{\left(\sum_{k\in U} x_k\right)^2} = \mathrm{E}\left[\frac{A_S + B_S}{\left(\sum_{k\in U} x_k\right)^2}\right],$$

we can construct an unbiased estimator of the variance

$$\widehat{\mathrm{var}}(\widehat{R}) = \widehat{R}^2(S) - \frac{A_S + B_S}{\left(\sum_{k\in U} x_k\right)^2}.$$

Exercise 3.15 *Midzuno's method*

Consider the following sampling design with unequal probabilities: in a population U of size $N \geq 3$, we select a unit with unequal probabilities $\alpha_k, k \in U$, where

$$\sum_{k\in U} \alpha_k = 1.$$

Next, among the $N - 1$ remaining units, we select in the sample $n - 1$ units according to a simple random design without replacement. The final sample is thus of fixed size n.

1. Calculate the first- and second-order inclusion probabilities for this design. Write the second-order inclusion probabilities as a function of the first-order probabilities.
2. Do the second-order inclusion probabilities satisfy the Sen-Yates-Grundy conditions?

3. How do we determine the α_k in order to select the units according to $\pi_k, k \in U$, fixed *a priori*?
4. Is this method a) without replacement, b) applicable for every vector of inclusion probabilities fixed *a priori*? Explain.

Solution

1. We condition with respect to the result of the first drawing. Unit k is selected with probability α_k

$$\pi_k = \alpha_k + (1 - \alpha_k)\frac{n-1}{N-1} = \alpha_k\frac{N-n}{N-1} + \frac{n-1}{N-1}.$$

We condition according to three possible occurrences in the first drawing: k is kept, or ℓ is kept, or neither k nor ℓ is kept.

$$\pi_{k\ell} = \alpha_k\frac{n-1}{N-1} + \alpha_\ell\frac{n-1}{N-1} + (1 - \alpha_k - \alpha_\ell)\frac{(n-1)(n-2)}{(N-1)(N-2)}$$

$$= (\alpha_k + \alpha_\ell)\frac{(n-1)(N-n)}{(N-1)(N-2)} + \frac{(n-1)(n-2)}{(N-1)(N-2)}.$$

The second-order inclusion probabilities are written:

$$\pi_{k\ell} = \left(\pi_k\frac{N-1}{N-n} - \frac{n-1}{N-n} + \pi_\ell\frac{N-1}{N-n} - \frac{n-1}{N-n}\right)\frac{(n-1)(N-n)}{(N-1)(N-2)}$$
$$+ \frac{(n-1)(n-2)}{(N-1)(N-2)}$$
$$= \pi_k\frac{(n-1)}{N-2} + \pi_\ell\frac{(n-1)}{N-2} - 2\frac{(n-1)}{N-1}\frac{(n-1)}{N-2} + \frac{(n-1)(n-2)}{(N-1)(N-2)}$$
$$= \frac{(n-1)}{N-2}\left(\pi_k + \pi_\ell - \frac{n}{N-1}\right).$$

2. The second-order inclusion probabilities satisfy the Sen-Yates-Grundy conditions. Indeed, we have

$$\pi_k\pi_\ell - \pi_{k\ell} = \left(\alpha_k\frac{N-n}{N-1} + \frac{n-1}{N-1}\right)\left(\alpha_\ell\frac{N-n}{N-1} + \frac{n-1}{N-1}\right)$$
$$-(\alpha_k + \alpha_\ell)\frac{(n-1)(N-n)}{(N-1)(N-2)} - \frac{(n-1)(n-2)}{(N-1)(N-2)}$$
$$= (\alpha_k + \alpha_\ell)\frac{(N-n)(n-1)}{(N-1)^2} + \frac{(n-1)^2}{(N-1)^2} + \alpha_k\alpha_\ell\frac{(N-n)^2}{(N-1)^2}$$
$$-(\alpha_k + \alpha_\ell)\frac{(n-1)(N-n)}{(N-1)(N-2)} - \frac{(n-1)(n-2)}{(N-1)(N-2)}$$
$$= (1 - \alpha_k - \alpha_\ell)\frac{(n-1)(N-n)}{(N-1)^2(N-2)} + \alpha_k\alpha_\ell\frac{(N-n)^2}{(N-1)^2} \geq 0.$$

3. We calculate α_k as a function of π_k by

$$\alpha_k = \pi_k \frac{N-1}{N-n} - \frac{n-1}{N-n}.$$

Clearly, we have $\alpha_k \leq 1$ for all k in U.

4. The method is of course without replacement but in order for it to be applicable, it is necessary and sufficient that

$$\alpha_k = \pi_k \frac{N-1}{N-n} - \frac{n-1}{N-n} \geq 0,$$

and therefore that

$$\pi_k \geq \frac{n-1}{N-1}, \text{ for all } k = 1, \ldots, N,$$

which is rarely the case (also see Exercise 3.14).

Remark: this method was proposed by Midzuno   (see on this topic Midzuno, 1952; Singh, 1975).

Exercise 3.16 *Brewer's method*

Consider a sampling design of fixed size n with unequal probabilities in a population U of size N whose first two order probabilities are denoted as π_k and $\pi_{k\ell}$. We denote $p(s)$ as the probability of selecting sample s. We say that a design $p^*(.)$ of size $n^* = N - n$ is the complement of $p(s)$ if $p^*(U\backslash s) = p(s)$, for all $s \subset U$.

1. Give the first- and second-order inclusion probabilities for the design $p^*(.)$ as a function of π_k and $\pi_{k\ell}$.
2. Show that if the Sen-Yates-Grundy conditions are satisfied for a sampling design, they are equally satisfied for the complementary design.
3. The Brewer method (see Brewer, 1975) can be written as a succession of splitting steps (see on this topic Deville and Tillé, 1998; Tillé, 2001, chapter 6). A splitting method consists of transforming in a random manner the vector of inclusion probabilities. At each step, the same procedure is applied to the non-integer inclusion probabilities: we randomly choose one of the vectors given in Figure 3.4 with a probability λ_j, where

$$\lambda_j = \left\{ \sum_{z \in U} \frac{\pi_z(n - \pi_z)}{1 - \pi_z} \right\}^{-1} \frac{\pi_j(n - \pi_j)}{1 - \pi_j}.$$

Give the splitting step of the complementary design of the Brewer method. Express this step as a function of the sample size and the inclusion probabilities of the complementary design.
4. Brewer's method consists of selecting at each step only one unit, that which corresponds to the coordinate equal to 1. What about the complementary design?

Fig. 3.4. Brewer's method shown as a technique of splitting into N parts: Exercise 3.16

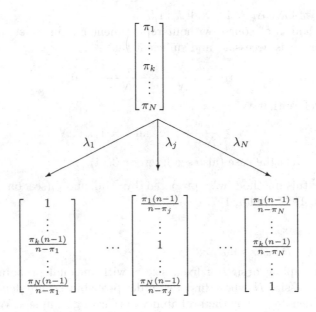

Solution

1. The first- and second-order inclusion probabilities for the complementary design are

$$\pi_k^* = \Pr(k \in U \backslash S) = \Pr(k \notin S) = 1 - \Pr(k \in S) = 1 - \pi_k,$$

$$\pi_{k\ell}^* = \Pr[k \notin S \cap \ell \notin S] = 1 - \Pr[k \in S \cup \ell \in S].$$

Thus
$$\pi_{k\ell}^* = 1 - \pi_k - \pi_\ell + \pi_{k\ell},$$

as
$$\Pr(A \cup B) = \Pr(A) + \Pr(B) - \Pr(A \cap B).$$

2. Since

$$\pi_k^* \pi_\ell^* - \pi_{k\ell}^* = (1 - \pi_k)(1 - \pi_\ell) - (1 - \pi_k - \pi_\ell + \pi_{k\ell}) = \pi_k \pi_\ell - \pi_{k\ell} \geq 0,$$

the Sen-Yates-Grundy conditions are also satisfied for the complementary design.

3. The Brewer method produces the vector split

$$\pi_k^{(j)} = \begin{cases} 1 & k = j \\ \pi_k \dfrac{n-1}{n-\pi_j} & k \neq j, \end{cases}$$

with λ_j such that

$$\sum_{j \in U} \lambda_j \pi_k^{(j)} = \pi_k.$$

The complementary design gives the following split:

$$\pi_k^{(j)*} = 1 - \pi_k^{(j)} = \begin{cases} 1-1 & k = j \\ 1 - \pi_k \dfrac{n-1}{n-\pi_j} & k \neq j. \end{cases}$$

Being

$$\pi_k^{(j)*} = \begin{cases} 0 & k = j \\ 1 - (1-\pi_k^*)\dfrac{N-n^*-1}{N-n^*-1+\pi_j^*} & k \neq j, \end{cases}$$

with the same λ_j. We indeed have $\sum_{j \in U} \lambda_j \pi_k^{(j)*} = 1 - \pi_k = \pi_k^*$.

4. We note that the method now consists of eliminating one unit at each step (eliminatory method), since $\pi_j^{(j)*} = 0$ for all $j = 1, \ldots, N$. Thus, in $n = N - n^*$ successive steps, the complementary design indeed gives a sample of size $n^* = N - n$, while respecting the set of inclusion probabilities π_k^* initially presented.

Exercise 3.17 *Sampling with replacement and comparison of means*
Consider $x_1, \ldots, x_k, \ldots, x_N$ as a family of any positive values.

1. Mention the sampling probabilities used in a sample proportional to sizes x_k with replacement.
2. What is the unbiased estimator $\widehat{N}$ of the population size N? What is its variance?
3. Using the fact that a variance is always positive, find a well-known mathematical inequality for a family of any positive values.

Solution

1. The sampling probability of individual k is

$$p_k = \frac{x_k}{X} \quad \text{with} \quad X = \sum_{k \in U} x_k,$$

and therefore

$$\sum_{k \in U} p_k = 1.$$

2. In a sample with replacement with unequal probabilities p_k we can estimate N without bias using the Hansen-Hurwitz estimator:

$$\widehat{N} = \frac{1}{n} \sum_{\alpha=1}^{n} \frac{1}{p_{k(\alpha)}},$$

where α is the sample number, $k(\alpha)$ is the identifier of the individual selected for drawing number α, and n is the size of the sample. We have:

$$\mathrm{var}(\widehat{N}) = \frac{1}{n} \sum_{k \in U} p_k \left(\frac{1}{p_k} - N\right)^2,$$

where k is indeed the identifier of the individual.

3. The variance of $\widehat{N}$ is

$$\mathrm{var}(\widehat{N}) = \frac{1}{n} \sum_{k \in U} p_k \left(\frac{1}{p_k^2} + N^2 - \frac{2N}{p_k}\right) = \frac{1}{n}\left[\sum_{k \in U} \frac{1}{p_k} - N^2\right].$$

In fact,

$$\mathrm{var}(\widehat{N}) \geq 0 \quad \Rightarrow \quad \sum_{k \in U} \frac{1}{p_k} \geq N^2 \quad \Rightarrow \quad \frac{1}{N}\sum_{k \in U} \frac{X}{x_k} \geq N.$$

Thus

$$\frac{1}{N} \sum_{k \in U} \frac{1}{x_k} \geq \frac{N}{\sum_{k \in U} x_k} \quad \Rightarrow \quad \overline{X} \geq \frac{1}{\frac{1}{N}\sum_{k \in U} 1/x_k}.$$

The arithmetic mean is greater than or equal to the harmonic mean: it is a well-known result, which is not obvious to obtain using a direct method.

Exercise 3.18 *Geometric mean and Poisson design*

In a Poisson design where the units are selected from a population U of size N with inclusion probabilities π_k, we want to estimate the geometric mean

$$\theta = \prod_{k \in U} y_k^{1/N}.$$

To do this, we propose to use the following estimator:

$$\widehat{\theta} = \prod_{k \in S} \frac{y_k^{1/N} - 1 + \pi_k}{\pi_k}.$$

1. Express $\widehat{\theta}$ by changing the product on the sample with a product on the population and by making use of the indicator variables I_k in the presence of units k in sample S. What problem can occur for $\widehat{\theta}$?
2. Using the expression given in 1., show that $\widehat{\theta}$ is unbiased for θ.
3. Give $\mathrm{E}\left(\widehat{\theta}^2\right)$ and deduce the exact variance of $\widehat{\theta}$.

Solution

1. The estimator can be written

$$\widehat{\theta} = \prod_{k \in U} \left(\frac{y_k^{1/N} - 1 + \pi_k}{\pi_k} \right)^{I_k}.$$

We notice that $\widehat{\theta}$ can be negative, especially if π_k and y_k are jointly small. This nuisance never occurs in the estimation of a total with the Horvitz-Thompson estimator.

2. Since the random variables I_k are independent:

$$\mathrm{E}\left(\widehat{\theta}\right) = \prod_{k \in U} \mathrm{E}\left\{ \left(\frac{y_k^{1/N} - 1 + \pi_k}{\pi_k} \right)^{I_k} \right\}$$

$$= \prod_{k \in U} \left\{ \left(\frac{y_k^{1/N} - 1 + \pi_k}{\pi_k} \right) \pi_k + (1 - \pi_k) \right\}$$

$$= \prod_{k \in U} y_k^{1/N} = \theta.$$

3. It remains to calculate

$$\mathrm{E}\left(\widehat{\theta}^2\right) = \prod_{k \in U} \mathrm{E}\left[\left(\frac{y_k^{1/N} - 1 + \pi_k}{\pi_k} \right)^{2I_k} \right]$$

$$= \prod_{k \in U} \left[\frac{\left(y_k^{1/N} - 1 + \pi_k \right)^2 + \pi_k - \pi_k^2}{\pi_k} \right],$$

to obtain

$$\mathrm{var}\left(\widehat{\theta}\right) = \mathrm{E}\left(\widehat{\theta}^2\right) - \left\{ \mathrm{E}\left(\widehat{\theta}\right) \right\}^2$$

$$= \prod_{k \in U} \left[\frac{\left(y_k^{1/N} - 1 + \pi_k \right)^2 + \pi_k - \pi_k^2}{\pi_k} \right] - \prod_{k \in U} y_k^{2/N}.$$

Exercise 3.19 *Sen-Yates-Grundy variance*

The goal of this exercise is to show that, when the sample size is fixed, the accuracy of a sample without replacement with unequal probabilities can be expressed under a 'pleasant' form, known as the Sen-Yates-Grundy variance.

1. If we denote π_k as the inclusion probability of individual k, N as the population size, and n as the fixed size of the sample, show that:

$$\sum_{k \in U} \pi_k = n.$$

2. If we denote $\pi_{k\ell}$ as the second-order inclusion probability of k and ℓ, show that, for all $k \in U$,

$$\sum_{\substack{\ell \in U \\ \ell \neq k}} \pi_{k\ell} = (n-1)\,\pi_k,$$

   (*hint*: use indicator variables).

3. Show that, for all $k \in U$,

$$\sum_{\substack{\ell \in U \\ \ell \neq k}} \pi_k \pi_\ell = \pi_k(n - \pi_k),$$

   and deduce that, for all $k \in U$,

$$\sum_{\substack{\ell \in U \\ \ell \neq k}} (\pi_k \pi_\ell - \pi_{k\ell}) = \pi_k(1 - \pi_k).$$

4. Put the accuracy of the Horvitz-Thompson estimator $\widehat{Y}$ in the form:

$$\mathrm{var}(\widehat{Y}) = \left[\sum_{k \in U} \left(\frac{y_k}{\pi_k} \right)^2 \sum_{\substack{\ell \in U \\ \ell \neq k}} (\pi_k \pi_\ell - \pi_{k\ell}) - \sum_{k \in U} \sum_{\substack{\ell \in U \\ \ell \neq k}} \frac{y_k}{\pi_k} \frac{y_\ell}{\pi_\ell} (\pi_k \pi_\ell - \pi_{k\ell}) \right].$$

   By showing:

$$\mathrm{var}(\widehat{Y}) = \frac{1}{2} \sum_{k \in U} \sum_{\substack{\ell \in U \\ \ell \neq k}} (\pi_k \pi_\ell - \pi_{k\ell}) \left(\frac{y_k}{\pi_k} - \frac{y_\ell}{\pi_\ell} \right)^2.$$

   What is the interest of this form?

Solution

1. Suppose

$$I_k = \begin{cases} 1 & \text{if } k \in S \\ 0 & \text{otherwise.} \end{cases}$$

The variable I_k therefore has a Bernoulli distribution, $\mathcal{B}(1, \pi_k)$. We know that:

$$\mathrm{E}(I_k) = \pi_k, \qquad \mathrm{var}(I_k) = \pi_k(1 - \pi_k).$$

Indeed,

$$\sum_{k \in U} I_k = n \quad \text{(by definition of } n\text{)}.$$

Therefore,

$$\mathrm{E}\left(\sum_{k \in U} I_k\right) = \mathrm{E}(n) = n, \text{ because } n \text{ is fixed.}$$

Finally,

$$\sum_{k \in U} \pi_k = \sum_{k \in U} \mathrm{E}(I_k) = n.$$

2. Since $\pi_{k\ell} = \mathrm{E}[I_k I_\ell]$, by fixing k we have

$$\sum_{\substack{\ell \in U \\ \ell \neq k}} \pi_{k\ell} = \mathrm{E}\left(\sum_{\substack{\ell \in U \\ \ell \neq k}} I_k I_\ell\right) = \mathrm{E}\left[I_k\left(\sum_{\substack{\ell \in U \\ \ell \neq k}} I_\ell\right)\right]$$

$$= \mathrm{E}\left(I_k(n - I_k)\right) = \mathrm{E}[nI_k - (I_k)^2] = n\pi_k - \pi_k,$$

as $(I_k)^2 = I_k$. We conclude that for all k in U:

$$\sum_{\substack{\ell \in U \\ \ell \neq k}} \pi_{k\ell} = (n - 1)\,\pi_k.$$

3. For all k,

$$\sum_{\substack{\ell \in U \\ \ell \neq k}} \pi_k \pi_\ell = \pi_k\left(\sum_{\substack{\ell \in U \\ \ell \neq k}} \pi_\ell\right) = \pi_k(n - \pi_k).$$

Thus,

$$\sum_{\substack{\ell \in U \\ \ell \neq k}} (\pi_k \pi_\ell - \pi_{k\ell}) = \pi_k(n - \pi_k) - (n - 1)\,\pi_k$$

$$= n\pi_k - \pi_k^2 - n\pi_k + \pi_k = \pi_k(1 - \pi_k).$$

4. The unbiased Horvitz-Thompson estimator is:

$$\widehat{Y} = \sum_{k \in S} \frac{y_k}{\pi_k} = \underbrace{\sum_{k \in U} \frac{y_k}{\pi_k}}_{\text{non-random}} \overbrace{\times I_k}^{\text{random}}.$$

Its variance is:

$$\mathrm{var}(\widehat{Y}) = \sum_{k \in U} \left(\frac{y_k}{\pi_k}\right)^2 \mathrm{var}(I_k) + \sum_{k \in U} \sum_{\substack{\ell \in U \\ \ell \neq k}} \frac{y_k y_\ell}{\pi_k \pi_\ell} \mathrm{cov}(I_k, I_\ell)$$

$$= \sum_{k \in U} \left(\frac{y_k}{\pi_k}\right)^2 \pi_k(1 - \pi_k) - \sum_{k \in U} \sum_{\substack{\ell \in U \\ \ell \neq k}} \frac{y_k y_\ell}{\pi_k \pi_\ell} (\pi_k \pi_\ell - \pi_{k\ell}).$$

By the expression from 3., we have

$$\mathrm{var}(\widehat{Y}) = \sum_{k \in U} \left(\frac{y_k}{\pi_k}\right)^2 \sum_{\substack{\ell \in U \\ \ell \neq k}} (\pi_k \pi_\ell - \pi_{k\ell}) - \sum_{k \in U} \sum_{\substack{\ell \in U \\ \ell \neq k}} \frac{y_k y_\ell}{\pi_k \pi_\ell} (\pi_k \pi_\ell - \pi_{k\ell})$$

$$= \frac{1}{2} \sum_{k \in U} \sum_{\substack{\ell \in U \\ \ell \neq k}} \left(\frac{y_k^2}{\pi_k^2} + \frac{y_\ell^2}{\pi_\ell^2} - 2\frac{y_k y_\ell}{\pi_k \pi_\ell}\right) (\pi_k \pi_\ell - \pi_{k\ell})$$

$$= \frac{1}{2} \sum_{k \in U} \sum_{\substack{\ell \in U \\ \ell \neq k}} \left(\frac{y_k}{\pi_k} - \frac{y_\ell}{\pi_\ell}\right)^2 (\pi_k \pi_\ell - \pi_{k\ell}).$$

This form of the variance is known as the Sen-Yates-Grundy variance, and has the advantage of highlighting the terms $\left(\frac{y_k}{\pi_k} - \frac{y_\ell}{\pi_\ell}\right)^2$. A 'good' sample with unequal probabilities is therefore like π_k as much as possible *proportional* to y_k.

Exercise 3.20 *Balanced design*

Consider U as a finite population and S as the random sample obtained from U using a design with inclusion probabilities π_k and $\pi_{k\ell}$ strictly positive. We assume that this design is balanced on a characteristic z, otherwise stated

$$\sum_{k \in S} \frac{z_k}{\pi_k} = \sum_{k \in U} z_k.$$

The total of the characteristic of interest y given by

$$Y = \sum_{k \in U} y_k$$

can be estimated without bias by

$$\widehat{Y}_\pi = \sum_{k \in S} \frac{y_k}{\pi_k}.$$

1. Show that, for all ℓ in U:

$$\sum_{\substack{k \in U \\ k \neq \ell}} \frac{z_k \pi_{k\ell}}{\pi_k} = \pi_\ell \sum_{k \in U} z_k - z_\ell. \tag{3.3}$$

2. What particular result do we obtain when $z_k = \pi_k, k \in U$?
3. Show that, for all $\ell \in U$,

$$\operatorname{var}\left(\widehat{Y}_\pi\right) = \frac{1}{2} \sum_{k \in U} \sum_{\substack{\ell \in U \\ \ell \neq k}} \left(\frac{y_k}{z_k} - \frac{y_\ell}{z_\ell}\right)^2 z_k z_\ell \frac{\pi_k \pi_\ell - \pi_{k\ell}}{\pi_k \pi_\ell}. \tag{3.4}$$

4. What result is generalised by Expression (3.4)?
5. Construct an unbiased estimator of the variance starting from Expression (3.4).

Solution

1. As the design is balanced we have, denoting I_ℓ as the indicator for unit ℓ being in S:

$$\sum_{k \in S} \frac{z_k}{\pi_k} I_\ell = \sum_{k \in U} z_k I_\ell,$$

and thus

$$E\left(\sum_{\substack{k \in U \\ k \neq \ell}} \frac{z_k}{\pi_k} I_k I_\ell + \frac{z_\ell I_\ell^2}{\pi_\ell}\right) = E\left(\sum_{k \in U} z_k I_\ell\right),$$

which gives

$$\sum_{\substack{k \in U \\ k \neq \ell}} \frac{z_k \pi_{k\ell}}{\pi_k} = \pi_\ell \sum_{k \in U} z_k - z_\ell. \tag{3.5}$$

2. If $z_k = \pi_k$, then the balanced design is written

$$\sum_{k \in S} 1 = \sum_{k \in U} \pi_k = n.$$

The balanced design is in that case of fixed sample size. Expression (3.5) becomes

$$\sum_{\substack{k \in U \\ k \neq \ell}} \pi_{k\ell} = \pi_\ell(n-1).$$

3.

$$\frac{1}{2} \sum_{k \in U} \sum_{\substack{\ell \in U \\ \ell \neq k}} \left(\frac{y_k}{z_k} - \frac{y_\ell}{z_\ell} \right)^2 z_k z_\ell \frac{\pi_k \pi_\ell - \pi_{k\ell}}{\pi_k \pi_\ell}$$

$$= \sum_{k \in U} \sum_{\substack{\ell \in U \\ \ell \neq k}} \frac{y_k^2}{z_k^2} z_k z_\ell \frac{\pi_k \pi_\ell - \pi_{k\ell}}{\pi_k \pi_\ell} - \sum_{k \in U} \sum_{\substack{\ell \in U \\ \ell \neq k}} \frac{y_k}{z_k} \frac{y_\ell}{z_\ell} z_k z_\ell \frac{\pi_k \pi_\ell - \pi_{k\ell}}{\pi_k \pi_\ell}$$

$$= \sum_{k \in U} \frac{y_k^2}{z_k \pi_k} \sum_{\substack{\ell \in U \\ \ell \neq k}} \left(z_\ell \pi_k - z_\ell \frac{\pi_{k\ell}}{\pi_\ell} \right) - \sum_{k \in U} \sum_{\substack{\ell \in U \\ \ell \neq k}} \frac{y_k y_\ell}{\pi_k \pi_\ell} \left(\pi_k \pi_\ell - \pi_{k\ell} \right).$$

In fact,

$$\sum_{\substack{\ell \in U \\ \ell \neq k}} \left(z_\ell \pi_k - z_\ell \frac{\pi_{k\ell}}{\pi_\ell} \right) = \pi_k(Z - z_k) - (\pi_k Z - z_k) = z_k(1 - \pi_k),$$

where

$$Z = \sum_{k \in U} z_k.$$

We therefore get

$$\frac{1}{2} \sum_{k \in U} \sum_{\substack{\ell \in U \\ \ell \neq k}} \left(\frac{y_k}{z_k} - \frac{y_\ell}{z_\ell} \right)^2 z_k z_\ell \frac{\pi_k \pi_\ell - \pi_{k\ell}}{\pi_k \pi_\ell}$$

$$= \sum_{k \in U} \frac{y_k^2}{\pi_k}(1 - \pi_k) - \sum_{k \in U} \sum_{\substack{\ell \in U \\ \ell \neq k}} \frac{y_k y_\ell}{\pi_k \pi_\ell} \left(\pi_k \pi_\ell - \pi_{k\ell} \right) = \text{var}(\widehat{Y}_\pi).$$

4. When $z_k = \pi_k$, we get the Sen-Yates-Grundy variance for designs of fixed size.

5. An unbiased estimator is

$$\widehat{\text{var}}(\widehat{Y}_\pi) = \frac{1}{2} \sum_{k \in S} \sum_{\substack{\ell \in S \\ \ell \neq k}} \left(\frac{y_k}{z_k} - \frac{y_\ell}{z_\ell} \right)^2 z_k z_\ell \frac{\pi_k \pi_\ell - \pi_{k\ell}}{\pi_k \pi_\ell \pi_{k\ell}}.$$

Exercise 3.21 *Design effect*

When we introduce complex sampling designs and we look to calculate the accuracy using a computer program, we get in general the calculation of a ratio called the *'design effect'*. This ratio is defined as the ratio of the variance of the estimator of the total $\widehat{Y}$ over the variance of the estimator that we would get if we would have performed a simple random sampling design of the same size n. We denote $\widehat{\overline{Y}}$ as the simple mean of the y_k for k in S.

1. Letting $\text{var}_p(\widehat{Y})$ be the true variance (possibly very complicated) obtained under the complex design (denoted p), give the expression of the design effect (henceforth denoted DEFF).

2. How are we going to naturally estimate DEFF(we denote $\widehat{\text{DEFF}}$ as the estimator)?
   We are henceforth limited to complex designs p with equal probabilities and of fixed size.

3. Under these conditions, how do we estimate without bias any 'true' total Y?

4. Calculate the expected value of the sample variance s_y^2 in the sample, under the design p (we denote this $E_p(s_y^2)$). We will express this as a function of $\text{var}_p(\widehat{\overline{Y}})$, S_y^2, n and N.

5. Considering the denominator of $\widehat{\text{DEFF}}$, show that its use introduces a bias that we express as a function of n, N and $\text{var}_p(\widehat{\overline{Y}})$. For this question, we assume that n is 'large'.

6. Deduce that the denominator of $\widehat{\text{DEFF}}$ has an expected value equal to the desired value multiplied by the factor:

$$1 - \frac{1-f}{n}\text{DEFF}.$$

Find the conclusions in the case where n is 'large'.

Solution

1. The *design effect* is:

$$\text{DEFF} = \frac{\text{var}_p(\widehat{Y})}{N^2\frac{1-f}{n}S_y^2}.$$

where S_y^2 is the variance of the y_k in the population.

2. The estimator is

$$\widehat{\text{DEFF}} = \widehat{\text{var}}_p(\widehat{Y})/N^2\frac{1-f}{n}s_y^2,$$

where

$$s_y^2 = \frac{1}{n-1}\sum_{k\in S}(y_k - \widehat{\overline{Y}})^2,$$

$\widehat{\text{var}}_p(\widehat{Y})$ is an unbiased estimator of $\text{var}_p(\widehat{Y})$, and $\overline{\widehat{Y}}$ is the simple mean in the sample.

3. We estimate without bias Y by

$$\widehat{Y} = \sum_{k \in S} \frac{y_k}{\pi_k}.$$

Now $\sum_{k \in U} \pi_k = n$ and π_k are constant, which implies that for all $k = 1, \dots, N$, $\pi_k = n/N$, and $\widehat{Y} = N\overline{\widehat{Y}}$, where

$$\overline{\widehat{Y}} = \frac{1}{n} \sum_{k \in S} y_k.$$

4. We have:

$$s_y^2 = \frac{1}{n-1} \sum_{k \in S} y_k^2 - \frac{n}{n-1} \overline{\widehat{Y}}^2,$$

and

$$E_p(s_y^2) = \frac{1}{n-1} \left(\sum_{k \in U} y_k^2 \frac{n}{N} \right) - \frac{n}{n-1} \left\{ \text{var}_p(\overline{\widehat{Y}}) + \left(E_p \overline{\widehat{Y}} \right)^2 \right\}.$$

Since the sampling is with equal probabilities, we have $E_p(\overline{\widehat{Y}}) = \overline{Y}$. Hence:

$$E_p(s_y^2) = \frac{n}{n-1} \left(\frac{1}{N} \sum_{k \in U} y_k^2 - (\overline{Y})^2 \right) - \frac{n}{n-1} \text{var}_p(\overline{\widehat{Y}})$$

$$= \frac{n}{n-1} \left(\frac{N-1}{N} S_y^2 - \text{var}_p(\overline{\widehat{Y}}) \right).$$

5. We have, in the case where n (and therefore N) is large

$$E_p \left(N^2 \frac{1-f}{n} s_y^2 \right) \approx N^2 \frac{1-f}{n} \left(S_y^2 - \text{var}_p(\overline{\widehat{Y}}) \right)$$

$$= N^2 \frac{1-f}{n} S_y^2 - \underbrace{\frac{1-f}{n} \text{var}_p(N\overline{\widehat{Y}})}_{\text{Bias}}.$$

6. The expected value is

$$E_p \left(N^2 \frac{1-f}{n} s_y^2 \right) \approx N^2 \frac{1-f}{n} S_y^2 - \frac{1-f}{n} \left(\text{DEFF} N^2 \frac{1-f}{n} S_y^2 \right)$$

$$= N^2 \frac{1-f}{n} S_y^2 \left(1 - \text{DEFF} \frac{1-f}{n} \right).$$

Of course, $N^2(1-f)S_y^2/n$ is the value that we try to approach (it is a question of the denominator of DEFF). DEFF is a real value which only rarely exceeds the value by a few: indeed, the sample designs are by construction built to restrict this. We very rarely find values greater than 10, for example. As a result, as soon as n is large, the factor $(1 - f)/n$ is very small, and the coefficient $1 - \mathrm{DEFF}(1 - f)/n$ is close to 1. The conclusion is that with sample designs with equal probabilities and of fixed size, the use of 'naive' computer programs based upon the calculation of s_y^2 is without any harmful consequence when n is large. On the other hand, in the contrary case (n of the order of 10, 20 or 30), the calculation of $\widehat{\mathrm{DEFF}}$ can turn out to be whimsical.

Exercise 3.22 *Rao-Blackwellisation*

In a finite population $U = \{1, \ldots, k, \ldots, N\}$, we select three units with replacement and with unequal probabilities $p_k, k \in U$, where

$$\sum_{k \in U} p_k = 1.$$

We let a_k be the random variable indicating the number of times that unit k is selected in the sample with replacement.

1. Give the distribution of the probability of vector $(a_1, \ldots, a_k, \ldots, a_N)'$.
2. Deduce $\mathrm{E}\,[a_k]$, $\mathrm{var}\,[a_k]$ and $\mathrm{cov}\,[a_k, a_\ell]$, $k \neq \ell$.
3. If S represents the random sample of distinct units, calculate

$$\Pr\,(k \notin S),$$

$$\Pr\,(S = \{k\}),$$

$$\Pr\,(S = \{k, \ell\})\,, k \neq \ell,$$

$$\Pr\,(S = \{k, \ell, m\})\,, k \neq \ell \neq m.$$

4. What is the inclusion probability $\pi_k = \Pr\,(k \in S)$?
5. Determine

$$\mathrm{E}\,[a_k \mid S = \{k\}]\,, \mathrm{E}\,[a_k \mid S = \{k, \ell\}]\,, k \neq \ell,$$

$$\mathrm{E}\,[a_k \mid S = \{k, \ell, m\}]\,, k \neq \ell \neq m.$$

6. Define the unbiased estimator of the total of a variable y_k using a summation of U and a_k (called the Hansen-Hurwitz estimator).
7. Define the estimator obtained by Rao-Blackwellising the Hansen-Hurwitz estimator by separately considering the cases $S = \{k\}$, $S = \{k, \ell\}$, and $S = \{k, \ell, m\}$. The Rao-Blackwellisation consists of calculating the expected value of the estimator conditional on the list of distinct units obtained, here being S (see Tillé, 2001, page 30).

Solution

1. The vector $(a_1, \ldots, a_k, \ldots, a_N)'$ has a multinomial distribution with exponent 3 and parameters $p_1, \ldots, p_k, \ldots, p_N$:

$$\Pr\left(a_k = r_k, k \in U\right) = \frac{3!}{r_1! \ldots r_k! \ldots r_N!} \prod_{k \in U} p_k^{r_k}.$$

and a_k following a binomial distribution $\mathcal{B}(3, p_k)$.

2. $\mathrm{E}\left[a_k\right] = 3p_k$, $\mathrm{var}\left[a_k\right] = 3p_k\left(1 - p_k\right)$ and $\mathrm{cov}\left[a_k, a_\ell\right] = -3p_k p_\ell$.

3. Since the three samples are independent, we have

$$\Pr\left(k \notin S\right) = \left(1 - p_k\right)^3,$$

$$\Pr\left(S = \{k\}\right) = p_k^3,$$

$$\begin{aligned}
\Pr\left(S = \{k, \ell\}\right) &= \Pr(a_k = 2 \cap a_\ell = 1) + \Pr(a_k = 1 \cap a_\ell = 2) \\
&= 3p_k^2 p_\ell + 3p_k p_\ell^2 \\
&= 3p_k p_\ell \left(p_k + p_\ell\right), k \neq \ell,
\end{aligned}$$

$$\Pr\left(S = \{k, \ell, m\}\right) = \Pr(a_k = a_\ell = a_m = 1) = 6p_k p_\ell p_m, k \neq \ell \neq m.$$

4. The inclusion probability is

$$\pi_k = \Pr\left(k \in S\right) = 1 - \Pr\left(k \notin S\right) = 1 - \left(1 - p_k\right)^3.$$

5. The conditional expectations are

$$\mathrm{E}\left[a_k \mid S = \{k\}\right] = \mathrm{E}\left[a_k \mid a_k = 3\right] = 3, k \in U,$$

$$\begin{aligned}
&\mathrm{E}\left[a_k \mid S = \{k, \ell\}\right] \\
&= \sum_{a_k = 1, 2} a_k \Pr\left(a_k \mid S = \{k, \ell\}\right) \text{ (null probability if } a_k = 0 \text{ or } a_k = 3) \\
&= 1 \frac{\Pr\left(a_k = 1 \cap S = \{k, \ell\}\right)}{\Pr\left(S = \{k, \ell\}\right)} + 2 \frac{\Pr\left(a_k = 2 \cap S = \{k, \ell\}\right)}{\Pr\left(S = \{k, \ell\}\right)} \\
&= \frac{2 \times \Pr\left[a_k = 2 \text{ and } a_l = 1\right] + 1 \times \Pr\left[a_k = 1 \text{ and } a_l = 2\right]}{\Pr\left[S = \{k, \ell\}\right]} \\
&= \frac{2 \times 3p_k^2 p_\ell + 1 \times 3p_k p_\ell^2}{3p_k^2 p_\ell + 3p_k p_\ell^2} = \frac{2p_k + p_\ell}{p_k + p_\ell}, k \neq \ell \in U,
\end{aligned}$$

$$\mathrm{E}\left[a_k \mid S = \{k, \ell, m\}\right] = 1, k \neq \ell \neq m \in U.$$

6. The unbiased Hansen-Hurwitz estimator is given by

$$\widehat{Y}_{HH} = \frac{1}{3} \sum_{i \in U} \frac{y_i a_i}{p_i}.$$

The estimator is unbiased, because for all i in U, we have $\mathrm{E}(a_i) = 3p_i$.

7. Finally, the Rao-Blackwellised estimator is obtained by calculating the conditional expectation of S, since S indicates the distinct units obtained in the initial sample of size 3.

$$\widehat{Y}_{RB} = \frac{1}{3} \sum_{i \in U} \frac{y_i}{p_i} \mathrm{E}\left[a_i \mid S\right].$$

In a general way, $\mathrm{E}\left[a_i \mid S\right] = 0$ if i does not appear in S. Question 5 gives the other values of $\mathrm{E}\left[a_i \mid S\right]$.
If $S = \{k\}$,

$$\widehat{Y}_{RB} = \frac{y_k}{p_k}.$$

If $S = \{k, \ell\}$,

$$\widehat{Y}_{RB} = \frac{y_k}{3p_k} \frac{2p_k + p_\ell}{p_k + p_\ell} + \frac{y_\ell}{3p_\ell} \frac{p_k + 2p_\ell}{p_k + p_\ell} = \frac{1}{3}\left(\frac{y_k}{p_k} + \frac{y_\ell}{p_\ell} + \frac{y_k + y_\ell}{p_k + p_\ell}\right).$$

If $S = \{k, \ell, m\}$,

$$\widehat{Y}_{RB} = \frac{1}{3}\left(\frac{y_k}{p_k} + \frac{y_\ell}{p_\ell} + \frac{y_m}{p_m}\right).$$

The Rao-Blackwellised estimator has a smaller variance than the Hansen-Hurwitz estimator, because

$$\mathrm{var}(\widehat{Y}_{HH}) = \mathrm{var}\mathrm{E}(\widehat{Y}_{HH} \mid S) + \mathrm{Evar}(\widehat{Y}_{HH} \mid S)$$
$$= \mathrm{var}(\widehat{Y}_{RB}) + \mathrm{Evar}(\widehat{Y}_{HH} \mid S)$$
$$\geq \mathrm{var}(\widehat{Y}_{RB}).$$

In a design with replacement, it is thus always theoretically possible to improve the Hansen-Hurwitz estimator. Unfortunately, the calculations turn out to be too complicated when $n > 3$.

Exercise 3.23 *Null second-order probabilities*

In a complex design of fixed size and with equal probabilities, show that the classical estimator $\widehat{\mathrm{var}}(\widehat{Y})$ underestimates on average the true variance $\mathrm{var}(\widehat{Y})$ as soon as there exists at least one pair (k, ℓ) whose second-order inclusion probability $\pi_{k\ell}$ is null. Express the bias as a function of y_k.

Solution

Since

$$\widehat{\mathrm{var}}(\widehat{Y}) = \frac{1}{2}\sum_{k\in S}\sum_{\substack{\ell\in S\\ \ell\neq k}} \frac{\pi_k\pi_\ell - \pi_{k\ell}}{\pi_{k\ell}}\left(\frac{y_k}{\pi_k} - \frac{y_\ell}{\pi_\ell}\right)^2$$

$$= \frac{1}{2}\sum_{k\in U}\sum_{\substack{\ell\in U\\ \ell\neq k\\ \pi_{k\ell}>0}} \frac{\pi_k\pi_\ell - \pi_{k\ell}}{\pi_{k\ell}}\left(\frac{y_k}{\pi_k} - \frac{y_\ell}{\pi_\ell}\right)^2 I_k I_\ell,$$

denoting I_k as the indicator for the inclusion in S, the expectation is

$$\mathrm{E}(\widehat{\mathrm{var}}(\widehat{Y})) = \frac{1}{2}\sum_{k\in U}\sum_{\substack{\ell\in U\\ \ell\neq k}} (\pi_k\pi_\ell - \pi_{k\ell})\left(\frac{y_k}{\pi_k} - \frac{y_\ell}{\pi_\ell}\right)^2 I\{\pi_{k\ell} > 0\}$$

$$= \mathrm{var}(\widehat{Y}) - \frac{1}{2}\sum_{k\in U}\sum_{\substack{\ell\in U\\ \ell\neq k}} (\pi_k\pi_\ell - \pi_{k\ell})\left(\frac{y_k}{\pi_k} - \frac{y_\ell}{\pi_\ell}\right)^2 I\{\pi_{k\ell} = 0\},$$

where $I\{A\}$ is 1 if A is true and 0 otherwise. The bias of $\widehat{\mathrm{var}}(\widehat{Y})$ is therefore:

$$\mathrm{E}(\widehat{\mathrm{var}}(\widehat{Y})) - \mathrm{var}(\widehat{Y}) = -\frac{1}{2}\sum_{k\in U}\sum_{\substack{\ell\in U\\ \ell\neq k}} \pi_k\pi_\ell\left(\frac{y_k}{\pi_k} - \frac{y_\ell}{\pi_\ell}\right)^2 I\{\pi_{k\ell} = 0\}.$$

This bias is strictly negative as soon as a $\pi_{k\ell}$ is null. If the probabilities π_k are all equal, the bias is:

$$\mathrm{B}(\widehat{Y}) = -\frac{1}{2}\sum_{k\in U}\sum_{\substack{\ell\in U\\ \ell\neq k}} (y_k - y_\ell)^2 I\{\pi_{k\ell} = 0\}.$$

In all cases, it is therefore harmful to use an algorithm imposing $\pi_{k\ell} = 0$ as soon as y_k and y_ℓ are very different.

Exercise 3.24 *Hájek's ratio*

The object of this exercise is to determine certain conditions in which the Hájek ratio is less efficient than the classical Horvitz-Thompson estimator. We consider that the size of the sample is large and that the sample is of fixed size.

1. Recall, for the estimation of a total Y, the variance expressions of the two estimators in question.

2. We can always write, for all $k \in U$,

$$y_k = \alpha + \beta x_k + u_k \quad \alpha, \ \beta \in \mathbb{R},$$

where α and β are the regression coefficients ('true' and unknown) for y on x, $\pi_k = n x_k / X$, x_k is a size variable, and the units are consequently selected proportionally to the size. Furthermore, we assume that u_k is 'small', that is to say that x 'explains well' y.

Under these conditions, what happens to the variance expressions for the two estimators? (Reminder: the terms in u_k are numerically negligible.)

3. What is (approximately) the ratio of the two variances?

4. In conclusion, under the conditions of a strong linear correlation between x and y (that is, u_k is small), when can we 'qualitatively' consider that the Horvitz-Thompson estimator is preferable to Hájek's ratio?

Solution

1. Let $\widehat{Y}$ be the Horvitz-Thompson estimator and $\widehat{Y}_H$ be the Hájek ratio. We must compare

$$\text{var}(\widehat{Y}) \quad \text{and} \quad \text{var}(\widehat{Y}_H) = \text{var}\left(N \frac{\widehat{Y}}{\widehat{N}}\right).$$

By linearisation, since the sample size is large, we have

$$\text{var}\left(\frac{\widehat{Y}}{\widehat{N}}\right) \approx \text{var}\left(\frac{1}{N}\left(\widehat{Y} - \overline{Y}\widehat{N}\right)\right) = \frac{1}{N^2}\, \text{var}(\widehat{Z}),$$

where

$$z_k = y_k - \overline{Y} \Rightarrow \text{var}(\widehat{Y}_H) \approx \text{var}(\widehat{Z}).$$

Thus

$$\text{var}(\widehat{Y}) = \frac{-1}{2} \sum_{k \in U} \sum_{\substack{\ell \in U \\ \ell \neq k}} \Delta_{k\ell} \left(\frac{y_k}{\pi_k} - \frac{y_\ell}{\pi_\ell}\right)^2,$$

where $\Delta_{k\ell} = \pi_{k\ell} - \pi_k \pi_\ell$, and

$$\text{var}(\widehat{Y}_H) \approx -\frac{1}{2} \sum_{k \in U} \sum_{\substack{\ell \in U \\ \ell \neq k}} \Delta_{k\ell} \left(\frac{y_k - \overline{Y}}{\pi_k} - \frac{y_\ell - \overline{Y}}{\pi_\ell}\right)^2.$$

2. If we can write

$$y_k = \alpha + \beta x_k + u_k,$$

then

$$\overline{Y} = \alpha + \beta \overline{X} + \overline{U},$$

where α and β are the 'true' regression coefficients and $\overline{U} = 0$. Then, considering that the u_k are small:

$$\left(\frac{y_k}{\pi_k} - \frac{y_\ell}{\pi_\ell}\right)^2 \approx \alpha^2 \left(\frac{1}{\pi_k} - \frac{1}{\pi_\ell}\right)^2,$$

and

$$\left(\frac{y_k - \overline{Y}}{\pi_k} - \frac{y_\ell - \overline{Y}}{\pi_\ell}\right)^2 \approx \beta^2 \overline{X}^2 \left(\frac{1}{\pi_k} - \frac{1}{\pi_\ell}\right)^2.$$

3. We therefore have:

$$\frac{\operatorname{var}(\widehat{Y})}{\operatorname{var}(\widehat{Y}_H)} \approx \left(\frac{\alpha}{\beta\overline{X}}\right)^2.$$

4. The estimator $\widehat{Y}$ is preferable to $\widehat{Y}_H$ if and only if

$$\left(\frac{\alpha}{\beta\overline{X}}\right)^2 < 1 \quad \Leftrightarrow \quad |\alpha| < |\beta|\,\overline{X}. \tag{3.6}$$

In fact,

$$\beta = \frac{S_{xy}}{S_x^2}, \quad \text{and} \quad \alpha = \overline{Y} - \beta\overline{X}.$$

By (3.6), we get

$$|\overline{Y} - \beta\overline{X}| < |\beta|\,\overline{X}.$$

Let us suppose that $\overline{Y} > 0$.

- Case 1: $S_{xy} > 0$: Expression (3.6) is equivalent to

$$-\beta\overline{X} < \overline{Y} - \beta\overline{X} < \beta\overline{X} \Leftrightarrow \overline{Y} < 2\beta\overline{X} \Leftrightarrow S_x^2\overline{Y} < 2S_{xy}\overline{X},$$

which is still equivalent to

$$1 \geq \rho > \frac{1}{2}\frac{\operatorname{CV}_x}{\operatorname{CV}_y},$$

where CV_x is the coefficient of variation of the x_k. This condition requires in particular $\operatorname{CV}_x < 2\operatorname{CV}_y$.

- Case 2: $S_{xy} \leq 0$: (3.6) is never satisfied.

Conclusion:
$\widehat{Y}$ is only preferable to $N\widehat{Y}/\widehat{N}$ under the following conditions:

$$S_{xy} > 0, \quad \operatorname{CV}_x < 2\operatorname{CV}_y \quad \text{and} \quad \rho > \frac{1}{2}\frac{\operatorname{CV}_x}{\operatorname{CV}_y}.$$

In practice, and qualitatively, as we are immediately under the conditions where ρ is close to 1 (u_k small), we retain that $\widehat{Y}$ is a serious competitor to Hájek's ratio when the correlation between x and y is positive and that CV_x is noticeably smaller than $2CV_y$.

These calculations are approximate, but they show that the ratio is not systematically better than $\widehat{Y}$: if the relationship between x and y is purely linear ($\alpha = 0$), it is certain that the ratio is not interesting. If it is strongly linear, the result is similar: the ratio is only interesting if there is a constant term α 'sufficiently large'.

When we select a sample with inclusion probabilities proportional to size x_k, we are placed exactly in a hypothesis where β is large and α rather small, that is, in a situation where $\widehat{Y}$ can happen to be greatly better than the Hájek ratio.

Exercise 3.25 *Weighting and estimation of the population size*

We consider a variable y measured on the observation units of a population U of size N. Letting a be a fixed value and known and y_k be the value of y for unit k in U, we construct the variable

$$z_k = y_k + a.$$

Let Y and Z denote respectively the totals of variables y and z in U. We select from U a random sample S using any design.

1. Recall the expressions for the linear estimators $\widehat{Z}$ of Z and $\widehat{Y}$ of Y.
2. Consider the estimator $\widehat{Y}_a = \widehat{Z} - a \times N$. Give the relations that the sampling weights must satisfy so that the estimators $\widehat{Y}$ and $\widehat{Y}_a$ are identical for all values of a.
3. We consider a complex design in which the weights are not random. If the estimator is unbiased, what are these weights?
4. With the estimator defined in 3., how do we write the relations from 2.? How do we obtain them?

Solution

1. The linear estimators are

$$\widehat{Z} = \sum_{k \in S} w_k(S) z_k, \quad \text{and} \quad \widehat{Y} = \sum_{k \in S} w_k(S) y_k,$$

where $w_k(S)$ is the weight for unit k. This weight, in general, depends on k but also on sample S.

2. The estimator $\widehat{Y}_a$ is

$$\widehat{Y}_a = \sum_{k \in S} w_k(S) z_k - aN = \sum_{k \in S} w_k(S) y_k + a \sum_{k \in S} w_k(S) - aN.$$

In order that $\widehat{Y}_a$ and $\widehat{Y}$ be equal for all a, it is necessary and sufficient that, for all S

$$\sum_{k \in S} w_k(S) = N. \tag{3.7}$$

There are as many equations as there are samples S possible.

3. Consider the estimator

$$\widehat{Y}_w = \sum_{k \in S} w_k y_k,$$

where the weights w_k are not random, meaning that they do not depend on S. The expected value is (denoting I_k as the indicator for the inclusion in S):

$$\mathrm{E}(\widehat{Y}_w) = \sum_{k \in S} w_k y_k \mathrm{E}(I_k) = \sum_{k \in U} y_k w_k \pi_k.$$

So that $\mathrm{E}(\widehat{Y}_w) = Y$, it is necessary and therefore sufficient that $w_k \pi_k = 1$, for all $k \in U$, meaning that $w_k = 1/\pi_k$. We then encounter the classical Horvitz-Thompson estimator.

4. If $w_k = 1/\pi_k$, then Condition (3.7) becomes

$$\sum_{k \in S} \frac{1}{\pi_k} = N.$$

This relation can be obtained if we use a sample balanced on the variable with a value of 1 everywhere on U.

Exercise 3.26 *Poisson sampling*

When we undertake a sample with unequal probabilities, in the large majority of cases, we use sampling methods of fixed size. Nevertheless, there exist very simple algorithms allowing for samples with unequal probabilities but conferring on the sample a size variable. We are interested here in one of these algorithms (called 'Poisson sampling').

Method: Having a population of size N, we hold a lottery for each individual, independent from one individual to another: if we have a set of N values $\pi_1, \ldots, \pi_k, \ldots, \pi_N$ such that $0 < \pi_k < 1$, we generate a set of N independent risks $u_1, \ldots, u_k, \ldots, u_N$ under the uniform distribution over [0,1], and we retain individual k if and only if $u_k \leq \pi_k$.

1. a) What is the inclusion probability for each individual k?

b) What is the expected value (denoted ν) of the sample size? What is its variance?

c) What is the probability that the sample has a size at least equal to 1?

2. We then use the estimator of the total:

$$\widehat{Y} = \sum_{k \in S} \frac{y_k}{\pi_k},$$

where S indicates the final sample selected at the outcome of the N lotteries (we assume that this sample has a size at least equal to 1).

a) Verify that $\widehat{Y}$ is unbiased for the true total Y.

b) What is the true variance of $\widehat{Y}$? How can we estimate it without bias? (We always suppose that the sample has a nonzero size.)

c) What is the second-order inclusion probability $\pi_{k\ell}$?

3. We assume here that the average size of the sample is a value that we fixed (denoted ν) and that it is sufficiently small so that the inclusion probabilities that we are manipulating are all less than 1.

a) What values π_k are we interested in *selecting* to obtain an optimal accuracy? We will specify in having ν 'sufficiently small' so as to not have problems with the calculation.

b) In the case where $\nu \ll N$, what is the optimal accuracy? We will express this as a function of N, ν, $\overline{Y}$ and CV, where CV represents the true coefficient of variation of the y_k in the population.

c) Still in the case where ν is 'sufficiently small', compare with a simple random sample of size ν (consider the magnitude of CV in a 'real situation').

4. Finally, what is the advantage of this method? Interpret this.

Solution

1. a) Clearly,
$$\Pr[k \in S] = \Pr[u_k \le \pi_k] = F(\pi_k) = \pi_k,$$

where $F(.)$ is the cumulative distribution function of the uniform distribution $U_{[0,1]}$, being $F(x) = x$ over $[0,1]$.

b) The sample size n is a random variable taking the values in
$$\{0, 1, \ldots, N\}.$$

If
$$I_k = \begin{cases} 1 \text{ if } k \in S \\ 0 \text{ if } k \notin S, \end{cases}$$

then
$$n = \sum_{k \in U} I_k, \quad \nu = \mathrm{E}(n) = \sum_{k \in U} \mathrm{E}(I_k) = \sum_{k \in U} \pi_k,$$

and

$$\text{var}(n) = \sum_{k \in U} \text{var}(I_k),$$

because the samples are *independent* from one individual to another (since the u_k are as well). The variance of n is:

$$\text{var}(n) = \sum_{k \in U} \pi_k(1 - \pi_k),$$

because the I_k follow a Bernoulli distribution: $I_k \sim \mathcal{B}(1, \pi_k)$.
c) $\Pr[n = 0] = (1 - \pi_1)(1 - \pi_2)(1 - \pi_3) \dots (1 - \pi_N)$.
   Therefore

$$\Pr[n \geq 1] = 1 - \prod_{k \in U}(1 - \pi_k).$$

2. a) We have

$$\widehat{Y} = \sum_{k \in U} \frac{y_k}{\pi_k} I_k \quad \Rightarrow \quad \text{E}(\widehat{Y}) = \sum_{k \in U} \frac{y_k}{\pi_k} \text{E}(I_k) = \sum_{k \in U} y_k.$$

   b) Due to the independence of samples, the variance of the sum is the sum of the variances,

$$\text{var}(\widehat{Y}) = \sum_{k \in U} \left(\frac{y_k}{\pi_k}\right)^2 \text{var}(I_k) = \sum_{k \in U} \frac{y_k^2}{\pi_k^2} \pi_k(1 - \pi_k) = \sum_{k \in U} \frac{1 - \pi_k}{\pi_k} y_k^2,$$

and

$$\widehat{\text{var}}(\widehat{Y}) = \sum_{k \in S} \frac{1 - \pi_k}{\pi_k^2} y_k^2 \text{ (instantly verifiable)}.$$

   c) Obvious: $\pi_{k\ell} = \pi_k \times \pi_\ell$ (due to the independence of samples).
   Note: The Sen-Yates-Grundy variance form is *no longer* valid here, as the sample is no longer of fixed size.
3. a) The problem to solve is:

$$\min_{\pi_k} \sum_{k \in U} \frac{1 - \pi_k}{\pi_k} y_k^2,$$

subject to

$$\sum_{k \in U} \pi_k = \nu \text{ and } 0 < \pi_k \leq 1.$$

We 'forget' for the moment the inequality constraints. This problem is equivalent to

$$\min_{\pi_k} \sum_{k \in U} \frac{y_k^2}{\pi_k}, \text{ subject to } \sum_{k \in U} \pi_k = \nu.$$

The Lagrangian function is:

$$\mathcal{L} = \sum_{k \in U} \frac{y_k^2}{\pi_k} - \mu \left(\sum_{k \in U} \pi_k - \nu \right),$$

and

$$\frac{\partial \mathcal{L}}{\partial \pi_k} = 0 \quad \Rightarrow \quad -\frac{y_k^2}{\pi_k^2} - \mu = 0.$$

The π_k must be proportional to y_k: $\pi_k = \lambda y_k$, with

$$\lambda = \frac{\nu}{\sum_{k \in U} y_k}.$$

Therefore,

$$\pi_k = \nu \frac{y_k}{\sum_{k \in U} y_k}, \quad \text{for all } k = 1, 2, \ldots, N.$$

To not have problems in calculation, it is necessary and sufficient that:

$$\nu \frac{y_k}{\sum_{\ell \in U} y_\ell} \le 1, \quad k = 1, 2, \ldots, N,$$

that is:

$$\nu \le \frac{\sum_{k \in U} y_k}{\max_{1 \le k \le N} (y_k)} = N \frac{\overline{Y}}{\max_{1 \le k \le N} (y_k)}.$$

In this case, the inequality constraints defined in the first optimisation problem do not have to be taken into account. Except for the pathological case where there exist some large values y_k, it is an 'easy' condition to obtain from the moment where $\nu \ll N$: this last inequality corresponds to that for which we understood to be 'sufficiently small'.

b) We let $\underset{\min}{V}$ be the optimal accuracy:

$$\underset{\min}{V} = \sum_{k \in U} \frac{1 - \pi_k}{\pi_k} y_k^2,$$

where

$$\pi_k = \nu \frac{y_k}{Y}, \quad \text{and} \quad Y = \sum_{k \in U} y_k.$$

Thus,

$$\underset{\min}{V} = \sum_{k \in U} \frac{y_k^2}{\nu y_k / Y} - \left(\sum_{k \in U} y_k^2 \right) = \frac{1}{\nu} \left(\sum_{k \in U} y_k \right)^2 - \left(\sum_{k \in U} y_k^2 \right).$$

We have:

$$NS_y^2 \approx \left(\sum_{k\in U} y_k^2\right) - \frac{\left(\sum_{k\in U} y_k\right)^2}{N},$$

which implies that

$$\underset{\min}{V} \approx \frac{1}{\nu} Y^2 - \left[NS_y^2 + \frac{Y^2}{N}\right] = \left(\frac{1}{\nu} - \frac{1}{N}\right) Y^2 - NS_y^2.$$

Let us set $f = \nu/N$, then

$$\underset{\min}{V} \approx \frac{1-f}{\nu} Y^2 - NS_y^2 = Y^2 \left[\frac{1-f}{\nu} - N\left(\frac{S_y}{Y}\right)^2\right]$$

$$= Y^2 \left[\frac{1-f}{\nu} - \frac{CV^2}{N}\right],$$

where CV is the (true) coefficient of variation of the y_k:

$$CV = \frac{S_y}{\overline{Y}},$$

with $\overline{Y} = Y/N$. Now, since we are placed in the situation $\nu \ll N$, we have $f \ll 1$. Thus:

$$\underset{\min}{V} \approx Y^2 \left[\frac{1}{\nu} - \frac{CV^2}{N}\right].$$

That is, *in fine*:

$$\underset{\min}{V}(\widehat{Y}) \approx N^2 \left(\frac{1}{\nu} - \frac{CV^2}{N}\right) \overline{Y}^2.$$

c) Furthermore:

$$\underset{SRS}{V} = \underset{SRS}{\mathrm{var}}(\widehat{Y}) = N^2 \frac{1-f}{\nu} S_y^2,$$

where $f = \nu/N$, that is

$$\underset{SRS}{\mathrm{var}}(\widehat{Y}) \approx N^2 \frac{1}{\nu} S_y^2,$$

as f is very small compared to 1 considering the hypothesis made upon ν.

In a 'real situation' the CV are *small*, in general noticeably smaller than 1. It is rare that we have $S_y > \overline{Y}$ with a distribution of income, or for physical heights of individuals for example, but even if $S_y > \overline{Y}$, we would anyhow have to have:

$$\frac{1}{\nu} \gg \frac{\mathrm{CV}^2}{N} \quad \Leftrightarrow \quad \frac{N}{\nu} \gg \mathrm{CV}^2,$$

which is indeed our 'starting hypothesis', being ν very small compared to N. Therefore

$$\underset{\min}{V} = \underset{\min}{V}(\widehat{Y}) \approx N^2 \frac{1}{\nu} \overline{Y}^2.$$

In conclusion, we have

$$\frac{\underset{\mathrm{SRS}}{V}}{\underset{\min}{V}} \approx \left(\frac{S_y}{\overline{Y}}\right)^2 = \mathrm{CV}^2.$$

Under these conditions, we would have to have *a priori*

$$\underset{\mathrm{SRS}}{V} \leq \underset{\min}{V}.$$

4. From the accuracy viewpoint, that Poisson sampling truly has little advantage, it is the contrary! Indeed, in the previous case, its variance had been calculated with the optimal π_k and, in spite of that, it is greater than that for simple random sampling. The problem is that each lottery brings its share to the variance and that all the variances *add up*. Even if we had y_k/π_k rigorously constant, it would even so be necessary to 'collect' the variability due to the size variable for the sample, and it is already an important source of inaccuracy. From the point of view of necessary and sufficient *information* to perform the sampling, it is clear that the Poisson sample uses a minimum amount of information, as it is unnecessary to have information about the individuals *other* than k when we prepare to hold the lottery on k. This is a kind of 'minimal information' sampling (but we pay for it with a degraded accuracy). In comparison, we indeed see that in 'classical' sampling, we have $\pi_k = nx_k/X$, and it is thus necessary *to know* X (auxiliary information), being the x_k for *all* the individuals before we can begin the algorithm. This last case, which corresponds for example to systematic sampling, is thus more demanding in terms of information.

Exercise 3.27 *Quota method*

We are interested in a quota sampling design based upon two qualitative variables x_1 and x_2 (see, for an introduction to the method and the vocabulary, Ardilly, 1994). We denote $\overline{Y}_{ij}$ as the true mean of the variable of interest y in the sub-population U_{ij} of size N_{ij} intersecting the modes i of x_1 and j of x_2. The sample S again intersecting U_{ij} is denoted S_{ij} and its size is n_{ij}. For every individual k of U_{ij}, we define $\varepsilon_k = y_k - \overline{Y}_{ij}$. We traditionally use the true simple mean $\widehat{\overline{Y}}$ in S as the estimator of the true mean $\overline{Y}$.

1. Write $\widehat{\overline{Y}}$ as a function of $\overline{Y}_{ij}$, n_{ij} and ε_k.
2. Empirical practice leads the interviewer to select individual k with a probability π_k (unknown). Calculate the bias of $\widehat{\overline{Y}}$, conditional on n_{ij}. We will express this while using the covariances between y and π in the subpopulations U_{ij}.
3. What can we fear, in practice?
4. Under what favourable conditions is the conditional bias null?
   We will be successively placed in the following cases:
   a) intersecting quotas,
   b) marginal quotas.

Solution

1. We have

$$\widehat{\overline{Y}} = \frac{1}{n} \sum_{k \in S} y_k = \sum_i \sum_j \frac{n_{ij}}{n} \widehat{\overline{Y}}_{ij},$$

where

$$\widehat{\overline{Y}}_{ij} = \frac{1}{n_{ij}} \sum_{k \in S_{ij}} y_k = \overline{Y}_{ij} + \frac{1}{n_{ij}} \sum_{k \in S_{ij}} \varepsilon_k.$$

2. We conditionally justify on n_{ij} (for which the distribution is complex):

$$\mathrm{E}(\widehat{\overline{Y}}) = \sum_i \sum_j \frac{n_{ij}}{n} \mathrm{E}(\widehat{\overline{Y}}_{ij}).$$

In fact,

$$\mathrm{E}(\widehat{\overline{Y}}_{ij}) = \overline{Y}_{ij} + \frac{1}{n_{ij}} \sum_{k \in U_{ij}} \varepsilon_k \pi_k.$$

Furthermore, if we denote C_{ij} as the covariance in U_{ij} between y and π, we have:

$$C_{ij} = \frac{1}{N_{ij}} \sum_{k \in U_{ij}} (y_k - \overline{Y}_{ij})(\pi_k - \overline{\pi}_{ij}) = \frac{1}{N_{ij}} \sum_{k \in U_{ij}} \varepsilon_k(\pi_k - \overline{\pi}_{ij}),$$

where

$$\overline{\pi}_{ij} = \frac{1}{N_{ij}} \sum_{k \in U_{ij}} \pi_k.$$

Since

$$\sum_{k \in U_{ij}} \varepsilon_k = 0,$$

we have

$$C_{ij} = \frac{1}{N_{ij}} \sum_{k \in U_{ij}} \varepsilon_k \pi_k.$$

Finally,

$$E(\widehat{\overline{Y}}_{ij}) = \overline{Y}_{ij} + \frac{N_{ij}}{n_{ij}}C_{ij},$$

and

$$E(\widehat{\overline{Y}} - \overline{Y}) = \left(\sum_i \sum_j \frac{n_{ij}}{n}\overline{Y}_{ij} - \overline{Y}\right) + \frac{N}{n}\sum_i \sum_j \frac{N_{ij}}{N}C_{ij}$$

$$= \sum_i \sum_j \left(\frac{n_{ij}}{n} - \frac{N_{ij}}{N}\right)\overline{Y}_{ij} + \frac{N}{n}\sum_i \sum_j \frac{N_{ij}}{N}C_{ij}.$$

3. The bias is composed of two terms:

$$A = \sum_i \sum_j \left(\frac{n_{ij}}{n} - \frac{N_{ij}}{N}\right)\overline{Y}_{ij}, \qquad B = \frac{N}{n}\sum_i \sum_j \frac{N_{ij}}{N}C_{ij}.$$

That suggests the following comments:

- The term A, linked in a complex way to π_k through n_{ij}, is not zero unless, for all (i, j) we have $n_{ij}/n = N_{ij}/N$. This case corresponds to the intersecting quotas, which can only be brought into use if we know the N_{ij}. In the case of marginal quotas, even if n becomes very large, we cannot really count on a convergence of A towards zero, because we can imagine that certain categories U_{ij} are left frequently underrepresented while the marginal quotas remain satisfied.
- The term B does not have any particular reason to be zero (also see 4.), as in a general manner, we very well imagine that in practice the empirical methods produce a correlation between y and π: the interviewer could indeed select an individual k with a probability π_k linked to the value of y_k. If we take the example of a survey on the duration of work, π_k will be most probably negatively correlated with y_k, because a person who works a great deal will be more difficult to contact. This term is moreover still less sensitive to the size n than is A: we can even say that it only depends a little on n, as through this analogy with probabilistic samples, we can think that π_k varies *a priori* like n ($\pi_k = n/N$ in simple random sampling, $\pi_k = nX_i/X$ in sampling proportional to size, for example). By this analogy, we can suppose that it signifies that C_{ij} varies like n (more or less), and therefore that B does not practically depend on n. This persistence of a bias *a priori* (unmeasurable) is generally presented as a weakness of the quota method.

4. a) Case of intersecting quotas:
    The bias is reduced to the term B. In order for it to be zero, it is necessary and sufficient that the C_{ij} are all zero. In practice, two cases appear favourable:

- either y_k is constant in U_{ij}: it is indeed to approach this context that we find in practice the quota variables x_1 and x_2 which best explain y,
- or π_k is constant in U_{ij}: to approach this context, we conceive collection instructions in such a way that sampling carried out by the interviewer is the most 'uniform' possible.

b) Case of marginal quotas:

The conclusions in case (a) still hold, but it is necessary to add the condition $A = 0$. In a conditional approach, the favourable case is that for the additive decomposition of $\overline{Y}_{ij}$ of type:

$$\overline{Y}_{ij} = a_i + b_j \text{ for all } i, \text{ for all } j.$$

Indeed:

$$A = \sum_i \sum_j \left(\frac{n_{ij}}{n} - \frac{N_{ij}}{N} \right)(a_i + b_j)$$

$$= \sum_i \left(\sum_j \frac{n_{ij}}{n} - \sum_j \frac{N_{ij}}{N} \right) a_i + \sum_j \left(\sum_i \frac{n_{ij}}{n} - \sum_i \frac{N_{ij}}{N} \right) b_j.$$

Satisfying the marginal quotas sets,

$$\text{for all } j \ \sum_i \frac{n_{ij}}{n} = \frac{n_{.j}}{n} = \frac{N_{.j}}{N} = \sum_i \frac{N_{ij}}{N},$$

$$\text{and for all } i \ \sum_j \frac{n_{ij}}{n} = \sum_j \frac{N_{ij}}{N}, \text{ that is } A = 0.$$

In conclusion, to protect against bias that is of great importance in the framework of the marginal quota method (it is a question of a frequently encountered design), we can proceed by simultaneously ensuring:

- selection probabilities π_k as invariable as possible;
- a choice of quota variables that explains well the variable of interest, according to a model of type (for the two quota variables)

$$y_k = a_i + b_j + \varepsilon_k$$

with ε_k small.

Exercise 3.28 *Successive balancing*

We select a sample S_1 with probabilities of selection $\pi_{k,1}$ from a population U of size N. This sample is balanced (see, for an overview of balanced sampling, Deville and Tillé, 2004; Tillé, 2001, Chapter 8) on two variables: $\pi_{k,1}$ and an auxiliary variable x_k. Sometime later on, we select from the complement $U \backslash S_1$ a second sample S_2, with equal probabilities and of fixed size. We balance S_2 on x_k.

1. Write the balancing equations. What can we say about the size of S_1?
2. What is the selection probability π_k in the overall sample $S = S_1 \cup S_2$?
3. Is the overall sample S balanced on x_k?
4. Examine the particular case where $\pi_{k,1}$ is constant.

Solution

1. The balancing of S_1 leads to two equations:

$$\sum_{k \in S_1} \frac{\pi_{k,1}}{\pi_{k,1}} = \sum_{k \in U} \pi_{k,1}, \qquad (3.8)$$

$$\sum_{k \in S_1} \frac{x_k}{\pi_{k,1}} = \sum_{k \in U} x_k. \qquad (3.9)$$

The left-hand term of (3.8) represents the size of S_1, which is consequently constant. The balancing of S_2 is carried out in $U \backslash S_1$ and leads to the following equation:

$$\sum_{k \in S_2} \frac{x_k}{\pi_{k,2}} = \sum_{k \in U \backslash S_1} x_k, \qquad (3.10)$$

where $\pi_{k,2}$ is the selection probability of k in S_2. Since S_2 is of fixed size n_2 and with equal probabilities, we have

$$\pi_{k,2} = \frac{n_2}{N - n_1} \text{ for all } k \text{ of } U \backslash S_1.$$

Hence

$$\frac{1}{n_2} \sum_{k \in S_2} x_k = \frac{\sum_{k \in U} x_k - \sum_{k \in S_1} x_k}{N - n_1}. \qquad (3.11)$$

2. We denote $p(s_1)$ as the probability of selecting s_1.

$$\pi_k = Pr(k \in S) = \sum_{s_1} Pr(k \in S|s_1)p(s_1)$$

$$= \sum_{s_1 \ni k} 1 \times p(s_1) + \sum_{s_1 \not\ni k} Pr(k \in S|s_1)p(s_1).$$

In fact, by definition $\pi_{k,1} = \sum_{s_1 \ni k} p(s_1)$ and, for s_1 such that $k \notin s_1$,

$$Pr(k \in S|s_1) = Pr(k \in S_2) = \frac{n_2}{N - n_1}.$$

Therefore,

$$\pi_k = \pi_{k,1} + \frac{n_2}{N - n_1} \sum_{s_1 \not\ni k} p(s_1) = \pi_{k,1} + \frac{n_2}{N - n_1}(1 - \pi_{k,1}).$$

3. Balancing S on x_k would correspond to the equality, for all S:

$$\sum_{k \in S} \frac{x_k}{\pi_k} = \sum_{k \in U} x_k,$$

such that

$$\sum_{k \in S_1} \frac{x_k}{\pi_k} + \sum_{k \in S_2} \frac{x_k}{\pi_k} = \sum_{k \in U} x_k. \tag{3.12}$$

Obviously, considering the relation which links π_k to $\pi_{k,1}$, we cannot use Equations (3.9) and (3.11) to get this balance, in any general context where $\pi_{k,1}$ is unspecified.

4. If $\pi_{k,1}$ is constant, since S_1 is of fixed size n_1, we inevitably have $\pi_{k,1} = n_1/N$. In this particular case, we get $\pi_k = (n_1+n_2)/N$. But (3.9) simplifies to:

$$\frac{1}{n_1} \sum_{k \in S_1} x_k = \frac{1}{N} \sum_{k \in U} x_k.$$

Likewise, (3.11) becomes:

$$\frac{1}{n_2} \sum_{k \in S_2} x_k = \frac{1}{N - n_1} \left(1 - \frac{n_1}{N}\right) \sum_{k \in U} x_k = \frac{1}{N} \sum_{k \in U} x_k.$$

Along the way, we notice that this last equality signifies that S_2 is balanced on x_k in U (and no longer only in $U \backslash S_1$). Finally, the left-hand side of (3.12) becomes:

$$\frac{N}{n_1 + n_2} \left(\sum_{k \in S_1} x_k + \sum_{k \in S_2} x_k \right) = \frac{N}{n_1 + n_2} \left(\frac{n_1}{N} \sum_{k \in U} x_k + \frac{n_2}{N} \sum_{k \in U} x_k \right)$$

$$= \sum_{k \in U} x_k.$$

There is indeed a balancing of S on x_k.

Exercise 3.29 *Absence of a sampling frame*

This exercise deals with estimation in a context of an absence of an exhaustive sampling frame of individuals. More precisely, it is about introducing a method of estimation in a survey of homeless people who frequent a given shelter in a given city. The shelter does not have any list of names other than from day to day.

1. What statistical problem are we going to naturally encounter if we are content to place a team of interviewers for a given day at the shelter?
2. We decide to observe the population for a period of T consecutive days (t represents the day, t varies from 1 to T, for example describing a complete month). We consider that a homeless person frequents the shelter at most one time each day, and we denote U_t as the population having frequented the shelter on day t. Under these conditions, what is the population of interest $\tilde{U}_T$? What is the unit of observation and what is the sampling unit? What technical difficulty are we going to face during this phase of estimation?

3. We are interested in a variable y_k that does not depend on time (example: age at the end of the study). We denote r_k as the total number of visits made to the shelter by individual k of $\widetilde{U}_T$ during the course of the period T $(r_k = 1, 2, \ldots, T)$. Express the total

$$Y = \sum_{\widetilde{U}_T} y_k$$

as a function of the sums of y_k/r_k on the populations U_t.

4. If, on day t, individual k frequenting the shelter is selected with probability π_k^t, how do we estimate Y without bias? (We denote $\widehat{Y}$ as the estimator and s_t as the sample for day t.)

5. What is the variance of $\widehat{Y}$? (The sample is of fixed size every day t, and the samples are independent from one day to the next.) How must we choose the sampling probabilities π_k^t?

6. Write $\widehat{Y}$ in the form of a linear estimator involving a sum on S_T, the overall sample obtained during the period T, with

$$S_T = \bigcup_{t=1}^{T} s_t.$$

Does $\widehat{Y}$ depend on the inclusion probability in S_T?

7. In practice, where does the difficulty lie with estimation for this survey?

Solution

1. There exists a considerable 'conditional' bias on the selected day: indeed, we claim to estimate a parameter on the entire population for which certain individuals cannot be surveyed. The inference is only valid for the homeless population having frequented the shelter on day t (the other homeless individuals have an inclusion probability of zero), which is surely not the result we are looking for. If we consider that the day is 'randomly' selected, the bias can disappear if an adequate weighting scheme is used and if we begin from the hypothesis that over the course of period T each homeless person frequents the shelter at least once. On the other hand, the variance is strong if the characteristics of frequenting depend appreciably on the selected day (we can imagine that weather conditions are a deciding factor, for example: according to whether we select a very cold day or a mild day instead, we probably have survey units with a rather different profile).

2. The inference focuses in this case on:

$$\widetilde{U}_T = \bigcup_{t=1}^{T} U_t.$$

We obviously get a better coverage of the population in a precarious situation with $\widetilde{U}_T$ than for any U_t. Obviously, this coverage is improved when T increases because there exist people who only visit the shelter occasionally. It is, however, very difficult (impossible?) to find a totally satisfying concept for the population in this context, starting from the moment where we are interested in a population that is naturally unstable in time (like all human population everywhere, but this is especially marked in this sensitive domain): conversely if T is large (one year for example), the 'social' sense of $\widetilde{U}_T$ such as the collection of punctual populations U_t which are evolving becomes questionable. The observation unit is the homeless person as long as he is part of $\widetilde{U}_T$. The sampling unit is instead the visit made by the individual in the shelter, on a given day. The difficulty is due to the fact that $U_t \cap U_{t'} \neq \emptyset$. An individual can thus be selected through several visits (at the most, he can be selected each day t if he frequents the shelter every day). This multiplicity of visits constitutes a particular technical difficulty for estimation, as a homeless individual who frequents the shelter often has a greater chance of being selected than an individual who seldom visits. It is then necessary to find adequate weighting.

3. We have:

$$\sum_{t=1}^{T} \sum_{k \in U_t} y_k = \sum_{k \in \widetilde{U}_T} r_k y_k,$$

as on T days, y_k is encountered r_k times as a member of the left-hand side. Therefore,

$$Y = \sum_{k \in \widetilde{U}_T} y_k = \sum_{t=1}^{T} \sum_{k \in U_t} \frac{y_k}{r_k}.$$

4. The fundamental contribution of the previous rewriting is due to the fact that the sampling is effectively practical for the units of U_t (in fact for the visits, but a visit refers to a single individual on a given day), and not for those of $\widetilde{U}_T$ (population constructed from U_t but that does not directly identify the sampling units). We estimate without bias

$$\sum_{k \in U_t} \frac{y_k}{r_k} \qquad \text{by} \qquad \sum_{k \in s_t} \frac{y_k}{\pi_k^t r_k}.$$

Therefore:

$$\widehat{Y} = \sum_{t=1}^{T} \sum_{k \in s_t} \frac{y_k}{\pi_k^t r_k}$$

estimates Y without bias, where s_t designates the selected sample over the course of day t.

5. The variance is:

$$\mathrm{var}(\widehat{Y}) = \sum_{t=1}^{T} \mathrm{var}\left(\sum_{k \in s_t} \frac{y_k / r_k}{\pi_k^t} \right),$$

because the samples are independent from one day to another. Thus

$$\text{var}(\widehat{Y}) = \sum_{t=1}^{T} \underbrace{\sum_{i \neq j}}_{U_t} (\pi_i^t \pi_j^t - \pi_{i,j}^t) \left(\frac{y_i}{r_i \pi_i^t} - \frac{y_j}{r_j \pi_j^t} \right)^2.$$

This is the classical expression, applied on the individual variable y_i/r_i. We are interested in having π_i^t as proportional as possible to y_i/r_i: what is original here is the presence of the factor $1/r_i$. If it is possible, that is if in practice we have at our disposal *a priori* the information r_i (or some information which is more or less proportional to it), we then more likely select the individuals where r_i is small, being the homeless people who only rarely frequent the shelter.

6. The estimator is written

$$\widehat{Y} = \sum_{k \in S_T} \left(\sum_{t \in \mathcal{S}_k} \frac{1}{\pi_k^t} \right) \frac{y_k}{r_k},$$

where

$$\mathcal{S}_k = \{t = 1, \ldots, T \text{ such that } k \in s_t\}.$$

Therefore

$$\widehat{Y} = \sum_{k \in S_T} w_k y_k \quad \text{where} \quad w_k = \frac{1}{r_k} \left(\sum_{t \in \mathcal{S}_k} \frac{1}{\pi_k^t} \right).$$

The inclusion probability of k in S_T is:

$$\Pr[k \in S_T] = \Pr \left[\bigcup_{t=1}^{T} \{k \in s_t\} \right].$$

The events $\{k \in s_t\}$ are not disjoint, and the probability of their union is expressed in a complicated way as a function of $\Pr[k \in s_t] = \pi_k^t$ and is not involved in $\widehat{Y}$ (which recalls that the sample is not undertaken in a 'simple' manner in $\widetilde{U}_T$).

7. In concrete terms, for all k, the difficulty consists in obtaining a value for r_k for the past period: if it is short enough, it is still possible by questioning the survey subject. On the other hand, obtaining r_k in a sufficiently reliable manner on the set of periods T (and therefore partly on the future) is practically impossible. We thus estimate r_k, for example after looking at a short enough time period before the interview (several days) and then by using the rules of three to estimate the frequency on the set of periods, under the hypothesis of a relatively stable behaviour of frequenting the shelter over time.

4

Stratification

4.1 Definition

Consider a population U split into H parts U_h, called 'strata', such that

$$\bigcup_{h=1}^{H} U_h = U, \quad \text{and} \quad U_h \cap U_i = \emptyset,$$

for all (h, i) with $h \neq i$. A design is called stratified if in each stratum U_h we select a random sample S_h of fixed size, and that the sample selection in each stratum is taken independently of the selection done in all other strata (see Figure 4.1).

Fig. 4.1. Stratified design

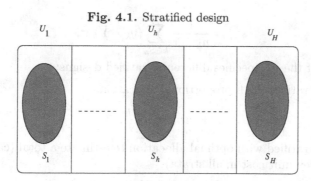

4.2 Estimation and variance

We furthermore assume throughout this chapter that the designs are simple without replacement within each stratum. The population size U_h is denoted

N_h and the sample size S_h is denoted n_h, where $h = 1, ..., H$. Since the inclusion probability is $\pi_k = n_h/N_h$, for all $k \in U_h$, the Horvitz-Thompson estimator of the total becomes

$$\widehat{Y}_\pi = \sum_{k \in S} \frac{y_k}{\pi_k} = \sum_{h=1}^{H} \frac{N_h}{n_h} \sum_{k \in S_h} y_k = \sum_{h=1}^{H} N_h \widehat{\overline{Y}}_h,$$

where $\widehat{\overline{Y}}_h$ is the unbiased mean estimator for stratum h:

$$\widehat{\overline{Y}}_h = \frac{1}{n_h} \sum_{k \in S_h} y_k.$$

The variance of $\widehat{Y}_\pi$ is

$$\mathrm{var}(\widehat{Y}_\pi) = \sum_{h=1}^{H} N_h^2 \frac{N_h - n_h}{N_h} \frac{S_{yh}^2}{n_h},$$

where

$$S_{yh}^2 = \frac{1}{N_h - 1} \sum_{k \in U_h} (y_k - \overline{Y}_h)^2,$$

and

$$\overline{Y}_h = \frac{1}{N_h} \sum_{k \in U_h} y_k.$$

The variance can be estimated by

$$\widehat{\mathrm{var}}(\widehat{Y}_\pi) = \sum_{h=1}^{H} N_h^2 \frac{N_h - n_h}{N_h} \frac{s_{yh}^2}{n_h},$$

where

$$s_{yh}^2 = \frac{1}{n_h - 1} \sum_{k \in S_h} (y_k - \widehat{\overline{Y}}_h)^2.$$

The choice of the n_h specifies different stratified designs:

- designs stratified with proportional allocation,

$$n_h = n \frac{N_h}{N}; \tag{4.1}$$

- designs stratified with optimal allocation to estimate a total (case of identical survey unit cost in all strata),

$$n_h = n \frac{N_h S_{yh}}{\sum_{i=1}^{H} N_i S_{yi}}. \tag{4.2}$$

Expressions (4.1) and (4.2) do not generally give an integer value for n_h; it is therefore necessary to turn to a rounding procedure. Furthermore, Expression (4.2) sometimes leads to having $n_h > N_h$. In this case, we take a census in the strata where this problem exists, and we restart the calculation of n_h for the remaining strata.

EXERCISES

Exercise 4.1 *Awkward stratification*

Given a population $U = \{1, 2, 3, 4\}$ and $y_1 = y_2 = 0, y_3 = 1, y_4 = -1$, the values taken by the characteristic y.

1. Calculate the variance of the mean estimator for a simple random design without replacement of size $n = 2$.
2. Calculate the variance of the mean estimator for a stratified random design for which only one unit is selected per stratum and the strata are given by $U_1 = \{1, 2\}$ and $U_2 = \{3, 4\}$.

Solution

1. The mean of y is zero. Indeed,

$$\overline{Y} = \frac{1}{N} \sum_{k \in U} y_k = \frac{1}{4}(0 + 0 + 1 - 1) = 0.$$

The population variance is

$$S_y^2 = \frac{1}{N-1} \sum_{k \in U} (y_k - \overline{Y})^2 = \frac{1}{4-1} \left\{ 0^2 + 0^2 + 1^2 + (-1)^2 \right\} = \frac{2}{3}.$$

We thus have

$$\operatorname{var}\left(\widehat{\overline{Y}}\right) = \frac{N-n}{N} \frac{S_y^2}{n} = \frac{4-2}{4} \times \frac{1}{2} \times \frac{2}{3} = \frac{1}{6}.$$

2. For the stratified design, we start by calculating the parameters within the strata

$$\overline{Y}_1 = \frac{1}{N_1} \sum_{k \in U_1} y_k = \frac{1}{2}(0 + 0) = 0,$$

$$\overline{Y}_2 = \frac{1}{N_2} \sum_{k \in U_2} y_k = \frac{1}{2}(1 - 1) = 0,$$

$$S_{y1}^2 = \frac{1}{N_1 - 1} \sum_{k \in U_1} (y_k - \overline{Y}_1)^2 = \frac{1}{1}(0^2 + 0^2) = 0,$$

and

$$S_{y2}^2 = \frac{1}{N_2 - 1} \sum_{k \in U_2} (y_k - \overline{Y}_2)^2 = \frac{1}{1} \left\{ 1^2 + (-1)^2 \right\} = 2.$$

The variance of the Horvitz-Thompson estimator is, with regards to a proportional allocation,

$$\operatorname{var}\left(\widehat{\overline{Y}}_\pi\right) = \frac{N-n}{nN} \sum_{h=1}^{2} \frac{N_h}{N} S_{yh}^2 = \frac{4-2}{2 \times 4^2}(2 \times 2 + 2 \times 0) = \frac{1}{4}.$$

We therefore see that the variance for a stratified design is larger than for the simple design, despite proportional allocation. This surprising result recalls that stratification does not lead to a systematic improvement in accuracy; it is due to the fact that, in this example, the inter-strata variance is zero and that the population size is small.

Exercise 4.2 *Strata according to income*

Among the 7500 employees of a company, we wish to know the proportion P of them that owns at least one vehicle. For each individual in the sampling frame, we have the value of his income. We then decide to construct three strata in the population: individuals with low income (stratum 1), with medium income (stratum 2), and with high income (stratum 3). We denote:

N_h = the stratum size h,

n_h = the sample size in stratum h (simple random sampling),

p_h = the estimator of the proportion of individuals in stratum h owning at least one vehicle.

The results are given in Table 4.1.

Table 4.1. Employees according to income: Exercise 4.2

	h=1	h=2	h=3
N_h	3500	2000	2000
n_h	500	300	200
p_h	0.13	0.45	0.50

1. What estimator $\widehat{P}$ of P do you propose? What can we say about its bias?
2. Calculate the accuracy of $\widehat{P}$, and give a 95% confidence interval for P.
3. Do you consider the stratification criteria to be adequate?

Solution

1. The Horvitz-Thompson estimator for the stratified design is given by

$$\widehat{P} = \sum_{h=1}^{3} \frac{N_h p_h}{N} = \frac{1}{7500}(3500 \times 0.13 + 2000 \times 0.45 + 2000 \times 0.50)$$
$$= 0.314.$$

This estimator is unbiased.

2. As the estimated variance is

$$\widehat{\text{var}}\left(\widehat{P}\right) = \frac{1}{N^2}\sum_{h=1}^{3} N_h^2 \frac{N_h - n_h}{N_h(n_h - 1)}p_h(1 - p_h) = (0.013)^2,$$

the 95% confidence interval for P is given by

$$\text{CI}(0.95) = [0.314 - 0.026 \; ; \; 0.314 + 0.026].$$

The normal distribution can be used without hesitation, because n is large.
3. The stratification criteria is adequate, as income is strongly correlated to owning a vehicle.

Exercise 4.3 *Strata of elephants*

A circus director has 100 elephants classified into two categories: 'males' and 'females'. The director wants to estimate the total weight of his herd because he wants to cross a river by boat. However, the previous year, this same circus director had all the elephants of the herd weighed and had obtained the results presented in Table 4.2 (averages are expressed in tonnes).

Table 4.2. Average weights and variances by stratum: Exercise 4.3

	Size N_h	Means $\overline{Y}_h$	Variances S_{yh}^2
Males	60	6	4
Females	40	4	2.25

1. Calculate the variance in the population for the variable 'elephant weight' for the previous year.
2. The director assumes from now on that the variances of the weights do not noticeably change from one year to another (this type of hypothesis here remains very reasonable and commonly occurs in practice when we repeat surveys in time). If the director conducts a simple random sample survey without replacement of 10 elephants, what is the variance of the estimator for the total weight of the herd?
3. If the director conducts a stratified sample survey with proportional allocation of 10 elephants, what is the variance of the estimator for the total weight of the herd?
4. If the director conducts an optimal stratified sample survey of 10 elephants, what are the sample sizes in each of the two strata and what is the variance of the estimator for the total?

Solution

1. The mean weight of an elephant in the population is

$$\overline{Y} = \frac{1}{N}\left(N_1\overline{Y}_1 + N_2\overline{Y}_2\right) = \frac{1}{100}(60 \times 6 + 40 \times 4) = \frac{360 + 160}{100} = 5.2.$$

The uncorrected variances are

$$\sigma_{y1}^2 = \frac{N_1 - 1}{N_1}S_{y1}^2 = \frac{60 - 1}{60} \times 4 = 3.9333,$$

$$\sigma_{y2}^2 = \frac{N_2 - 1}{N_2}S_{y2}^2 = \frac{40 - 1}{40} \times 2.25 = 2.19375.$$

We can then calculate the total variance (equation called 'analysis of variance')

$$\sigma_y^2 = \frac{1}{N}\left\{N_1\sigma_{y1}^2 + N_2\sigma_{y2}^2\right\} + \frac{1}{N}\left\{N_1(\overline{Y} - \overline{Y}_1)^2 + N_2(\overline{Y} - \overline{Y}_2)^2\right\}$$

$$= \frac{1}{100}\left\{60 \times 3.9333 + 40 \times 2.19375\right\}$$

$$+ \frac{1}{100}\left\{60 \times (6 - 5.2)^2 + 40 \times (4 - 5.2)^2\right\}$$

$$= 4.1975.$$

Therefore,

$$S_y^2 = \sigma_y^2\frac{N}{N - 1} = 4.1975 \times \frac{100}{100 - 1} = 4.2399.$$

2. The variance of the estimator for the total weight of the herd in the case of a simple design without replacement is therefore

$$\mathrm{var}\left(\widehat{Y}_\pi\right) = \frac{N(N - n)}{n}S_y^2 = \frac{100 \times 90}{10} \times 4.2399 = 3815.91.$$

3. If we stratify with proportional allocation, we get

$$n_1 = \frac{N_1}{N}n = \frac{60}{100} \times 10 = 6, \quad \text{and} \quad n_2 = \frac{N_2}{N}n = \frac{40}{100} \times 10 = 4.$$

The variance of the estimator for the total is directly obtained

$$\mathrm{var}\left(\widehat{Y}_\pi\right) = \frac{N - n}{n}\sum_{h=1}^{2} N_h S_{yh}^2 = \frac{100 - 10}{10}\left\{60 \times 4 + 40 \times 2.25\right\} = 2970.$$

4. If we use an optimal allocation, we get

$$N_1 S_{y1} = 60 \times \sqrt{4} = 120 \quad \text{and} \quad N_2 S_{y2} = 40 \times \sqrt{2.25} = 60.$$

The sample sizes within the strata are therefore

$$n_1 = \frac{nN_1 S_{y1}}{N_1 S_{y1} + N_2 S_{y2}} = \frac{10 \times 120}{120 + 60} = 6.66667,$$

and

$$n_1 = \frac{nN_2 S_{y2}}{N_1 S_{y1} + N_2 S_{y2}} = \frac{10 \times 60}{120 + 60} = 3.33333.$$

By rounding to the nearest whole number, we get $n_1 = 7$ and $n_2 = 3$. The variance of the Horvitz-Thompson estimator for the total is thus the following:

$$\mathrm{var}\left(\widehat{Y}_\pi\right) = \sum_{h=1}^{2} N_h \frac{N_h - n_h}{n_h} S_{yh}^2$$

$$= 60 \times \frac{60 - 7}{7} \times 4 + 40 \times \frac{40 - 3}{3} \times 2.25$$

$$= 2927.14.$$

The gain in accuracy is therefore not very important with respect to the proportional stratification (well-known result: the two allocations in question only differ slightly, and the optimum is rather 'flat'). We therefore prefer to use proportional stratification, which is more simple to calculate and which has the determining advantage of not depending on a particular variable.

Exercise 4.4 *Strata according to age*

In a very large population composed of actual individuals, we are looking to estimate the mean age $\overline{Y}$. Given information on age groups, we stratify the population into three parts, and we select a sample using simple random sampling in each part.
We denote:

- N_h/N: the true weight of stratum h,
- $\widehat{\overline{Y}}_h$: the mean age calculated on the sample in stratum h,
- n_h: the allocation chosen in stratum h,
- S_{yh}^2: the population variance of ages in stratum h (note the squared term),
- C_h: the unit cost of surveying in stratum h.

Table 4.3 gives the useful data:

1. What is the unbiased stratified estimator of $\overline{Y}$? (We denote $\widehat{\overline{Y}}_\pi$ as this estimator.)
2. Is this estimator different from the simple mean calculated on the overall sample?

Table 4.3. Distribution of ages: Exercise 4.4

Stratum	N_h/N	$\overline{Y}_h$	S_{yh}^2	n_h	C_h
Less than 40	50%	25	16	40	1
Between 40 and 50	30%	45	10	20	1
Over 50	20%	58	20	40	4

3. Neglecting all the sampling rates, calculate the accuracy of $\widehat{\overline{Y}}_\pi$.
4. Calculate the proportional allocation and recall the expression of the estimator which follows (the total size of the sample is $n = 100$).
5. What is the accuracy obtained with the proportional allocation?
6. What is the gain in accuracy from using Neyman allocation instead of proportional allocation? (Use comparable situations.)

Solution

1. The estimator is given by

$$\widehat{\overline{Y}}_\pi = \sum_{h=1}^{3} \frac{N_h}{N}\,\widehat{\overline{Y}}_h = 0.50 \times 25 + 0.30 \times 45 + 0.20 \times 58 = 37.6 \text{ years.}$$

2. Yes, because

$$\frac{n_h}{n} \neq \frac{N_h}{N}.$$

3. Neglecting the sampling rate,

$$\mathrm{var}(\widehat{\overline{Y}}_\pi) = \sum_{h=1}^{3} \left(\frac{N_h}{N}\right)^2 \frac{S_{yh}^2}{n_h} = (0.5)^2 \times \frac{16}{40} + (0.3)^2 \times \frac{10}{20} + (0.2)^2 \times \frac{20}{40}$$
$$= 0.165 \approx (0.41)^2.$$

4. The proportional allocation leads to

$$n_h = n\frac{N_h}{N}.$$

Therefore,

$$n_1 = 100 \times 50\% = 50, \quad n_2 = 30, \quad \text{and} \quad n_3 = 20.$$

The unbiased estimator is the simple mean in the sample

$$\widehat{\overline{Y}} = \sum_{h=1}^{3} \frac{n_h}{n}\widehat{\overline{Y}}_h.$$

5. The variance is

$$\mathrm{var}_{\mathrm{prop}}(\widehat{\overline{Y}}) \approx \frac{S^2_{\mathrm{intra}}}{n} \ \ \text{(neglecting the sampling rate)} ,$$

where

$$S^2_{\mathrm{intra}} = \sum_{h=1}^{3} \frac{N_h}{N} S^2_{yh} = 0.50 \times 16 + 0.30 \times 10 + 0.20 \times 20 = 15.$$

Thus,

$$\mathrm{var}_{\mathrm{prop}}(\widehat{\overline{Y}}) = 0.150 \approx (0.39)^2.$$

We therefore improve the accuracy with respect to the initial allocation.

6. We obviously reason on a constant cost: with 100 interviewers, the cost is, with proportional allocation:

$$50 \times 1 + 30 \times 1 + 20 \times 4 = 160.$$

We have, for a Neyman allocation

$$n_h = \frac{N_h S_{yh}}{\sqrt{C_h \lambda}},$$

with λ calculated in a way such that the total cost is 160, being $n_1 + n_2 + 4n_3 = 160$. We find (rounding to the nearest whole number): $n_1 = 68, n_2 = 32$, and $n_3 = 15$. With this allocation, using the general formula from 3., we get the minimum variance $\mathrm{var}_{\mathrm{opti}}(\widehat{\overline{Y}}_\pi) = 0.140 \approx (0.37)^2$, which is a gain in the order of 5% for the standard deviation as compared to proportional allocation.

Exercise 4.5 *Strata of businesses*

We want to estimate average sales related to a population of businesses. The businesses are *a priori* listed in three classes by sales. The data are presented in Table 4.4. We want to select a sample of 111 businesses. Having confidence in

Table 4.4. Distribution of sales: Exercise 4.5

Sales in millions of Euros	Number of businesses
0 to 1	1000
1 to 10	100
10 to 100	10

the expert assessments and in the absence of any other information, we assume that the distribution of sales is uniform within each class: give the variances of the mean estimator of sales for a stratified design with proportional allocation and for a stratified design with optimal (or Neyman) allocation.

Solution

A random variable X is called uniform over an interval $[a, b]$ with $b > a$ if its density function is given by

$$f(x) = \begin{cases} (b-a)^{-1} & \text{if } a \leq x \leq b \\ 0 & \text{otherwise.} \end{cases}$$

We can thus calculate the expected value and the variance of X:

$$E(X) = \int_a^b \frac{x}{b-a} dx = \frac{1}{2(b-a)} \left[x^2 \right]_a^b = \frac{1}{2(b-a)} (b^2 - a^2) = \frac{b+a}{2}.$$

$$\begin{aligned} \text{var}(X) &= \int_a^b \frac{x^2}{b-a} dx - \left(\frac{b+a}{2} \right)^2 \\ &= \frac{1}{3(b-a)} \left[x^3 \right]_a^b - \left(\frac{b+a}{2} \right)^2 \\ &= \frac{1}{3(b-a)} (b^3 - a^3) - \left(\frac{b+a}{2} \right)^2 \\ &= \frac{a^2 + ab + b^2}{3} - \frac{a^2 + 2ab + b^2}{4} \\ &= \frac{(b-a)^2}{12}. \end{aligned}$$

The standard error of a uniform variable is therefore proportional to the length of the interval $[a, b]$:

$$\sqrt{\text{var}(X)} = \frac{b-a}{2\sqrt{3}}.$$

We can thus complete Table 4.4 with the population variances within each stratum and we get Table 4.5. The corrected variances are therefore

Table 4.5. Distribution of sales and population variances: Exercise 4.5

Sales in millions of Euros	Number of businesses	σ_{yh}^2
0 to 1	1000	1/12
1 to 10	100	81/12
10 to 100	10	8100/12

$$S_{y1}^2 = \frac{1}{12} \times \frac{1000}{999} = 0.0834168,$$

$$S_{y2}^2 = \frac{81}{12} \times \frac{100}{99} = 6.81818,$$

$$S_{y3}^2 = \frac{8100}{12} \times \frac{10}{9} = 750.$$

1. Stratification with proportional allocation

$$\text{var}\left(\widehat{\overline{Y}}_\pi\right) = \frac{N-n}{nN} \sum_{h=1}^{3} \frac{N_h}{N} S_{yh}^2$$

$$= \frac{1110 - 111}{111 \times 1110^2} \times \{1000 \times 0.0834168 + 100 \times 6.81818 + 10 \times 750\}$$

$$\approx 0.0604.$$

We easily prove that the largest stratum is the one that contributes the most to this variance (it creates roughly 91% of the total variance).

2. Optimal stratification

We calculate the products of the standard errors S_{yh} and the stratum sizes

$$N_1 S_{y1} = 1000 \times \sqrt{0.0834168} = 288.82,$$

$$N_2 S_{y2} = 100 \times \sqrt{6.81818} = 261.116,$$

$$N_3 S_{y3} = 10 \times \sqrt{750} = 273.861,$$

which gives the optimal allocation:

$$n_1 = \frac{nN_1 S_{y1}}{\sum_{h=1}^{3} N_h S_{yh}} = \frac{111 \times 288.82}{288.82 + 261.116 + 273.861} = 38.9161$$

$$n_2 = \frac{nN_2 S_{y2}}{\sum_{h=1}^{3} N_h S_{yh}} = \frac{111 \times 261.116}{288.82 + 261.116 + 273.861} = 35.1833$$

$$n_3 = \frac{nN_3 S_{y3}}{\sum_{h=1}^{3} N_h S_{yh}} = \frac{111 \times 273.861}{288.82 + 261.116 + 273.861} = 36.9006.$$

The sample size in the third stratum $n_3 = 36.9$ is larger than $N_3 = 10$. In this case, we select all units from the third stratum by setting $n_3 = N_3 = 10$, and it remains to select (in an optimal manner) 101 units among the 1100 units from strata 1 and 2. Thus, we have

$$n_1 = \frac{101 \times N_1 S_{y1}}{N_1 S_{y1} + N_2 S_{y2}} = \frac{101 \times 288.82}{288.82 + 261.116} = 53.0439 \approx 53,$$

$$n_2 = \frac{101 \times N_2 S_{y2}}{N_1 S_{y1} + N_2 S_{y2}} = \frac{101 \times 261.116}{288.82 + 261.116} = 47.9561 \approx 48.$$

The optimal distribution is thus $(n_1 = 53, n_2 = 48, n_3 = 10)$. It remains to calculate the variance of the mean estimator

$$\text{var}\left(\widehat{\overline{Y}}_\pi\right) = \sum_{h=1}^{3} \frac{N_h^2}{N^2} \frac{N_h - n_h}{N_h n_h} S_{yh}^2$$

$$= \frac{1000^2}{1110^2} \frac{1000 - 53}{1000 \times 53} \times 0.0834168 + \frac{100^2}{1110^2} \frac{100 - 48}{100 \times 48} \times 6.81818 + 0$$

$$= 0.0018.$$

We note that it is much more interesting to use an optimal than a proportional allocation: the gain essentially follows from exhaustive sampling in the stratum with the largest sales.

Exercise 4.6 *Stratification and unequal probabilities*

When we have available auxiliary information, we try to use it to improve the accuracy of estimators. When this individual information is quantitative, we particularly think of two types of concurrent sampling designs:

- stratified samples,
- samples with unequal probabilities.

It is not possible, *a priori* and without further specifying the context, to say that one of the two methods is better than the other. What follows has the objective of showing that, in certain cases, we arrive all the same at determining which of these two methods has to be used.

We consider a population U of size N partitioned into H classes. We assume that in class U_h of size $N_h, h = 1, \ldots, H$, we can rewrite the variable attached to individual k, being y_{hk}, in the following form using the auxiliary information x:

$$y_{hk} = \beta x_h + e_{hk},$$

with, for all h,

$$\sum_{k \in U_h} e_{hk} = 0, \quad \text{and} \quad \frac{1}{N_h} \sum_{k \in U_h} e_{hk}^2 = a x_h^g.$$

The individuals are therefore found by the indicator (hk).

Here, β is an unknown positive value, a and g are known positive values and x is an auxiliary variable known everywhere, with the notation x_h signifying that all individuals of class U_h take the same value of x.

1. Recall the expression of usual estimators for the mean $\overline{Y}$, as well as their respective variances, in the following three cases (the sample is always of size n):

   - stratified sampling with proportional allocation ($\widehat{\overline{Y}}_{\text{prop}}$),
   - stratified sampling with Neyman optimal allocation ($\widehat{\overline{Y}}_{\text{opti}}$),
   - sampling with probabilities proportional to x_h, *with replacement* ($\widehat{\overline{Y}}_{\text{pps}}$).

   To simplify matters, we always ignore the sampling rates, and we assume $N_h \gg 1$.

2. Using the rewritten form of y_{hk}, express the variances coming from the previous section by only using the quantities a and n as well as the true means (known) for variables of type x_k^α, where α is a real value. We denote:

$$\overline{X^{(\alpha)}} = \frac{1}{N} \sum_{h=1}^{H} \sum_{k \in U_h} x_h^\alpha = \frac{1}{N} \sum_{k \in U} x_k^\alpha.$$

The three variances must be made using easily comparable forms.

3. Compare the three sampling designs and specify, in particular, under which condition sampling proportional to size (with replacement) is more efficient than stratification by proportional allocation.

Solution

1. We denote $\widehat{\overline{Y}}_h$ as the simple mean of the y_k in the sample of stratum U_h, and $\widehat{\overline{Y}}$ as the simple mean in the total sample.

- $\widehat{\overline{Y}}_{\text{prop}} = \sum_{h=1}^{H} \frac{N_h}{N} \widehat{\overline{Y}}_h = \widehat{\overline{Y}}$ where $\frac{N_h}{N} = \frac{n_h}{n}$,

$$\underset{\text{prop}}{V} = \text{var}(\widehat{\overline{Y}}_{\text{prop}}) = \frac{1}{n} \left(\sum_{h=1}^{H} \frac{N_h}{N} S_{yh}^2 \right),$$

with

$$S_{yh}^2 = \frac{1}{N_h - 1} \sum_{k \in U_h} (y_{hk} - \overline{Y}_h)^2 \approx \frac{1}{N_h} \sum_{k \in U_h} e_{hk}^2 = a x_h^g,$$

as $\overline{Y}_h = \beta x_h$.

- $\widehat{\overline{Y}}_{\text{opti}} = \sum_{h=1}^{H} \frac{N_h}{N} \widehat{\overline{Y}}_h$ where $n_h = n \dfrac{N_h S_{yh}}{\sum_{h=1}^{H} N_h S_{yh}}$,

$$\underset{\text{opti}}{V} = \text{var}(\widehat{\overline{Y}}_{\text{opti}}) = \frac{1}{nN^2} \left(\sum_{h=1}^{H} N_h S_{yh} \right)^2.$$

- $\widehat{\overline{Y}}_{\text{pps}} = \dfrac{1}{N} \sum_{j=1}^{n} \dfrac{y_{hk_j}}{n p_{hk_j}}$ (sampling with replacement),

where (hk_j) is the label of the unit selected in the hth stratum at the jth trial, and p_{hk} is the probability of selecting individual hk in each drawing, given by

$$p_{hk} = \frac{x_h}{X}, \quad \text{where} \quad X = \sum_{h=1}^{H} \sum_{k \in U_h} x_h = \sum_{h=1}^{H} N_h x_h.$$

The probability p_{hk} only depends on h. We denote p_h as the common value of all individuals of U_h.

$$\underset{\text{pps}}{V} = \text{var}(\widehat{\overline{Y}}_{\text{pps}}) = \frac{1}{nN^2} \sum_{h=1}^{H} \sum_{k \in U_h} p_h \left(\frac{y_{hk}}{p_h} - Y \right)^2,$$

where Y is the true total

$$Y = \sum_{h=1}^{H} N_h \overline{Y}_h.$$

2. a) With proportional allocation, the variance is

$$\underset{\text{prop}}{V} = \frac{1}{n} \sum_{h=1}^{H} \frac{N_h}{N} a x_h^g = \frac{a}{n} \frac{1}{N} \left(\sum_{h=1}^{H} N_h x_h^g \right).$$

Indeed,

$$\sum_{h=1}^{H} N_h x_h^g = \sum_{h=1}^{H} \sum_{k \in U_h} x_h^g,$$

thus

$$\underset{\text{prop}}{V} = \frac{a}{n} \overline{X^{(g)}}.$$

We note that $\overline{X^{(g)}}$ is the true mean of the x_h^g.
b) With optimal allocation, the variance is

$$\underset{\text{opti}}{V} = \frac{a}{nN^2} \left(\sum_{h=1}^{H} N_h x_h^{g/2} \right)^2 = \frac{a}{nN^2} \left(\sum_{h=1}^{H} \sum_{k \in U_h} x_h^{g/2} \right)^2$$

$$= \frac{a}{n} \left[\frac{1}{N} \sum_{h=1}^{H} \sum_{k \in U_h} x_h^{g/2} \right]^2 = \frac{a}{n} \left[\overline{X^{(g/2)}} \right]^2.$$

Here, $\overline{X^{(g/2)}}$ is the true mean of the $\sqrt{x_h^g}$.
c) For the design with unequal probabilities, the variance satisfies

$$nN^2 \underset{\text{pps}}{V} = \sum_{h=1}^{H} \sum_{k \in U_h} p_h \left(\frac{y_{hk}}{p_h} - Y \right)^2 = \sum_{h=1}^{H} \sum_{k \in U_h} \frac{x_h}{X} \left(X \frac{y_{hk}}{x_h} - Y \right)^2,$$

where $p_h = x_h/X$. Indeed, $\overline{Y}_h = \beta x_h$, therefore

$$Y = \sum_{h=1}^{H} N_h \overline{Y}_h = \beta \sum_{h=1}^{H} N_h x_h = \beta X,$$

which gives

$$nN^2 \underset{\text{pps}}{V} = \sum_{h=1}^{H} \sum_{k \in U_h} \frac{x_h}{X} X^2 \left(\frac{y_{hk}}{x_h} - \beta \right)^2$$

$$= X \sum_{h=1}^{H} \sum_{k \in U_h} x_h \frac{1}{x_h^2} (y_{hk} - \beta x_h)^2$$

$$= X \sum_{h=1}^{H} \sum_{k \in U_h} \frac{1}{x_h} \sum_{k \in U_h} (y_{hk} - \overline{Y}_h)^2$$

$$= X \sum_{h=1}^{H} \frac{N_h}{x_h} \left[\frac{1}{N_h} \sum_{k \in U_h} (y_{hk} - \overline{Y}_h)^2 \right].$$

We obtain approximately

$$nN^2 \underset{\text{pps}}{V} \approx X \sum_{h=1}^{H} \frac{N_h}{x_h} S_{yh}^2 = Xa \sum_{h=1}^{H} N_h x_h^{g-1} = XaN \overline{X^{(g-1)}}.$$

Finally,

$$\underset{\text{pps}}{V} = \frac{a}{n} \frac{X}{N} \overline{X^{(g-1)}},$$

and as

$$\frac{X}{N} = \overline{X},$$

we get

$$\underset{\text{pps}}{V} = \frac{a}{n} \overline{X} \, \overline{X^{(g-1)}}.$$

Here, $\overline{X^{(g-1)}}$ is the true mean of the x_h^{g-1}.

3. We obtained:

$$\begin{cases} \underset{\text{prop}}{V} = \frac{a}{n} \overline{X^{(g)}} \\[2mm] \underset{\text{opti}}{V} = \frac{a}{n} [\overline{X^{(g/2)}}]^2 \\[2mm] \underset{\text{pps}}{V} = \frac{a}{n} \overline{X} \, \overline{X^{(g-1)}}. \end{cases}$$

The problem therefore is to rank, as a function of g, the three expressions:

$$\overline{X^{(g)}}, \quad [\overline{X^{(g/2)}}]^2 \quad \text{and} \quad \overline{X} \, \overline{X^{(g-1)}}.$$

We notice that a completely disappears in this comparison process.

a) Without hesitation, we can say $\underset{\text{prop}}{V} \geq \underset{\text{opti}}{V}$, because $\underset{\text{opti}}{V}$ corresponds to the optimal method.

Thus, we must have $\overline{X^{(g)}} \geq [\overline{X^{(g/2)}}]^2$. It is well understood to be true, but we can eventually convince doubtful readers: the idea is to write the empirical means as *expected values* of discrete random variables, and to use the 'well-known' properties on the expected values. We know that for every real random variable X, we have: $E(X)^2 \geq (EX)^2$ (as $var(X) \geq 0$). Let us apply that for the variable $X^{g/2}$

$$E(X^g) = E([X^{g/2}]^2) \geq [EX^{g/2}]^2.$$

Indeed

$$E(X^g) = \frac{1}{N} \sum_{h=1}^{H} \sum_{k \in U_h} x_h^g = \overline{X^{(g)}}$$

and

$$EX^{g/2} = \frac{1}{N} \sum_{h=1}^{H} \sum_{k \in U_h} x_h^{g/2} = \overline{X^{(g/2)}},$$

which is to say:

$$\overline{X^{(g)}} \geq [\overline{X^{(g/2)}}]^2.$$

b) Let us consider two real random variables, some X and Y. We have:

$$cov(X,\ Y) = EXY - (EX)\,(EY).$$

Therefore

$$cov(X^{g-1},\ X) = EX^g - (EX)\,(EX^{g-1}) = \overline{X^{(g)}} - \overline{X}\ \overline{X^{(g-1)}}.$$

Indeed

$$g \geq 1 \quad \Leftrightarrow \quad cov(X^{g-1},\ X) \geq 0.$$

In fact, X and X^{g-1} vary 'along the same lines' if and only if

$$g - 1 \geq 0.$$

Therefore

$$g \geq 1 \quad \Leftrightarrow \quad \overline{X^{(g)}} \geq \overline{X}\,\overline{X^{(g-1)}}.$$

c) Let us change methods and return to the (well-known) Schwarz inequality: given two vectors $\mathbf{a}$ and $\mathbf{b}$ of $\mathbb{R}^H$ with coordinates

$$\mathbf{a} = \left(\sqrt{N_h x_h} \right)_{1 \leq h \leq H} \text{ and } \mathbf{b} = \left(\sqrt{N_h x_h^{g-1}} \right)_{1 \leq h \leq H}, \text{ in } \mathbb{R}^H.$$

We know that:

$$|\ \mathbf{ab}\ | \leq \|\ \mathbf{a}\ \| \times \|\ \mathbf{b}\ \|,$$

being:

$$\sum_{h=1}^{H} (N_h x_h^{1/2} x_h^{g-1/2}) \leq \sqrt{\sum_{h=1}^{H} N_h x_h} \sqrt{\sum_{h=1}^{H} N_h x_h^{g-1}},$$

thus

$$\left[\sum_{h=1}^{H} \frac{N_h}{N} (x_h)^{g/2}\right]^2 \leq \left(\sum_{h=1}^{H} \frac{N_h}{N} x_h\right)\left(\sum_{h=1}^{H} \frac{N_h}{N} x_h^{g-1}\right).$$

Finally

$$\left[\overline{X^{(g/2)}}\right]^2 \leq \overline{X}\,\overline{X^{(g-1)}},$$

for all g.

We can conclude by distinguishing two cases:

Case 1:

$$0 < g < 1 \;:\; \left[\overline{X^{(g/2)}}\right]^2 \leq \overline{X^{(g)}} < \overline{X}\,\overline{X^{(g-1)}}$$

$$\Leftrightarrow \quad V_{\text{pps}} > V_{\text{prop}} \geq V_{\text{opti}}$$

Case 2:

$$g \geq 1 \;:\; \left[\overline{X^{(g/2)}}\right]^2 \leq \overline{X}\,\overline{X^{(g-1)}} \leq \overline{X^{(g)}}$$

$$\Leftrightarrow \quad V_{\text{prop}} \geq V_{\text{pps}} \geq V_{\text{opti}}$$

Stratified sampling with optimal allocation is always the most efficient; on the other hand everything depends on g in ranking stratified sampling with proportional allocation and sampling proportional to size.

Exercise 4.7 *Strata of doctors*

In a large city, we are studying the mean number of patients that a doctor sees during a working day. We begin with the *a priori* idea that the more experience a doctor has, the more clients she or he has. That leads us to classify the population of doctors into 3 groups: the 'beginners' (class 1), the 'intermediates' (class 2) and the 'experienced' (class 3). Furthermore, we assume that we know, from the sampling frame of doctors, the class of each one (1 or 2 or 3). Thus, we list 500 doctors in class 1, 1 000 in class 2 and 2 500 in class 3. Using simple random sampling, we select 200 doctors in each class. We then calculate, in each class, the mean number of patients by day and by sampled doctor: 10 in class 1, then 15 in class 2 and 20 in class 3.

We finally calculate the variances of the number of patients by doctor in each of the three samples and we find respectively 4 (class 1), 7 (class 2), and 10 (class 3).

1. What do we call this sample design? Justify *a priori* its usage.
2. How do you estimate the mean number of patients treated by day and by doctor?
3. Give a 95% confidence interval for the 'true' mean number of patients treated by doctor and by day.
4. If you had a constraint on the total number of doctors to survey (being 600), would you proceed as shown above?
5. What is the gain in estimated variance obtained with a proportional allocation in comparison with simple random sampling (of size 600)?
6. Would this gain have been numerically different if we had naively estimated the population variance S_y^2 by the simple sample variance s_y^2 calculated on the whole sample?

Solution

1. It is stratified sampling. The three groups defined are supposedly *a priori* relatively 'intra' homogeneous; that is, the number of patients is well explained by the experience of the doctor.
2. The mean estimator is:

$$\widehat{\overline{Y}}_\pi = \sum_{h=1}^{3} \frac{N_h}{N}\, \widehat{\overline{Y}}_h = \frac{500}{4\,000} \times 10 + \frac{1\,000}{4\,000} \times 15 + \frac{2\,500}{4\,000} \times 20 = 17.5.$$

3. The number of doctors selected per stratum ($n_h = 200$) is sufficiently large so that we consider that each $\widehat{\overline{Y}}_h$ follows a normal distribution, and therefore that the linear combination $\widehat{\overline{Y}}$ follows a normal distribution as well (the $\widehat{\overline{Y}}_h$ are independent).

$$\widehat{\mathrm{var}}(\widehat{\overline{Y}}) = \sum_{h=1}^{3} \left(\frac{N_h}{N}\right)^2 \left(1 - \frac{n_h}{N_h}\right) \frac{s_{yh}^2}{n_h}$$

$$= \left(\frac{500}{4\,000}\right)^2 \left(1 - \frac{200}{500}\right) \frac{4}{200} + \left(\frac{1\,000}{4\,000}\right)^2 \left(1 - \frac{200}{1\,000}\right) \frac{7}{200}$$

$$+ \left(\frac{2\,500}{4\,000}\right)^2 \left(1 - \frac{200}{2\,500}\right) \frac{10}{200}$$

$$\approx 19.9 \times 10^{-3}.$$

Therefore

$$2 \times \sqrt{\widehat{\mathrm{var}}(\widehat{\overline{Y}})} \approx 0.282 \quad \text{and} \quad \overline{Y} \in [17.5 \pm 0.28],$$

95 times out of 100.

4. Everything depends on the information which we have *a priori* on the population variances by stratum for the variable 'number of patients'. In the absence of such information, we choose a proportional allocation, which assures a better accuracy than that for simple random sampling:

$$n_h = 600 \times \frac{N_h}{N},$$

$$n_1 = 75, \qquad n_2 = 150, \qquad \text{and} \qquad n_3 = 375.$$

If we have an estimation *a priori* of standard errors S_{yh} (previous survey on the same subject, preliminary sampling), we choose an optimal Neyman allocation. For example, if we have to again carry out a survey, we use the s_{yh} estimated by the previous survey, being: n_h proportional to $N_h s_{yh}$.

$$N_1 s_{y1} = 1\,000, \qquad N_2 s_{y2} = 2\,646, \qquad N_3 s_{y3} = 7\,906.$$

That is:

$$n_1 = 52, \qquad n_2 = 137, \qquad n_3 = 411.$$

5. The difficulty consists of estimating the true *overall* population variance S_y^2 starting from the stratified sample with 200 doctors selected per stratum. Using the decomposition formula of the variance, we have:

$$S_y^2 \approx \sum_{h=1}^{3} \frac{N_h}{N} S_{yh}^2 + \sum_{h=1}^{3} \frac{N_h}{N} (\overline{Y}_h - \overline{Y})^2.$$

We know that $E(s_{yh}^2) = S_{yh}^2$ (simple sampling in each stratum). Furthermore, it is natural to be interested in the expected value of

$$A = \sum_{h=1}^{3} \frac{N_h}{N} (\widehat{\overline{Y}}_h - \widehat{\overline{Y}}_\pi)^2 = \sum_{h=1}^{3} \frac{N_h}{N} \widehat{\overline{Y}}_h^2 - \widehat{\overline{Y}}_\pi^2.$$

We have

$$E(A) = \sum_{h=1}^{3} \frac{N_h}{N} E(\widehat{\overline{Y}}_h^2) - E\left(\widehat{\overline{Y}}_\pi^2\right)$$

$$= \sum_{h=1}^{3} \frac{N_h}{N} \left(\mathrm{var}(\widehat{\overline{Y}}_h) + \overline{Y}_h^2\right) - \left(\mathrm{var}(\widehat{\overline{Y}}_\pi) + \overline{Y}^2\right)$$

$$= \sum_{h=1}^{3} \frac{N_h}{N} (\overline{Y}_h - \overline{Y})^2 + \sum_{h=1}^{3} \frac{N_h}{N} \mathrm{var}(\widehat{\overline{Y}}_h) - \mathrm{var}(\widehat{\overline{Y}}_\pi).$$

But

$$\widehat{\mathrm{var}}(\widehat{\overline{Y}}_h) = \left(1 - \frac{n_h}{N_h}\right) \frac{s_{yh}^2}{n_h} \qquad (n_h = 200)$$

estimates $\mathrm{var}(\widehat{\overline{Y}}_h)$ without bias.

In conclusion, gathering the unbiased estimators for each component of S_y^2, we got the unbiased estimator:

$$\widehat{S}_y^2 = \sum_{h=1}^{3} \frac{N_h}{N} s_{yh}^2 + \sum_{h=1}^{3} \frac{N_h}{N} (\widehat{\overline{Y}}_h - \widehat{\overline{Y}}_\pi)^2 - \sum_{h=1}^{3} \frac{N_h}{N} \widehat{\mathrm{var}}(\widehat{\overline{Y}}_h) + \widehat{\mathrm{var}}(\widehat{\overline{Y}}_\pi)$$

$$\approx 8.5 + 12.5 - 0.037 + 0.020$$

$$= 20.983.$$

The estimated variance with proportional allocation is:

$$\widehat{V}_{\mathrm{prop}} = \frac{1-f}{n} \widehat{S}_{\mathrm{intra}}^2 = \frac{1-f}{n} \left(\sum_{h=1}^{3} \frac{N_h}{N} s_{yh}^2 \right).$$

The variance that we would obtain with a simple random sample is therefore estimated by:

$$\widehat{V}_{\mathrm{SRS}} = \frac{1-f}{n} \widehat{S}_y^2.$$

The desired gain is:

$$\frac{\widehat{V}_{\mathrm{prop}}}{\widehat{V}_{\mathrm{SRS}}} = \frac{\left(\sum_{h=1}^{3} \frac{N_h}{N} s_{yh}^2 \right)}{\widehat{S}_y^2} \approx \frac{8.5}{21} = 40.5\%.$$

It is a substantial gain, ensuing from a quite strong inter-strata variance (that signifies that the strata are well constructed).

6. If we use s_y^2 to estimate S_y^2, we create a bias as under the stratified sample design effectively carried out, $\mathrm{E}(s_y^2) \neq S_y^2$. Numerically, if we denote $\widehat{\overline{Y}}$ as the simple mean of y_k in the overall sample,

$$s_y^2 = \frac{1}{n-1} \sum_{k \in S} (y_k - \widehat{\overline{Y}})^2$$

$$\approx \sum_{h=1}^{3} \frac{n_h}{n} s_{yh}^2 + \sum_{h=1}^{3} \frac{n_h}{n} (\widehat{\overline{Y}}_h - \widehat{\overline{Y}})^2 = 7 + 16.67 = 23.67.$$

We would therefore get a weaker variance relationship (slight overestimation of the gain).

Exercise 4.8 *Estimation of the population variance*

1. Give an unbiased estimator for the population variance σ_y^2 for a stratified survey with proportional allocation.
2. Show that the corrected sample variance s_y^2 is a biased estimator of σ_y^2 but that this bias approaches zero when n becomes very large.

Solution

Method 1.

In any stratified design, we have

$$
\pi_{k\ell} = \begin{cases}
\dfrac{n_h(n_h - 1)}{N_h(N_h - 1)} & \text{if } k, \ell \in U_h, k \neq \ell, \\[3mm]
\dfrac{n_h n_i}{N_h N_i} & \text{if } k \in U_h, \ell \in U_i, h \neq i.
\end{cases}
$$

The unbiased estimator of σ_y^2 is given by (see Exercise 2.7)

$$
\begin{aligned}
\hat{\sigma}_y^2 &= \frac{1}{2N^2} \sum_{k \in S} \sum_{\substack{\ell \in S \\ \ell \neq k}} \frac{(y_k - y_\ell)^2}{\pi_{k\ell}} \\[2mm]
&= \frac{1}{2N^2} \sum_{h=1}^{H} \sum_{k \in S_h} \sum_{\substack{\ell \in S_h \\ \ell \neq k}} (y_k - y_\ell)^2 \frac{N_h(N_h - 1)}{n_h(n_h - 1)} \\[2mm]
&\quad + \frac{1}{2N^2} \sum_{h=1}^{H} \sum_{\substack{i=1 \\ i \neq h}}^{H} \sum_{k \in S_h} \sum_{\ell \in S_i} (y_k - y_\ell)^2 \frac{N_h N_i}{n_h n_i} \\[2mm]
&= \frac{1}{2N^2} \sum_{h=1}^{H} \sum_{i=1}^{H} \sum_{k \in S_h} \sum_{\substack{\ell \in S_i \\ \ell \neq k}} (y_k - y_\ell)^2 \frac{N_h N_i}{n_h n_i} \\[2mm]
&\quad + \frac{1}{2N^2} \sum_{h=1}^{H} \sum_{k \in S_h} \sum_{\substack{\ell \in S_h \\ \ell \neq k}} (y_k - y_\ell)^2 \left\{ \frac{N_h(N_h - 1)}{n_h(n_h - 1)} - \frac{N_h^2}{n_h^2} \right\}.
\end{aligned}
$$

As the allocation is proportional

$$
\frac{N_h^2}{n_h^2} = \frac{N^2}{n^2},
$$

and that

$$
\frac{N_h(N_h - 1)}{n_h(n_h - 1)} - \frac{N_h^2}{n_h^2} = \frac{N_h(N_h - n_h)}{n_h^2(n_h - 1)} = \frac{N - n}{n} \frac{N_h}{n_h(n_h - 1)},
$$

we get

$$\hat{\sigma}_y^2 = \frac{1}{2n^2} \sum_{h=1}^{H} \sum_{i=1}^{H} \sum_{k \in S_h} \sum_{\substack{\ell \in S_i \\ \ell \neq k}} (y_k - y_\ell)^2$$

$$+ \frac{1}{2N^2} \sum_{h=1}^{H} \sum_{k \in S_h} \sum_{\substack{\ell \in S_h \\ \ell \neq k}} (y_k - y_\ell)^2 \frac{N-n}{n} \frac{N_h}{n_h(n_h-1)}$$

$$= \frac{1}{2n^2} \sum_{k \in S} \sum_{\substack{\ell \in S \\ \ell \neq k}} (y_k - y_\ell)^2 + \frac{N-n}{nN^2} \sum_{h=1}^{H} N_h \frac{1}{2n_h(n_h-1)} \sum_{k \in S_h} \sum_{\substack{\ell \in S_h \\ \ell \neq k}} (y_k - y_\ell)^2$$

$$= s_y^2 \frac{n-1}{n} + \frac{N-n}{nN^2} \sum_{h=1}^{H} N_h s_{yh}^2$$

$$= s_y^2 \frac{n-1}{n} + \widehat{\text{var}}(\widehat{\overline{Y}}_{\text{prop}}),$$

where $\widehat{\overline{Y}}_{\text{prop}}$ is the unbiased estimator of $\overline{Y}$.

Method 2.
Due to the proportional allocation, the unbiased estimator of $\overline{Y}$ is $\widehat{\overline{Y}}_{\text{prop}}$, the simple mean in the sample. We therefore have:

$$E(\widehat{\overline{Y}}_{\text{prop}}^2) = \text{var}(\widehat{\overline{Y}}_{\text{prop}}) + [E(\widehat{\overline{Y}}_{\text{prop}})]^2 = E[\widehat{\text{var}}(\widehat{\overline{Y}}_{\text{prop}})] + \overline{Y}^2,$$

where $\widehat{\text{var}}(\widehat{\overline{Y}}_{\text{prop}})$ estimates $\text{var}(\widehat{\overline{Y}}_{\text{prop}})$ without bias. We know that with such an allocation,

$$\widehat{\text{var}}(\widehat{\overline{Y}}_{\text{prop}}) = \frac{N-n}{Nn} \sum_{h=1}^{H} \frac{N_h}{N} s_{yh}^2.$$

Furthermore,

$$\sigma_y^2 = \frac{1}{N} \sum_{k \in U} y_k^2 - \overline{Y}^2.$$

If we let

$$w_h = \frac{1}{n_h} \sum_{k \in S_h} y_k^2,$$

we have:

$$E\left(\sum_{h=1}^{H} \frac{N_h}{N} w_h \right) = \sum_{h=1}^{H} \frac{N_h}{N} \frac{1}{N_h} \sum_{k \in U_h} y_k^2 = \frac{1}{N} \sum_{k \in U} y_k^2,$$

and therefore

$$\sigma_y^2 = \mathrm{E}\left(\sum_{h=1}^{H} \frac{N_h}{N} w_h\right) - \mathrm{E}\left(\widehat{\overline{Y}}_{\text{prop}}^2 - \widehat{\text{var}}(\widehat{\overline{Y}}_{\text{prop}})\right).$$

An unbiased estimator of σ_y^2 is therefore:

$$\widehat{\sigma}_y^2 = \sum_{h=1}^{H} \frac{N_h}{N} w_h - \widehat{\overline{Y}}_{\text{prop}}^2 + \widehat{\text{var}}(\widehat{\overline{Y}}_{\text{prop}}) = \sum_{h=1}^{H} \frac{n_h}{n} w_h - \widehat{\overline{Y}}_{\text{prop}}^2 + \widehat{\text{var}}(\widehat{\overline{Y}}_{\text{prop}})$$

$$= \frac{1}{n} \sum_{k \in S} (y_k - \widehat{\overline{Y}}_{\text{prop}})^2 + \widehat{\text{var}}(\widehat{\overline{Y}}_{\text{prop}}) = \frac{n-1}{n} s_y^2 + \frac{N-n}{Nn} \sum_{h=1}^{H} \frac{N_h}{N} s_{yh}^2.$$

2. We have
$$\mathrm{E}\left(\widehat{\sigma}_y^2\right) = \sigma_y^2 = \mathrm{E}\left(s_y^2\right) \frac{n-1}{n} + \text{var}(\widehat{\overline{Y}}_{\text{prop}}).$$

Therefore

$$\mathrm{E}\left(s_y^2\right) = \frac{n}{n-1}\left\{\sigma_y^2 - \text{var}(\widehat{\overline{Y}}_{\text{prop}})\right\}$$

$$= \sigma_y^2 + \frac{\sigma_y^2}{n-1} - \text{var}(\widehat{\overline{Y}}_{\text{prop}})\frac{n}{n-1} = \sigma_y^2 + O\left(\frac{1}{n}\right).$$

As a reminder, we say that a function $f(n)$ of n is of order of magnitude $g(n)$ (denoted $f(n) = O(g(n))$) if and only if $f(n)/g(n)$ is restricted; that is to say, if there exists a quantity M such that, for all $n \in \mathbb{N}$, $|f(n)|/g(n) \leq M$. The bias is of $1/n$: it is very low if n is very large.

Exercise 4.9 *Expected value of the sample variance*

Consider the uncorrected sample variance in the sample:

$$v_y^2 = \frac{1}{n} \sum_{k \in S} \left(y_k - \widehat{\overline{Y}}\right)^2, \quad \text{where} \quad \widehat{\overline{Y}} = \frac{1}{n} \sum_{k \in S} y_k.$$

1. Give the expected value of v_y^2 for a stratified design with proportional allocation (we neglect the rounding problems which arise when calculating $n_h = nN_h/N$).
2. If v_y^2 is used to estimate
$$\sigma_y^2 = \frac{1}{N} \sum_{k \in U} \left(y_k - \overline{Y}\right)^2,$$

what is the bias of this estimator? Do we have a tendency to overestimate or underestimate σ_y^2?
3. What is the practical interest of the previous result?

Solution

1. *Method 1.*

$$E(v_y^2) = E\left[\frac{1}{n}\sum_{k\in S} y_k^2 - \widehat{\overline{Y}}^2\right] = E\left[\frac{1}{n}\sum_{k\in S} y_k^2\right] - \left[\mathrm{var}(\widehat{\overline{Y}}) + (E\widehat{\overline{Y}})^2\right].$$

A stratified design with proportional allocation is a design with equal probabilities and of fixed size: in this case, every calculated mean in the sample estimates without bias the mean defined in an identical manner in the population. Thus:

$$E\left(\frac{1}{n}\sum_{k\in S} y_k^2\right) = \frac{1}{N}\sum_{k\in U} y_k^2,$$

and $E(\widehat{\overline{Y}}) = \overline{Y}$, which implies that

$$E(v_y^2) = \left(\frac{1}{N}\sum_{k\in U} y_k^2 - \overline{Y}^2\right) - \mathrm{var}(\widehat{\overline{Y}}) = \sigma_y^2 - \mathrm{var}(\widehat{\overline{Y}}).$$

Method 2.
By the result from Exercise 2.7, we have

$$v_y^2 = \frac{1}{n}\sum_{k\in S}\left(y_k - \widehat{\overline{Y}}\right)^2$$

$$= \frac{1}{2n^2}\sum_{k\in S}\sum_{\substack{\ell\in S \\ \ell\neq k}} (y_k - y_\ell)^2 = \frac{1}{2n^2}\sum_{k\in U}\sum_{\substack{\ell\in U \\ \ell\neq k}} (y_k - y_\ell)^2 I_k I_\ell.$$

Separating the sums by stratum, we can write

$$v_y^2 = \frac{1}{2n^2}\sum_{h=1}^{H}\sum_{i=1}^{H}\sum_{k\in U_h}\sum_{\substack{\ell\in U_i \\ \ell\neq k}} (y_k - y_\ell)^2 I_k I_\ell$$

$$= \frac{1}{2n^2}\left[\sum_{h=1}^{H}\sum_{k\in U_h}\sum_{\substack{\ell\in U_h \\ \ell\neq k}} (y_k - y_\ell)^2 I_k I_\ell \right.$$

$$\left. + \sum_{h=1}^{H}\sum_{\substack{i=1 \\ i\neq h}}^{H}\sum_{k\in U_h}\sum_{\ell\in U_i} (y_k - y_\ell)^2 I_k I_\ell\right].$$

The expected value is

$$E(v_y^2) = \frac{1}{2n^2}\left[\sum_{h=1}^{H}\sum_{k\in U_h}\sum_{\substack{\ell\in U_h\\ \ell\neq k}}(y_k-y_\ell)^2\,E(I_kI_\ell)\right.$$

$$\left.+\sum_{h=1}^{H}\sum_{\substack{i=1\\ i\neq h}}^{H}\sum_{k\in U_h}\sum_{\ell\in U_i}(y_k-y_\ell)^2\,E(I_kI_\ell)\right].$$

Since

$$E(I_kI_\ell) = \frac{n_h(n_h-1)}{N_h(N_h-1)} = \frac{n(n_h-1)}{N(N_h-1)} = \frac{n^2}{N^2} - \frac{n(N-n)}{N^2(N_h-1)},$$

if $k\neq\ell\in U_h$, and that

$$E(I_kI_\ell) = \frac{n_h}{N_h}\frac{n_i}{N_i} = \frac{n^2}{N^2},$$

if $k\in U_h, \ell\in U_i, h\neq i$, we get

$$E(v_y^2) = \frac{1}{2n^2}\sum_{h=1}^{H}\sum_{k\in U_h}\sum_{\substack{\ell\in U_h\\ \ell\neq k}}(y_k-y_\ell)^2\left[\frac{n^2}{N^2} - \frac{n(N-n)}{N^2(N_h-1)}\right]$$

$$+\frac{1}{2n^2}\sum_{h=1}^{H}\sum_{\substack{i=1\\ i\neq h}}^{H}\sum_{k\in U_h}\sum_{\ell\in U_i}(y_k-y_\ell)^2\frac{n^2}{N^2}$$

$$=-\frac{1}{2n^2}\sum_{h=1}^{H}\sum_{k\in U_h}\sum_{\substack{\ell\in U_h\\ \ell\neq k}}(y_k-y_\ell)^2\frac{n(N-n)}{N^2(N_h-1)}$$

$$+\frac{1}{2n^2}\sum_{h=1}^{H}\sum_{i=1}^{H}\sum_{k\in U_h}\sum_{\ell\in U_i}(y_k-y_\ell)^2\frac{n^2}{N^2}$$

$$=\frac{1}{2N^2}\sum_{h=1}^{H}\sum_{i=1}^{H}\sum_{k\in U_h}\sum_{\ell\in U_i}(y_k-y_\ell)^2$$

$$-\frac{N-n}{N^2n}\sum_{h=1}^{H}N_h\frac{1}{2N_h(N_h-1)}\sum_{k\in U_h}\sum_{\substack{\ell\in U_h\\ \ell\neq k}}(y_k-y_\ell)^2$$

$$=\sigma_y^2 - \mathrm{var}(\widehat{\overline{Y}}).$$

2. The bias is

$$B(v_y^2) = E(v_y^2) - \sigma_y^2 = \sigma_y^2 - \mathrm{var}(\widehat{\overline{Y}}) - \sigma_y^2 = -\mathrm{var}(\widehat{\overline{Y}}) < 0.$$

The variance σ_y^2 is therefore underestimated.

3. The practical interest resides in the calculation of the estimated *design effect*, defined as the ratio of the variance estimated with the design used over the variance estimated by a random sample of the same size n: to estimate the population variance σ_y^2 in the denominator, certain software packages are going to naturally calculate v_y^2. The bias introduced, of order of magnitude $1/n$, is very low if n is large, and therefore the design effect thus estimated is correct, even if there is a theoretical overestimation.

Exercise 4.10 *Stratification and difference estimator*

Given a stratified design composed of H strata of size N_h. The objective is to estimate the population mean $\overline{Y}$ of a characteristic y. Denote $\overline{X}_h, h = 1, ..., H$ as the means in the strata (in the population) of an auxiliary characteristic x. The $\overline{X}_h$ are supposedly known and we propose to estimate $\overline{Y}$ using the following estimator:

$$\widehat{\overline{Y}}_D = \widehat{\overline{Y}}_\pi + \overline{X} - \widehat{\overline{X}}_\pi.$$

We undertake a simple random sample in each stratum.

1. Show that $\widehat{\overline{Y}}_D$ estimates $\overline{Y}$ without bias.
2. Give the variance of $\widehat{\overline{Y}}_D$.
3. What is the optimal allocation of the n_h in order to minimise the variance of $\widehat{\overline{Y}}_D$? We consider that the unit cost of the survey does not depend on the stratum.
4. In which favourable case is $\widehat{\overline{Y}}_D$ unquestionably preferable to $\widehat{\overline{Y}}_\pi$?

Solution

1. The estimator is unbiased. Indeed, since

$$\widehat{\overline{Y}}_\pi = \sum_{h=1}^{H} \frac{N_h}{N} \widehat{\overline{Y}}_h,$$

where $\widehat{\overline{Y}}_h$ indicates the simple mean of the y_k in the sample of stratum h,

$$E(\widehat{\overline{Y}}_D) = \overline{X} + E(\widehat{\overline{Y}}_\pi) - E(\widehat{\overline{X}}_\pi) = \overline{X} + \overline{Y} - \overline{X} = \overline{Y}.$$

2. Let $z_k = y_k - x_k$. We have

$$\widehat{\overline{Y}}_D = \overline{X} + \widehat{\overline{Z}}_\pi.$$

Therefore

$$\mathrm{var}(\widehat{\overline{Y}}_D) = \mathrm{var}(\widehat{\overline{Z}}_\pi) = \sum_{h=1}^{H} \left(\frac{N_h}{N}\right)^2 \left(1 - \frac{n_h}{N_h}\right) \frac{S_{zh}^2}{n_h},$$

where

$$S_{zh}^2 = \frac{1}{N_h - 1} \sum_{k \in U_h} (z_k - \widehat{\overline{Z}}_h)^2 = S_{yh}^2 + S_{xh}^2 - 2S_{xyh},$$

and

$$S_{xyh} = \frac{1}{N_h - 1} \sum_{k \in U_h} (x_k - \widehat{\overline{X}}_h)(y_k - \widehat{\overline{Y}}_h).$$

3. Letting $z_k = y_k - z_k$, the problem goes back to minimising $\mathrm{var}(\widehat{\overline{Z}}_\pi)$ subject to fixed sample size, which is written here $\sum_{h=1}^{H} n_h = n$. Indeed, the unit cost is the same in all of the strata, which gives

$$n_h = \frac{n N_h S_{zh}}{\sum_{\ell=1}^{H} N_\ell S_{z\ell}}.$$

In practice, we estimate *a priori* the S_{zh} and we round n_h to the nearest whole number, after having fixed n as a function of the overall budget which we have. It can happen that we get $n_h > N_h$ for certain h: in this case, we set $n_h = N_h$ and we perform the calculation again with the remaining strata.

4. As

$$\mathrm{var}(\widehat{\overline{Y}}_\pi) = \sum_{h=1}^{H} \left(\frac{N_h}{N}\right)^2 \left(1 - \frac{n_h}{N_h}\right) \frac{S_{yh}^2}{n_h},$$

and that the two estimators are unbiased, $\widehat{\overline{Y}}_D$ is indisputably preferable to $\widehat{\overline{Y}}_\pi$ when, for all h, $S_{yh}^2 > S_{zh}^2$, meaning, for all h,

$$\frac{S_{xyh}}{S_{xh}^2} > \frac{1}{2}.$$

This condition comes back to obtaining a regression line for y on x which, in each stratum, has a slope greater than $1/2$. This is particularly the case if we let $y = x$ (slope equal to 1): this result is natural, as then $\widehat{\overline{X}}_D = \overline{X}$ for whatever sample is selected. We say that the estimator $\widehat{\overline{Y}}_D$ is 'calibrated' on $\overline{X}$.

Exercise 4.11 *Optimality for a domain*

Consider a population U of size N partitioned into H strata denoted U_1, ..., U_h, ..., U_H, with respective sizes $N_1, ..., N_h, ..., N_H$. We also denote $\overline{Y}_1, ...,$ $\overline{Y}_h, ..., \overline{Y}_H$, as the H true means calculated within the strata. The sampling in each stratum is simple random. We have of course

$$\overline{Y} = \frac{1}{N} \sum_{h=1}^{H} N_h \overline{Y}_h.$$

The objective of the survey is to compare a particular stratum U_i to the total population: more specifically we want to estimate $D_i = \overline{Y}_i - \overline{Y}$.

1. Construct $\widehat{D}_{i\pi}$, the Horvitz-Thompson estimator of D_i for a stratified design with any allocation.
2. Give the variance of $\widehat{D}_{i\pi}$.
3. Give the optimal allocation minimising the variance of $\widehat{D}_{i\pi}$ for a fixed sample size n.
4. How does this allocation differ from the 'classical' optimal allocation?

Solution

1. Since

$$D_i = \overline{Y}_i \left(1 - \frac{N_i}{N}\right) - \frac{1}{N} \sum_{\substack{h=1 \\ h \neq i}}^{H} N_h \overline{Y}_h,$$

we have the unbiased estimator:

$$\widehat{D}_{i\pi} = \widehat{\overline{Y}}_i \left(1 - \frac{N_i}{N}\right) - \frac{1}{N} \sum_{\substack{h=1 \\ h \neq i}}^{H} N_h \widehat{\overline{Y}}_h,$$

where $\widehat{\overline{Y}}_h$ indicates the simple mean in the sample of stratum h.
2. The variance of $\widehat{D}_{i\pi}$ is:

$$\operatorname{var}\left(\widehat{D}_{i\pi}\right) = \left(1 - \frac{N_i}{N}\right)^2 \frac{N_i - n_i}{n_i N_i} S_{yi}^2 + \frac{1}{N^2} \sum_{\substack{h=1 \\ h \neq i}}^{H} N_h^2 \frac{N_h - n_h}{n_h N_h} S_{yh}^2.$$

3. Letting

$$z_k = \begin{cases} y_k(N/N_i - 1) & \text{if } k \in U_i \\ -y_k & \text{otherwise,} \end{cases}$$

we can write

$$\widehat{D}_{i\pi} = \sum_{h=1}^{H} \frac{N_h}{N} \widehat{\overline{Z}}_h.$$

The optimal allocation is given by the classical Neyman expression with a constant unit cost:

$$n_h = \frac{nN_hS_{zh}}{\sum_{j=1}^{H} N_jS_{zj}} \text{ (if } n_h \leq N_h),$$

where

$$S_{zh} = \begin{cases} S_{yi}(N/N_i - 1), & \text{if } h = i \\ S_{yh}, & \text{otherwise.} \end{cases}$$

As always, it is necessary to 'round' n_h after having estimated S_{zh} a priori (via S_{yh}).

4. In comparison to the classical optimal allocation, we 'overrepresent' stratum U_i by a factor of $(N/N_i - 1)$ whenever N_i is 'not too large' (more precisely, as soon as $N_i < N/2$). Otherwise, there is on the contrary 'underrepresentation' of stratum i and we again find exactly the Neyman allocation whenever $N_i = N/2$.

Exercise 4.12 *Optimality for a difference*

We wish to compare, using a sample survey, a metropolitan population with an overseas population. We assume that we know the variances of the variable y in both of the populations where we select a simple random sample without replacement. The objective is to estimate the difference:

$$D = \overline{Y}_1 - \overline{Y}_2,$$

where $\overline{Y}_1$ and $\overline{Y}_2$ are respectively the means of characteristic y in metropolitan France and in 'overseas entities' of France. We know furthermore that an overseas interview costs two times more than in metropolitan France.

1. Define your notation and give an unbiased estimator $\widehat{D}$ of D.
2. Give the variance of the estimator $\widehat{D}$. What criteria must be optimised to obtain the optimal allocation (to be determined) allowing to estimate at best D for a fixed cost C?
3. Give the variance of the optimal estimator obtained with the optimal allocation.

Solution

1. We denote C_1 as the cost of an interview in metropolitan France (population indicator 1) and $C_2 = 2C_1$ as the cost in overseas France (population indicator 2). We denote N_h as the population size h, n_h as the sample size in the population h, $\widehat{\overline{Y}}_h$ as the simple mean of the sample selected in the population h, and C as the total cost of the survey. We have:

$$\widehat{D} = \widehat{\overline{Y}}_1 - \widehat{\overline{Y}}_2.$$

2. Since the two surveys are independent, we must minimise

$$\mathrm{var}(\widehat{D}) = \mathrm{var}\left[\widehat{\overline{Y}}_1\right] + \mathrm{var}\left[\widehat{\overline{Y}}_2\right] = \frac{N_1 - n_1}{N_1 n_1} S_{y1}^2 + \frac{N_2 - n_2}{N_2 n_2} S_{y2}^2$$

subject to $n_1 C_1 + n_2 C_2 = C$. After some very simple calculations, we obtain

$$n_h = \frac{S_{yh}}{\sqrt{\lambda C_h}}, h = 1, 2,$$

where λ is the Lagrange multiplier, and therefore

$$n_h = \frac{S_{yh} C}{\sqrt{C_h} \left\{ \sqrt{C_1} \left(S_{y1} + S_{y2}\sqrt{2} \right) \right\}}, \text{ if } n_h \le N_h \text{ for } h = 1, 2.$$

3. We find, if $n_h \le N_h$ $(h = 1, 2)$,

$$\mathrm{var}(\widehat{D}) = \frac{C_1}{C} \left(S_{y1} + S_{y2}\sqrt{2} \right)^2 - \left(\frac{S_{y1}^2}{N_1} + \frac{S_{y2}^2}{N_2} \right).$$

Exercise 4.13 *Naive estimation*

Consider a population U of size N partitioned into H strata denoted U_1, ..., U_h, ..., U_H, of respective sizes $N_1, ..., N_h, ..., N_H$. We denote as well $\overline{Y}_1$, ..., $\overline{Y}_h$, ..., $\overline{Y}_H$, as the H means calculated within the strata. We have of course

$$\overline{Y} = \frac{1}{N} \sum_{h=1}^{H} N_h \overline{Y}_h.$$

In each stratum, we select a sample according to a simple random design without replacement of any size n_h, $h = 1, ..., H$. The samples are independent from one stratum to another. A young statistician proposes to estimate $\overline{Y}$ by

$$\widehat{\overline{Y}} = \frac{1}{n} \sum_{k \in S} y_k,$$

where $n = \sum_{h=1}^{H} n_h$.

1. Calculate $\mathrm{E}(\widehat{\overline{Y}})$, and deduce the bias of $\widehat{\overline{Y}}$.
2. Calculate the standard deviation of $\widehat{\overline{Y}}$, and deduce the bias ratio, defined as the ratio of the standard deviation over the bias.
3. Explain why it is not advised to use this estimator.

Solution

1. The expected value is

$$E(\widehat{\overline{Y}}) = \frac{1}{n} \sum_{k \in U} y_k \pi_k = \frac{1}{n} \sum_{h=1}^{H} \frac{n_h}{N_h} N_h \overline{Y}_h$$

$$= \frac{1}{n} \sum_{h=1}^{H} n_h \overline{Y}_h = \overline{Y} + \sum_{h=1}^{H} \left(\frac{n_h}{n} - \frac{N_h}{N} \right) \overline{Y}_h.$$

We deduce the bias:

$$B\left(\widehat{\overline{Y}}\right) = \sum_{h=1}^{H} \left(\frac{n_h}{n} - \frac{N_h}{N} \right) \overline{Y}_h.$$

2. We denote $\widehat{\overline{Y}}_h$ as the simple mean of the y_k in the sample of stratum h.

$$\mathrm{var}(\widehat{\overline{Y}}) = \mathrm{var}\left(\sum_{h=1}^{H} \frac{n_h}{n} \widehat{\overline{Y}}_h \right) = \sum_{h=1}^{H} \frac{n_h^2}{n^2} \mathrm{var}\left(\widehat{\overline{Y}}_h \right) = \sum_{h=1}^{H} \frac{n_h^2}{n^2} \frac{N_h - n_h}{N_h n_h} S_{yh}^2.$$

Therefore,

$$\sigma(\widehat{\overline{Y}}) = \left(\frac{1}{n} \sum_{h=1}^{H} \frac{n_h}{n} \frac{N_h - n_h}{N_h} S_{yh}^2 \right)^{1/2}.$$

3. The bias ratio is:

$$\mathrm{BR}\left(\widehat{\overline{Y}}\right) = \frac{B(\widehat{\overline{Y}})}{\sigma(\widehat{\overline{Y}})} = \frac{\sum_{h=1}^{H} \left(\frac{n_h}{n} - \frac{N_h}{N} \right) \overline{Y}_h}{\left(\frac{1}{n} \sum_{h=1}^{H} \frac{n_h}{n} \frac{N_h - n_h}{N_h} S_{yh}^2 \right)^{1/2}}.$$

We can consider the bias to be negligible when BR is small.
A priori the numerator does not systematically approach 0 when n increases (the convergence is only stochastic), while the denominator is always of magnitude $n^{-1/2}$, thus the bias ratio can be large when n is large. The estimator is thus banished whenever n_h/n differs from N_h/N, as we have another estimator (the unbiased 'classical' estimator) which does not have this unfortunate drawback.

Exercise 4.14 *Comparison of regions and optimality*

We perform a stratified survey on businesses in a country. The strata are regions and we study the variable 'sales' denoted y. In each stratum, we take a simple random sample. The objective is to compare the average sales of each region to that of the other regions. We use the following criteria to measure the pertinence of the sampling design.

$$W = \sum_{h=1}^{H} \sum_{\substack{\ell=1 \\ \ell \neq h}}^{H} \operatorname{var}\left\{\widehat{\overline{Y}}_h - \widehat{\overline{Y}}_\ell\right\},$$

where $\widehat{\overline{Y}}_h$ is the unbiased mean estimator of y in stratum h.

1. Show that W can equally be written

$$W = C \sum_{h=1}^{H} \frac{N_h - n_h}{N_h n_h} S_{yh}^2,$$

where C is a constant that does not depend on h. Give the value of C.
2. How do we choose the n_h while assuring a fixed sample size n?

Solution

1. In developing W, we have

$$W = \sum_{h=1}^{H} \sum_{\substack{\ell=1 \\ \ell \neq h}}^{H} \operatorname{var}\left(\widehat{\overline{Y}}_h - \widehat{\overline{Y}}_\ell\right)$$

$$= \sum_{h=1}^{H} \sum_{\substack{\ell=1 \\ \ell \neq h}}^{H} \left[\operatorname{var}\left(\widehat{\overline{Y}}_h\right) + \operatorname{var}\left(\widehat{\overline{Y}}_\ell\right)\right] \quad \text{(the } \widehat{\overline{Y}}_h \text{ are independent)}$$

$$= \sum_{h=1}^{H} \sum_{\ell=1}^{H} \left[\operatorname{var}\left(\widehat{\overline{Y}}_h\right) + \operatorname{var}\left(\widehat{\overline{Y}}_\ell\right)\right] - \sum_{h=1}^{H} 2\operatorname{var}\left(\widehat{\overline{Y}}_h\right)$$

$$= 2H \sum_{h=1}^{H} \operatorname{var}\left(\widehat{\overline{Y}}_h\right) - \sum_{h=1}^{H} 2\operatorname{var}\left(\widehat{\overline{Y}}_h\right)$$

$$= 2(H-1) \sum_{h=1}^{H} \operatorname{var}\left(\widehat{\overline{Y}}_h\right)$$

$$= 2(H-1) \sum_{h=1}^{H} \frac{N_h - n_h}{N_h n_h} S_{yh}^2,$$

which gives $C = 2(H-1)$.
2. We thus have:

$$W = C \sum_{h=1}^{H} \frac{S_{yh}^2}{n_h} + \text{constant}.$$

It remains to minimise $\sum_{h=1}^{H} S_{yh}^2 / n_h$ subject to $\sum_{h=1}^{H} n_h = n$. Deriving the Lagrangian function, we right away have

$$n_h = \frac{S_{yh}}{\sqrt{\lambda}},$$

where λ is the Lagrange multiplier, and therefore

$$n_h = \frac{n S_{yh}}{\sum_{\ell=1}^{H} S_{y\ell}} \quad \text{(if } n_h \leq N_h\text{)}.$$

Exercise 4.15 *Variance of a product*

Consider a population U of size N composed of two strata, U_1 and U_2 of sizes N_1 and N_2. We wish to estimate $\overline{Y}_1 \times \overline{Y}_2$, where $\overline{Y}_i$ represents the mean of characteristic y in U_i. In each stratum, we select (independently) a random sample. These samples denoted respectively e_1 and e_2 are selected according to two simple designs of respective fixed sizes n_1 and n_2.

1. Give the 'natural' estimator of $\overline{Y}_1 \times \overline{Y}_2$ and verify that it is unbiased.
2. Calculate its variance by expressing it as a function of the means $\overline{Y}_1$, $\overline{Y}_2$, and the corrected population variances calculated in the strata, denoted respectively S_1^2 and S_2^2.

Solution

1. We are going to naturally use $\widehat{\overline{Y}}_1 \times \widehat{\overline{Y}}_2$ where $\widehat{\overline{Y}}_i$ represents the simple mean of characteristic y in e_i. In fact $\widehat{\overline{Y}}_1$ and $\widehat{\overline{Y}}_2$ are independent, by construction. Therefore:

$$\mathrm{E}(\widehat{\overline{Y}}_1 \times \widehat{\overline{Y}}_2) = \mathrm{E}(\widehat{\overline{Y}}_1) \times \mathrm{E}(\widehat{\overline{Y}}_2) = \overline{Y}_1 \times \overline{Y}_2.$$

2. The variance is

$$\begin{aligned}
\mathrm{var}&\left(\widehat{\overline{Y}}_1 \times \widehat{\overline{Y}}_2\right) \\
&= \mathrm{E}\left(\widehat{\overline{Y}}_1^2 \times \widehat{\overline{Y}}_2^2\right) - \left\{\mathrm{E}\left(\widehat{\overline{Y}}_1 \times \widehat{\overline{Y}}_2\right)\right\}^2 \\
&= \mathrm{E}\left(\widehat{\overline{Y}}_1^2\right) \times \mathrm{E}\left(\widehat{\overline{Y}}_2^2\right) - \overline{Y}_1^2 \times \overline{Y}_2^2 \\
&= \left\{\mathrm{var}\left(\widehat{\overline{Y}}_1\right) + \overline{Y}_1^2\right\} \times \left\{\mathrm{var}\left(\widehat{\overline{Y}}_2\right) + \overline{Y}_2^2\right\} - \overline{Y}_1^2 \times \overline{Y}_2^2 \\
&= \left\{\left(\frac{1}{n_1} - \frac{1}{N_1}\right) S_1^2 + \overline{Y}_1^2\right\} \times \left\{\left(\frac{1}{n_2} - \frac{1}{N_2}\right) S_2^2 + \overline{Y}_2^2\right\} - \overline{Y}_1^2 \times \overline{Y}_2^2.
\end{aligned}$$

Exercise 4.16 *National and regional optimality*

We consider a stratified sample of individuals at a national scale, with each administrative region comprising a stratum. In each stratum, we select individuals through simple random sampling.

1. Recall the expression of Neyman allocation (indifferent costs) and express the accuracy of a *regional* simple mean as a function of the size of the region (the size of a region is the number of inhabitants which occupy it). What 'strange' occurrence can be detected concerning the quality of regional results?
2. Instead of minimising a 'national' variance, we use the following criterion:

$$\sum_{h=1}^{H} [(X_h)^{\alpha} \mathrm{CV}\,(\widehat{\overline{Y}}_h)]^2,$$

where:

- $\widehat{\overline{Y}}_h$ is the mean of y in the sample calculated within stratum h;
- X_h is some auxiliary information measuring the importance of the stratum (its population for example, or the total of a variable correlated to y);
- $\mathrm{CV}\,(\widehat{\overline{Y}}_h)$ is the coefficient of variation of $\widehat{\overline{Y}}_h$;
- α is a real and known fixed value, between 0 and 1.

a) Comment on the merits of such a criterion.
b) Express the criterion as a function of X_h, S_{yh}, $\overline{Y}_h$, n_h and N_h (traditional notations).
c) Minimise this criterion subject to the overall sample size equal to n. Deduce the optimal allocation.
d) With this allocation, what happens to the regional accuracy? In particular, we will measure this accuracy by the coefficient of variation (instead of the variance), neglecting the sampling rates.
e) Comment on the effect on the local accuracy (regional) of the following choices: $\alpha = 1$ and $X_h = N_h \overline{Y}_h$, then $\alpha = 0$ and finally $0 < \alpha < 1$.

Solution

1. The optimal allocation is given here by $n_h = \lambda N_h S_{yh}$, where λ is such that

$$\sum_{h=1}^{H} n_h = n.$$

If h represents a given region, we have

$$\mathrm{var}(\widehat{\overline{Y}}_h) = (1 - f_h)\,\frac{S_{yh}^2}{\lambda\,N_h S_{yh}} = \frac{1}{\lambda}\,(1 - f_h)\,\frac{S_{yh}}{N_h}.$$

Unfortunately, with this approach, the regions are treated in an unequal way: the smallest regions (N_h small) have the least precise results! The Neyman optimality is of an *overall* nature (here national): it is the best strategy to produce national results, but not regional results.

2. a) The CV $(\widehat{\overline{Y}}_h)$ is a measure of imprecision within the region h, and the X_h^α is a weight which puts into perspective this measure. The overall national quality criterion is obtained by weighting the regional qualities by the importance of the regions. This importance is measured by X_h^α (for example, if $X_h = N_h$, the most populated regions are going to have a larger importance in the quality criterion). But, quite cleverly, the exponent α comes *to moderate* the relative importance given to a region compared to the others.

   b) The square of the coefficient of variation is written

$$\mathrm{CV}^2(\widehat{\overline{Y}}_h) = \frac{\mathrm{var}(\widehat{\overline{Y}}_h)}{(\mathrm{E}\widehat{\overline{Y}}_h)^2} = \frac{(1-f_h)\frac{S_{yh}^2}{n_h}}{\overline{Y}_h^2} = \left(1 - \frac{n_h}{N_h}\right)\frac{1}{n_h}\left(\frac{S_{yh}}{\overline{Y}_h}\right)^2.$$

We get:

$$\mathrm{Criterion} = \sum_{h=1}^{H} X_h^{2\alpha}\left(\frac{1}{n_h} - \frac{1}{N_h}\right)\left(\frac{S_{yh}}{\overline{Y}_h}\right)^2$$

$$= \sum_{h=1}^{H}\left(\frac{X_h^\alpha S_{yh}}{\overline{Y}_h}\right)^2\frac{1}{n_h} + (\text{term without } n_h).$$

   c) Let

$$\Delta_h = \frac{X_h^\alpha S_{yh}}{\overline{Y}_h}.$$

Minimising the criterion comes back to minimising

$$\sum_{h=1}^{H} \Delta_h^2/n_h$$

subject to $\sum_{h=1}^{H} n_h = n$. We get:

$$-\frac{\Delta_h^2}{n_h^2} = \mathrm{Constant}.$$

The n_h must therefore be proportional to the Δ_h, or more precisely:

$$n_h = n\frac{\Delta_h}{\sum_{j=1}^{H}\Delta_j}.$$

d) The regional accuracy can be measured by:

$$\text{var}(\widehat{\overline{Y}}_h) = (1 - f_h)\ \frac{S_{yh}^2}{\dfrac{n}{\varDelta}\ \dfrac{X_h^\alpha S_{yh}}{\overline{Y}_h}},\ \ \text{where}\ \varDelta = \sum_{j=1}^{H} \varDelta_j,$$

that is,

$$\text{CV}^2(\widehat{\overline{Y}}_h) \approx \left(\frac{\varDelta}{n}\overline{Y}_h\ \frac{S_{yh}}{X_h^\alpha}\right) \frac{1}{\overline{Y}_h^2},$$

neglecting the sampling rates. $\text{CV}(\widehat{\overline{Y}}_h)$ is thus proportional to

$$\frac{1}{\sqrt{X_h^\alpha}} \times \sqrt{\frac{S_{yh}}{\overline{Y}_h}}.$$

e) • If $\alpha = 1$, $X_h = N_h \times \overline{Y}_h$. We then find the coefficient of variation attached to the Neyman *allocation*, which is not beneficial for the smaller regions (see 1.).

 • If $\alpha = 0$, we get a CV proportional to $\sqrt{S_{yh}/\overline{Y}_h}$. Indeed, $S_{yh}/\overline{Y}_h$ is the *true* coefficient of variation of y_k in region h. Except for particular circumstances, it *varies little* from one region to another. In this case, the regional accuracies (measured by the coefficient of variation) are absolutely comparable from a numerical point of view (the Limousin region is no more disadvantaged compared to the Ile-de-France region), but we lose in overall accuracy.

 • If $0 < \alpha < 1$, then we find ourselves in a *compromising* situation, which eventually allows to satisfy at the same time the national statisticians and the regional statisticians (for example, we compromise with $\alpha = 1/2$).

Exercise 4.17 *What is the design?*

In the population $U = \{1, 2, 3, 4, 5\}$, we consider the following sampling design:

$$p(\{1, 2, 4\}) = 1/6, \quad p(\{1, 2, 5\}) = 1/6, \quad p(\{1, 4, 5\}) = 1/6,$$

$$p(\{2, 3, 4\}) = 1/6, \quad p(\{2, 3, 5\}) = 1/6, \quad p(\{3, 4, 5\}) = 1/6.$$

Calculate the first- and second-order inclusion probabilities as well as the $\varDelta_{k\ell}$ (see Expression (1.1), page 3). Show that it is a matter of a stratified design.

Solution

The first-order inclusion probabilities are given by

$$\pi_1 = 1/2, \pi_2 = 2/3, \pi_3 = 1/2, \pi_4 = 2/3, \pi_5 = 2/3,$$

and the second-order inclusion probabilities by

$$\begin{pmatrix} - & \pi_{12} = 1/3 & \pi_{13} = 0 & \pi_{14} = 1/3 & \pi_{15} = 1/3 \\ \pi_{12} = 1/3 & - & \pi_{23} = 1/3 & \pi_{24} = 1/3 & \pi_{25} = 1/3 \\ \pi_{13} = 0 & \pi_{23} = 1/3 & - & \pi_{34} = 1/3 & \pi_{35} = 1/3 \\ \pi_{14} = 1/3 & \pi_{24} = 1/3 & \pi_{34} = 1/3 & - & \pi_{45} = 1/3 \\ \pi_{15} = 1/3 & \pi_{25} = 1/3 & \pi_{35} = 1/3 & \pi_{45} = 1/3 & - \end{pmatrix}.$$

Finally, the $\Delta_{k\ell} = \pi_{kl} - \pi_k \pi_l$ are given by

$$\begin{pmatrix} - & \Delta_{12} = 0 & \Delta_{13} = -1/4 & \Delta_{14} = 0 & \Delta_{15} = 0 \\ \Delta_{12} = 0 & - & \Delta_{23} = 0 & \Delta_{24} = -1/9 & \Delta_{25} = -1/9 \\ \Delta_{13} = -1/4 & \Delta_{23} = 0 & - & \Delta_{34} = 0 & \Delta_{35} = 0 \\ \Delta_{14} = 0 & \Delta_{24} = -1/9 & \Delta_{34} = 0 & - & \Delta_{45} = -1/9 \\ \Delta_{15} = 0 & \Delta_{25} = -1/9 & \Delta_{35} = 0 & \Delta_{45} = -1/9 & - \end{pmatrix}.$$

We see that a large number of $\Delta_{k\ell}$ are null, which is a sign of a stratified design. In fact, in a stratified design, if k and ℓ belong to any two different strata then $\pi_{k\ell} = \pi_k \pi_\ell$, that is $\Delta_{k\ell} = 0$. Anyway, if the design is stratified, two individuals k and ℓ such that $\Delta_{k\ell} \neq 0$ inevitably belong to the same stratum. Considering this principle, the two strata, if they exist, are inevitably:

$$\{1, 3\}, \{2, 4, 5\}.$$

If remains to verify that this stratified design corresponds well to the stated design. If, in the strata $\{1, 3\}$, we select a unit by simple random sampling (which explains that $\pi_{13} = 0$ and that $\pi_1 + \pi_3 = 1$) and if, in the strata $\{2, 4, 5\}$, independent from the previous selection, two units are selected by simple random sampling without replacement (where $\pi_2 + \pi_4 + \pi_5 = 2$), there are six possible samples S and we very well find the probabilities $p(s)$ previously stated.

5

Multi-stage Sampling

5.1 Definitions

We consider a partitioning of the population U into M parts, called primary units (PU). Each PU is itself partitioned into N_i parts, called secondary units (SU), identified by the pair (i, k), where k varies from 1 to N_i. The population of secondary units in PU i is denoted U_i. It is possible to repartition each SU and to iterate this process. We sample m PU (sample S) then, in general independently from one PU to another, we sample n_i SU in PU i if it is sampled (sample S_i): we say that we are faced with sampling of two stages. If this final stage is sampled exhaustively, the sampling is called 'cluster sampling'.

5.2 Estimator, variance decomposition, and variance

In a two-stage sampling design without replacement, if PU i is selected with inclusion probability π_i, and if SU (i, k) that it contains is selected with probability $\pi_{k|i}$, then we estimate the total

$$Y = \sum_{i=1}^{M} \sum_{k \in U_i} y_{i,k}$$

without bias by

$$\widehat{Y} = \sum_{i \in S} \sum_{k \in S_i} \frac{y_{i,k}}{\pi_i \pi_{k|i}}.$$

The variance $\text{var}(\widehat{Y})$ is the sum of two terms, knowing the 'inter-class' variance $\text{var}_1(\text{E}_{2|1}(\widehat{Y}))$ and the 'intra-class' variance $\text{E}_1(\text{var}_{2|1}(\widehat{Y}))$, where 1 and 2 are the indices representing the two successive sampling stages. In the case of a simple random sample at each stage, when n_i only depends on i, we show that:

$$\text{var}(\widehat{Y}) = M^2 \left(1 - \frac{m}{M}\right) \frac{S_T^2}{m} + \frac{M}{m} \sum_{i=1}^{M} N_i^2 \left(1 - \frac{n_i}{N_i}\right) \frac{S_{2,i}^2}{n_i},$$

where

$$S_T^2 = \frac{1}{M-1} \sum_{i=1}^{M} (Y_i - \overline{\overline{Y}})^2,$$

$$\overline{\overline{Y}} = \frac{1}{M} \sum_{i=1}^{M} Y_i,$$

and

$$S_{2,i}^2 = \frac{1}{N_i - 1} \sum_{k \in U_i} (y_{i,k} - \overline{Y}_i)^2,$$

with

$$\overline{Y}_i = \frac{Y_i}{N_i},$$

and

$$Y_i = \sum_{k \in U_i} y_{i,k}.$$

This variance can be estimated without bias by:

$$\widehat{\text{var}}(\widehat{Y}) = M^2 \left(1 - \frac{m}{M}\right) \frac{s_T^2}{m} + \frac{M}{m} \sum_{i \in S} N_i^2 \left(1 - \frac{n_i}{N_i}\right) \frac{s_{2,i}^2}{n_i},$$

where

$$s_T^2 = \frac{1}{m-1} \sum_{i \in S} (\widehat{Y}_i - \frac{1}{m} \sum_{i \in S} \widehat{Y}_i)^2,$$

and

$$s_{2,i}^2 = \frac{1}{n_i - 1} \sum_{k \in S_i} (y_{i,k} - \widehat{\overline{Y}}_i)^2,$$

with

$$\widehat{Y}_i = N_i \widehat{\overline{Y}}_i,$$

and

$$\widehat{\overline{Y}}_i = \frac{1}{n_i} \sum_{k \in S_i} y_{i,k}.$$

5.3 Specific case of sampling of PU with replacement

When the primary units are selected with replacement, we have a remarkable result. Denoting m as the sample size of PU, j as the order number of the drawing and i_j as the identifier of the PU selected at the jth drawing, and denoting:

- p_i as the sampling probability of PU i at the time of any drawing

$$\sum_{i=1}^{M} p_i = 1.$$

- $\widehat{Y}_i$ as the unbiased estimator of the true total Y_i (expression as a function of the sampling design within PU i).

We then estimate without bias the true total with the Hansen-Hurwitz estimator:

$$\widehat{Y}_{HH} = \frac{1}{m} \sum_{j=1}^{m} \frac{\widehat{Y}_{i_j}}{p_{i_j}},$$

and we estimate without bias its variance by:

$$\widehat{\mathrm{var}}\left(\widehat{Y}_{HH}\right) = \frac{1}{m(m-1)} \sum_{j=1}^{m} \left(\frac{\widehat{Y}_{i_j}}{p_{i_j}} - \widehat{Y}_{HH}\right)^2.$$

This very simple expression is valid for whatever sampling design used within the PU (we only require that $\widehat{Y}_i$ be unbiased for Y_i).

5.4 Cluster effect

We thus indicate the phenomenon conveying a certain 'similarity' among the individuals of the same PU, in comparison with the variable of interest y. We can formalise this by:

$$\rho = \frac{\sum_{i=1}^{M} \sum_{k=1}^{N_i} \sum_{\substack{\ell=1 \\ \ell \neq k}}^{N_i} (y_{i,k} - \overline{Y})(y_{i,\ell} - \overline{Y})}{\sum_{i=1}^{M} \sum_{k \in U_i} (y_{i,k} - \overline{Y})^2} \frac{1}{\overline{N} - 1},$$

where

$$\overline{N} = \frac{N}{M}.$$

With simple random sampling without replacement at each of the two stages and with the PU of same size, we show that

$$\mathrm{var}(\widehat{Y}) = N^2 \frac{S_y^2}{m\bar{n}} (1 + \rho(\bar{n} - 1))$$

as soon as $n_i = \bar{n}$ for all PU i (and that we neglect the sampling rate of PU). The cluster effect increases the variance, especially since $\bar{n}$ is large.

EXERCISES

Exercise 5.1 *Hard disk*

On a micro-computer hard disk, we count 400 files, each one consisting of exactly 50 records. To estimate the average number of characters per record, we decide to sample using simple random sampling 80 files, then 5 records in each file. We denote: $m = 80$ and $\overline{n} = 5$. After sampling we find:

- the sample variance of the estimators for the total number of characters per file, which is $s_T^2 = 905\ 000$;
- the mean of the m sample variances $s_{2,i}^2$ is equal to 805, where $s_{2,i}^2$ represents the variance for the number of characters per record in file i.

1. How do we estimate without bias the mean number $\overline{Y}$ of characters per record?
2. How do we estimate without bias the accuracy of the previous estimator?
3. Give a 95% confidence interval for $\overline{Y}$.

Solution

1. We denote $y_{i,k}$ as the number of characters in record k of file i. We have

$$\overline{Y} = \frac{1}{N} \sum_{i=1}^{M} \sum_{k \in U_i} y_{i,k} = \frac{1}{N} \sum_{i=1}^{M} \overline{N}\, \overline{Y}_i = \frac{1}{M} \sum_{i=1}^{M} \overline{Y}_i,$$

where
- $M = 400$ is the number of files (primary units),
- $\overline{N} = 50$ is the number of records per file,
- $N = M \times \overline{N} = 400 \times 50 = 20000$ is the total number of records,
- $\overline{Y}_i$ is the mean number of characters per record in file i,
- U_i is the set of identifiers for the records of file i.

We estimate $\overline{Y}$ without bias by

$$\widehat{\overline{Y}} = \frac{\widehat{Y}}{N} = \frac{1}{N} \sum_{i \in S_1} \frac{\widehat{Y}_i}{m/M},$$

where
- S_1 is the sample of files,
- $\widehat{Y}_i$ is the unbiased estimator of the total number of characters in file i

$$\widehat{Y}_i = \sum_{k \in S_i} \frac{y_{i,k}}{\overline{n}/\overline{N}} = \frac{\overline{N}}{\overline{n}} \sum_{k \in S_i} y_{i,k},$$

- S_i is the sample of records selected in file i.

We know that $\widehat{Y}_i = \overline{N} \times \widehat{\overline{Y}}_i$, where $\overline{Y}_i$ is the mean number of characters per record sampled in file i. We easily see that:

$$\widehat{\overline{Y}} = \frac{M}{Nm}\overline{N}\sum_{i\in S_1}\widehat{\overline{Y}}_i = \frac{1}{m}\sum_{i\in S_1}\widehat{\overline{Y}}_i = \frac{1}{m\bar{n}}\sum_{i\in S_1}\sum_{k\in S_i}y_{i,k},$$

which is the simple mean $\widehat{\overline{Y}}$ calculated on the sample of the $m \times \bar{n} = 400$ selected records. Arriving at this mean is natural if we observe that the sampling is of fixed size $m\bar{n}$.

2. This sampling design is of two stages, with primary units (the files) of constant size $\overline{N}$. In this case, we have:

$$\widehat{\mathrm{var}}(\overline{Y}) = \frac{1}{N^2}\widehat{\mathrm{var}}(\widehat{Y}) = \frac{1}{\overline{N}^2}\frac{1-\frac{m}{M}}{m}s_T^2 + \frac{m}{M}\frac{1}{m\bar{n}}\left(1-\frac{\bar{n}}{\overline{N}}\right)\left(\frac{1}{m}\sum_{i\in S_1}s_{2,i}^2\right),$$

which gives

$$\widehat{\mathrm{var}}(\widehat{\overline{Y}}) = \frac{1-80/400}{80}\times\frac{905\ 000}{(50)^2} + \frac{80}{400}\times\frac{1}{80\times 5}\left(1-\frac{5}{50}\right)\times 805$$

$$= \frac{14\ 480}{4\ 000} + \frac{1\ 449}{4\ 000}$$

$$\approx 3.98.$$

Note: In this design of two stages, the quantity $14\ 480/4\ 000$ overestimates the INTER-class variance and $1\ 449/4\ 000$ underestimates the INTRA-class variance (see Ardilly, 1994, page 101).

3. Taking into account the sampling sizes, we can consider that $\widehat{\overline{Y}}$ follows (approximately) a normal distribution. Then

$$\overline{Y} \in \left[\widehat{\overline{Y}} - 1.96\sqrt{\widehat{\mathrm{var}}(\widehat{\overline{Y}})}\ ;\ \ \widehat{\overline{Y}} + 1.96\sqrt{\widehat{\mathrm{var}}(\widehat{\overline{Y}})}\right] = \left[\widehat{\overline{Y}} - 3.9; \widehat{\overline{Y}} + 3.9\right],$$

95 times out of 100.

Exercise 5.2 *Selection of blocks*

The objective is to estimate the mean income of households in a district of a city consisting of 60 blocks of houses (of variable size). For this, we select three blocks using simple random sampling without replacement and we interview all households which live there. Furthermore, we know that 5000 households reside in this district. The result of the survey is given in Table 5.1.

1. Estimate the mean income $\widehat{\overline{Y}}_\pi$ and the total income $\widehat{Y}_\pi$ of the households in the district using the Horvitz-Thompson estimator.

Table 5.1. Table of three selected blocks: Exercise 5.2

Block number	Number of households in the block	Total household income in the block
1	120	2100
2	100	2000
3	80	1500

2. Estimate without bias the variance of the Horvitz-Thompson mean estimator.
3. Estimate the mean income $\widehat{\overline{Y}}_H$ of the households in the district using the Hájek ratio, and compare with the estimation from 1. Was the direction of the change predictable?

Solution

It is cluster sampling where the clusters are selected with equal probabilities with $M = 60, m = 3, N = 5000$. The inclusion probabilities are given by:

$$\pi_i = \frac{m}{M} = \frac{3}{60} = \frac{1}{20}.$$

The population total (known) in cluster i is $N_i \overline{Y}_i$.

1. We denote S as the sample of selected clusters. The Horvitz-Thompson mean estimator is defined by:

$$\widehat{\overline{Y}}_\pi = \frac{1}{N} \sum_{i \in S} \frac{N_i \overline{Y}_i}{\frac{m}{M}} = \frac{M}{N} \frac{1}{m} \sum_{i \in S} N_i \overline{Y}_i$$

$$= \frac{1}{5000} \left(\frac{1500}{1/20} + \frac{2000}{1/20} + \frac{2100}{1/20} \right) = 22.4.$$

The Horvitz-Thompson estimator of the total is:

$$\widehat{Y}_\pi = N \widehat{\overline{Y}}_\pi = 5000 \times 22.4 = 112\,000.$$

2. Since the sampling is simple random in the population of clusters, we have

$$\widehat{\mathrm{var}}(\widehat{\overline{Y}}_\pi) = \left(\frac{M}{N} \right)^2 \left(1 - \frac{m}{M} \right) \frac{1}{m} \frac{1}{m-1} \sum_{i \in S} \left(N_i \overline{Y}_i - \frac{N}{M} \widehat{\overline{Y}}_\pi \right)^2$$

$$= \frac{M - m}{m - 1} \frac{M}{m} \sum_{i \in S} \left(\frac{\overline{Y}_i N_i}{N} - \frac{\widehat{\overline{Y}}_\pi}{M} \right)^2$$

$$= \frac{60 - 3}{3 - 1} \times \frac{60}{3} \times \left\{ \left(\frac{1500}{5000} - \frac{22.4}{60} \right)^2 + \left(\frac{2000}{5000} - \frac{22.4}{60} \right)^2 \right.$$

$$\left. + \left(\frac{2100}{5000} - \frac{22.4}{60} \right)^2 \right\} \approx 4.7.$$

3. We denote $\widehat{N}_\pi$ as the unbiased estimator of N, being

$$\widehat{N}_\pi = \sum_{i \in S} \frac{N_i}{m/M},$$

in that case

$$\widehat{\overline{Y}}_H = \frac{\widehat{\overline{Y}}_\pi}{\widehat{N}_\pi} = \frac{\frac{1500}{1/20} + \frac{2000}{1/20} + \frac{2100}{1/20}}{\frac{120}{1/20} + \frac{100}{1/20} + \frac{80}{1/20}} = \frac{5600 \times 20}{300 \times 20} = 18.7.$$

Therefore $\widehat{\overline{Y}}_H < \widehat{\overline{Y}}_\pi$. The three blocks forming S are obviously 'too large' on average: their mean size is 100 households while in the entire population the mean block size is $5000/60 \approx 83.3$ households. Under these conditions, since the total income of a block is well explained by its size, it is logical to get an estimate $\widehat{\overline{Y}}_\pi$ of $\overline{Y}$ that is 'too large'. The usage of $\widehat{\overline{Y}}_H$ corrects this effect and decreases the estimate.

Exercise 5.3 *Inter-cluster variance*

Consider a simple random sample of clusters. We suppose that all clusters are of the same size. Recall the expression of the Horvitz-Thompson estimator. Give an expression of its variance as a function of the inter-cluster population variance.

Solution

With a simple random sample of clusters i of size N_i and of mean $\overline{Y}_i$, we have:

$$\widehat{\overline{Y}}_\pi = \frac{M}{N} \frac{1}{m} \sum_{i \in S} N_i \overline{Y}_i.$$

If all clusters are of the same size, we have

$$\frac{N_i}{N} = \frac{1}{M}, i = 1, ..., M.$$

That is, finally,

$$\widehat{\overline{Y}}_\pi = \frac{1}{m} \sum_{i \in S} \overline{Y}_i.$$

We will observe that it is the simple mean of y_k in the overall sample. The variance of the Horvitz-Thompson mean estimator is written in the case of clusters of size N_i:

$$\text{var}\left(\widehat{\overline{Y}}_\pi\right) = \left(\frac{M}{N}\right)^2 \frac{M-m}{mM} \frac{1}{M-1} \sum_{i=1}^{M} \left(N_i \overline{Y}_i - \frac{1}{M} \sum_{i=1}^{M} N_i \overline{Y}_i\right)^2$$

$$= \frac{M-m}{M-1} \frac{M}{m} \sum_{i=1}^{M} \left(\frac{\overline{Y}_i N_i}{N} - \frac{\overline{Y}}{M}\right)^2.$$

We therefore obtain, in the present case:

$$\mathrm{var}\left(\widehat{\overline{Y}}_\pi\right) = \frac{M-m}{M-1}\frac{M}{m}\sum_{i=1}^{M}\left(\frac{\overline{Y}_i}{M} - \frac{\overline{Y}}{M}\right)^2$$

$$= \frac{M-m}{M-1}\frac{1}{m}\sum_{i=1}^{M}\frac{1}{M}\left(\overline{Y}_i - \overline{Y}\right)^2$$

$$= \frac{M-m}{M-1}\frac{1}{m}\sum_{i=1}^{M}\frac{N_i}{N}\left(\overline{Y}_i - \overline{Y}\right)^2$$

$$= \frac{M-m}{M-1}\frac{\sigma_{\text{inter}}^2}{m}.$$

The variance of the estimator essentially depends on the size of the sample of clusters and on the inter-cluster population variance σ_{inter}^2. Contrary to the stratification, we thus have complete interest in constructing clusters for which the means are very close to one another. Note that, in the exclusive case of a simple random sample of clusters of equal size, we immediately deduce the variance of the unbiased estimator of the mean $\widehat{\overline{Y}}_\pi$, as this is the simple mean of values $\overline{Y}_i$, that is:

$$\mathrm{var}(\widehat{\overline{Y}}_\pi) = \left(1 - \frac{m}{M}\right)\frac{S^2(\overline{Y}_i)}{m},$$

where

$$S^2(\overline{Y}_i) = \frac{1}{M-1}\sum_{i=1}^{M}(\overline{Y}_i - \overline{Y})^2.$$

Exercise 5.4 *Clusters of patients*

A statistician wishes to carry out a survey on the quality of health care in the cardiology services of hospitals. For that, he selects by simple random sampling 100 hospitals among the 1 000 hospitals listed and then, in each of the selected hospitals, he collects the opinions of all the cardiology patients.

1. What do we call this sampling design and what is its reason for existence?
2. We consider that each cardiology unit is comprised of exactly 50 beds and that the 95% confidence interval on the true proportion P of dissatisfied patients is:

$$P \in [0.10 \pm 0.018],$$

(that signifies in particular that, in the sample, 10% of patients are dissatisfied with the quality of care). How do you estimate the cluster effect? (Start by estimating S_y^2.)

3. How would the accuracy of the statistician's survey on satisfaction evolve if, all at once, he sampled twice the number of hospitals but in each selected hospital he only collected his data on half of the cardiology units? (Say that the units are systematically divided by an aisle and that our statistician is exclusively interested in the 25 beds that are situated to the right of the aisle)?
4. Comment on this result in comparison to that given by the first sample design.

Solution

1. It is cluster sampling. It is justified by a search for savings in terms of budget.
2. We recall that a true proportion P is estimated without bias using a proportion in the sample $\widehat{P}$ whenever all the clusters have the same size (it is the case here, with the common size being 50). If $\widehat{\rho}$ is the estimated cluster effect, we have:

$$\widehat{\text{var}}(\widehat{P}) = (1 - f)\frac{\widehat{S}_y^2}{m\overline{N}}\left[1 + \widehat{\rho}(\overline{N} - 1)\right],$$

where $\widehat{S}_y^2$ is a 'good' estimator of S_y^2, because the clusters are of equal size $\overline{N}$. Indeed

$$2 \times \sqrt{\widehat{\text{var}}(\widehat{P})} = 0.018 \Rightarrow \widehat{\text{var}}(\widehat{P}) = 8.1 \times 10^{-5}.$$

Furthermore, $f = \frac{100}{1\,000}, m = 100$, and $\overline{N} = 50$. The problem remains to estimate S_y^2. We saw in Exercise 3.21 that the sample variance s_y^2 is a biased estimator of S_y^2 when the design is complex (which is the case here), but that this bias varies by $1/n$ if the design is of fixed size and with equal probabilities. Here, $n = 5000$, these conditions are satisfied and this bias is therefore totally negligible. That is,

$$\widehat{S}_y^2 = s_y^2 = \frac{1}{m\overline{N} - 1}\sum_{k \in S^*}(y_k - \widehat{\overline{Y}})^2,$$

where S^* is the sample of $m\overline{N}$ patients and y_k is 1 if patient k is dissatisfied, and 0 otherwise ($\widehat{\overline{Y}}$ is the mean of y_k on S^*). According to the decomposition formula for variance, denoting S as the sample of hospitals (the other notations are standard):

$$s_y^2 \approx \sum_{i \in S}\frac{\overline{N}}{n}s_{2,i}^2 + \sum_{i \in S}\frac{\overline{N}}{n}(\widehat{\overline{Y}}_i - \widehat{\overline{Y}})^2 = \frac{1}{m}\sum_{i \in S}s_{2,i}^2 + \frac{1}{m}\sum_{i \in S}(P_i - \widehat{P})^2,$$

where P_i is the true proportion of dissatisfied patients in hospital i. Here, $s_{2,i}^2 \approx P_i(1 - P_i)$. Therefore

$$s_y^2 \approx \frac{1}{m} \sum_{i \in S}[P_i - P_i^2 + (P_i - \widehat{P})^2] = \frac{1}{m} \sum_{i \in S}[P_i - 2P_i\widehat{P} + \widehat{P}^2] = \widehat{P}(1 - \widehat{P}),$$

that is, $s_y^2 \approx 0.1 \times 0.9 = 0.09$. Thus, $\widehat{\rho} = 4/49 \approx 0.08$. The estimator $\widehat{\rho}$ is biased for the true cluster effect ρ (unknown), but its bias is weak (on $1/n$).

3. To perform this type of simulation, we consider that the cluster effect does not change. It is mathematically false since it depends on the composition of clusters, but numerically it is a matter of an indicator of similarity which is, by construction, a little sensitive to the delimitation of clusters. We then obtain:

$$\widehat{\text{var}}' = \left(1 - \frac{200}{1\,000}\right) \times \frac{0.1 \times 0.9}{200 \times 25}\,[1 + 0.08(25 - 1)] \approx 4.2 \times 10^{-5}.$$

4. The variance goes from 8.1×10^{-5} to 4.2×10^{-5}, which is a decrease in standard deviation (and therefore in confidence interval length) of 28%, which conforms to the theory: it is preferable, with a constant final sample size and from the lone point of view of accuracy, to select more primary units (hospitals) and fewer secondary units (beds). In compensation, the second method is more expensive. In practice, the choice of method takes into account both the cost and the accuracy.

Exercise 5.5 *Clusters of households and size*

To estimate the average number $\overline{Y}$ of people per household in a given country, we carry out a two-stage sampling design:

- 1st stage: Random sampling with replacement of $m = 4$ villages among $M = 400$ proportional to size. The size of a village is the number of households it has. Thus, for each of the four independent selections, a village is selected with a probability proportional to its size.
- 2nd stage: Simple random sampling of n_i households among N_i if village i is selected.

The data are presented in Table 5.2.

$\widehat{\overline{Y}}_i$ is the mean number of people per household in village i, according to the sample.

The total number of households in the country is $N = 10\,000$.

1. a) What is the selection probability p_i for each of the four villages selected? (The selection probability is the probability a village has of being selected at the time of *each* of the four independent selections successively done under the same conditions.)

Table 5.2. Number of people per household: Exercise 5.5

i	N_i	$\overline{\widehat{Y}}_i$
1	20	5.25
2	23	5.50
3	25	4.50
4	18	5.00

b) Calculate $\Pr(i \notin S)$ as a function of $(1 - p_i)$. Deduce the inclusion probability $\pi_i = \Pr(i \in S)$ as a function of p_i. Examine the case where p_i is small.

2. What is the expression of $\overline{Y}$ (true value) and what is its unbiased estimator?

3. Estimate the variance of this estimator. What interest do we have in using sampling *with* replacement at the 1st stage?

Solution

1. a) The basic selection probability *with* replacement is proportional to the size N_i and is thus $p_i = N_i/N$, which gives

$$p_1 = \frac{20}{10\ 000}, \quad p_2 = \frac{23}{10\ 000}, \quad p_3 = \frac{25}{10\ 000}, \quad p_4 = \frac{18}{10\ 000}.$$

b) The probability that village i is not in the sample is:

$\Pr(i \notin S)$

$= \Pr\left[(i \text{ not selected in } 1^{\text{st}} \text{ trial}) \cap (i \text{ not selected in } 2^{\text{nd}} \text{ trial}) \cap \right.$

$\qquad \left. (i \text{ not selected in } 3^{\text{rd}} \text{ trial}) \cap (i \text{ not selected in } 4^{\text{th}} \text{ trial})\right]$

$= \displaystyle\prod_{\alpha=1,2,3,4} \Pr(i \text{ not selected in } \alpha^{\text{th}} \text{ trial})$

$= (1 - p_i)^4,$

which gives the inclusion probability

$$\pi_i = \Pr(i \in S) = 1 - \Pr(i \notin S) = 1 - (1 - p_i)^4 \quad \text{for all } i.$$

If we assume that p_i is small, then $\pi_i \approx 1 - (1 - 4p_i) = 4p_i$.
Note: we 'nearly' find the π_i from sampling *without* replacement, since in this case:

$$\pi_i = m\frac{N_i}{N} = mp_i \quad \text{here with} \quad m = 4.$$

2. The mean number of people per household is defined by:

$$\overline{Y} = \frac{\text{total number of people}}{\text{total number of households}}.$$

Denote $y_{i,k}$ as the number of people in household k of village i. The true mean is:

$$\overline{Y} = \frac{1}{N} \sum_{i=1}^{M} \sum_{k \in U_i} y_{i,k}.$$

There exists two unbiased estimators of the total $Y = \overline{Y}N$.
- the Hansen-Hurwitz estimator

$$\widehat{Y}_{HH} = \frac{1}{m} \sum_{i=1}^{m} \frac{\widehat{Y}_i}{p_i}$$

  (with an abuse of notation, i here indicates at the same time an identifier and a sampling number),
- the Horvitz-Thompson estimator

$$\widehat{Y}_{\pi} = \sum_{i \in S} \frac{\widehat{Y}_i}{\pi_i} = \sum_{i \in S} \frac{\widehat{Y}_i}{1 - (1 - p_i)^4},$$

  where S is the sample of villages with distinct identifiers *in fine* selected (therefore S does not have a fixed size).

These two estimators are approximately equal if the p_i are very small. The estimator $\widehat{Y}_i$ is an unbiased estimator of the total in village i

$$Y_i = \sum_{k \in U_i} y_{i,k}.$$

The estimator of Y_i is $\widehat{Y}_i = N_i \widehat{\overline{Y}}_i$, where $\widehat{\overline{Y}}_i$ is the simple mean calculated in the sample selected in village i. Finally, if we use the first estimator to estimate the mean $\overline{Y}$:

$$\widehat{\overline{Y}}_{HH} = \frac{1}{mN} \sum_{i=1}^{m} \frac{N_i \widehat{\overline{Y}}_i}{p_i},$$

with $p_i = N_i/N$, which gives

$$\widehat{\overline{Y}}_{HH} = \frac{1}{m} \sum_{i=1}^{m} \widehat{\overline{Y}}_i.$$

Numerical application: $\widehat{\overline{Y}}_{HH} \approx 5.06$.

3. We know that the unbiased estimator of $\text{var}(\widehat{Y}_{HH})$ is:

$$\widehat{\text{var}}(\widehat{Y}_{HH}) = \frac{1}{m(m-1)} \sum_{i=1}^{m} \left(\frac{\widehat{Y}_i}{p_i} - \widehat{Y}_{HH} \right)^2.$$

Indeed

$$\frac{\widehat{Y}_i}{p_i} = \frac{N_i \widehat{\overline{Y}}_i}{N_i/N} = N\widehat{\overline{Y}}_i \quad \text{and} \quad \widehat{Y}_{HH} = N\widehat{\overline{Y}}_{HH},$$

and thus

$$\widehat{\text{var}}(\widehat{\overline{Y}}_{HH}) = \frac{1}{N^2} \widehat{\text{var}}(\widehat{Y}_{HH}) = \frac{1}{m(m-1)} \sum_{i=1}^{m} (\widehat{\overline{Y}}_i - \widehat{\overline{Y}}_{HH})^2.$$

Numerical application:

$$\widehat{\text{var}}(\widehat{\overline{Y}}_{HH})$$
$$= \frac{1}{4 \times 3} [(5.25 - 5.06)^2 + (5.50 - 5.06)^2 + (4.50 - 5.06)^2 + (5 - 5.06)^2]$$
$$\approx 0.045.$$

The mean number of people per household is therefore known, 95 times out of 100, at nearly 0.42 individuals (if we make the assumption of a normal distribution).

Interest: The formula for estimating the accuracy $\widehat{\text{var}}$ is very simple (the true variance itself is complicated). This result is remarkable, as it is valid as soon as:

- the sampling of primary units is carried out with unequal probabilities and *with* replacement;
- any sampling of secondary units is used, with a single constraint nevertheless: $\widehat{Y}_i$ estimates Y_i without bias.

Exercise 5.6 *Which design?*

Consider the population $\{1, 2, 3, 4, 5, 6, 7, 8, 9\}$ and the following sample design:

$$p(\{1,2\}) = 1/6, \quad p(\{1,3\}) = 1/6, \quad p(\{2,3\}) = 1/6,$$
$$p(\{4,5\}) = 1/12, \quad p(\{4,6\}) = 1/12, \quad p(\{5,6\}) = 1/12,$$
$$p(\{7,8\}) = 1/12, \quad p(\{7,9\}) = 1/12, \quad p(\{8,9\}) = 1/12.$$

1. Give the first-order inclusion probabilities.
2. Is this design simple, stratified, clustered, two-stage or none of these particular designs? Justify your response.

Solution

1. The first-order probabilities are:

$$\pi_1 = \frac{1}{3}, \pi_2 = \frac{1}{3}, \pi_3 = \frac{1}{3}, \pi_4 = \frac{1}{6}, \pi_5 = \frac{1}{6}, \pi_6 = \frac{1}{6}, \pi_7 = \frac{1}{6}, \pi_8 = \frac{1}{6}, \pi_9 = \frac{1}{6}.$$

2. We see that the design is clearly developed as a function of the following partition of the population: $\{1, 2, 3\}, \{4, 5, 6\}, \{7, 8, 9\}$. The design consists of choosing one of the three parts with respective probabilities of $1/2$, $1/4$ and $1/4$. Next, in the selected part, we perform a simple sampling without replacement of size 2 among the three individuals (probability $1/3$ of selecting each of the possible samples). It is therefore a two-stage sample, where the primary units (the parts) are selected with unequal probabilities and the secondary units are selected according to simple random sampling without replacement of size $n = 2$.

Exercise 5.7 *Clusters of households*

1. A survey is carried out from a simple random sample of 90 clusters of 40 households each. The clusters are selected using simple random sampling at the rate $f = 1/300$. To improve the accuracy of the results, a statistician proposes to reduce by half the size of the clusters by selecting twice as many of them. What gain in accuracy can we hope for, 'all other things being equal'?
2. For an estimated proportion $\hat{P} = 0.1$, the actual survey produces a 95% confidence interval CI $= [0.1 \pm 0.014]$. Calculate the confidence interval that we obtain to estimate the same proportion with the new survey technique (we neglect the sampling rates).

Solution

1. If we estimate $\overline{Y}$ with $\widehat{\overline{Y}}$, we have, in the two-stage design proposed:

$$\text{var}[\widehat{\overline{Y}}] = \text{var}\left[\frac{\hat{Y}}{N}\right] = (1 - f)\,\frac{S_y^2}{m\overline{n}}\,[1 + \rho(\overline{n} - 1)],$$

where
- m = number of clusters selected ($f = m/M$),
- $\overline{n}$ = number of households per cluster (constant),
- ρ = 'intra-cluster' correlation coefficient, also called 'cluster effect',
- S_y^2 = true total variance in the population.

Note: This expression is true, as the sizes of the clusters are constant $(\bar{n} = 40)$; otherwise there is an additional term due to the variance of the sizes (see Exercise 5.12).

- 1st case:

$$\left.\begin{array}{l} m = 90 \\ \bar{n} = 40 \end{array}\right\} \quad f = \frac{1}{300} \quad \rightarrow \quad \text{variance } \text{var}_1.$$

- 2nd case:

$$\left.\begin{array}{l} m = 180 \\ \bar{n} = 20 \end{array}\right\} \quad f = \frac{1}{300} \quad \rightarrow \quad \text{variance } \text{var}_2.$$

Now, if we compare the two cases, $m\bar{n} = $ constant. Hence

$$\frac{\text{var}_2}{\text{var}_1} = \frac{1 + \rho(20 - 1)}{1 + \rho(40 - 1)} = \frac{1 + 19\rho}{1 + 39\rho} < 1.$$

This ratio measures, as a function of ρ, the expected gain in accuracy, 'all other things being equal'.

- *Note:*

We assume that ρ does not vary when the clusters go from 40 to 20 households. Strictly speaking, this is inaccurate, but in practice we consider that the modifications are slight and therefore that ρ is quite 'bearable' (*a priori*, we would instead have ρ decreasing if the size of the clusters increase, because the intra-cluster homogeneity would then have to decrease).

2. The proportion $\widehat{P}$ is in fact only a particular mean $\overline{Y}$ where y is an indicator variable. The variance expression from 1. is thus valid by adapting S_y^2 in the context of the indicator variables. Since the clusters are of identical size, $\bar{n} = 40$, we have $\widehat{P} = \overline{Y} = p$, the proportion in the sample. We are going to use (see Exercise 5.4 to justify the estimator of S_y^2):

$$\widehat{\text{var}}_1 = \frac{p(1-p)}{\bar{n}m}[1 + \rho(\bar{n} - 1)],$$

where $(1 - f)$ is close to 1, which gives

$$\widehat{\text{var}}_1 = \frac{0.1 \times 0.9}{90 \times 40}[1 + \rho \times 39] = \left(\frac{0.014}{2}\right)^2,$$

and therefore $\rho = 0.0246$. Hence

$$\widehat{\text{var}}_2 = \frac{0.1 \times 0.9}{180 \times 20}[1 + \rho \times 19],$$

and therefore

$$\frac{\widehat{\text{var}}_2}{\widehat{\text{var}}_1} = 0.75.$$

We deduce the new confidence interval estimated for the true proportion P:

$$\left[\hat{P} \pm 2 \times \sqrt{\widehat{\text{var}}_2}\right] = \left[0.1 \pm \sqrt{0.75} \times 0.014\right] = [0.1 \pm 0.012].$$

Exercise 5.8 *Bank clients*

A bank has 39 800 clients in its computer files, divided into 3 980 branches each managing exactly 10 clients. We wish to estimate the proportion of clients for whom the bank has granted a loan. For this, we sample, using simple random sampling (SRS), 40 branches (sample S), and we list, in each branch i, A_i clients having a loan. The data coming from the survey are:

$$\sum_{i \in S} A_i = 185, \quad \text{and} \quad \sum_{i \in S} A_i^2 = 1\,263.$$

1. What do we call this type of sampling?
2. Give the expression of the parameter to estimate and its unbiased estimator.
3. Estimate without bias the variance of this estimator, and provide a 95% confidence interval.
4. Calculate the design effect (DEFF), defined as a ratio measuring the loss in estimated variance obtained in comparison to a simple random sample of the same size.
5. Calculate the intra-cluster correlation coefficient ρ.
6. Estimate the accuracy that we would get by sampling (still using simple random sampling) 80 branches and 5 clients per selected branch.
7. We have a total budget $\overline{C}$ to proceed with the estimation (this budget corresponds to the cost of a simple random sampling of 400 clients). In concrete terms, in the first place we retrieve, by post from the sampled branches, the account numbers of sampled clients in the branch (C_2: cost of sending a letter) then we review the central computer list of loans, client by client, by means of the collected accounts (C_1: cost of reading a record). Then, we add a fixed cost C_0, independent of the sampling method. Discuss the interest of selecting, at a fixed budget, either a simple random sampling (with the mailout of a letter per client sampled to retrieve his account number) or a two-stage sampling (with the mailout of a letter per branch). Prior to this, justify that with simple random sampling, there is no interest in trying to group the clients by branch before the mailout by post.

Solution

1. The sample design is cluster sampling: each branch is a cluster of clients.

2. The function of interest to estimate is the proportion:

$$P = \frac{\sum_{i=1}^{M} A_i}{M\overline{N}},$$

with $M = 3\,980$ (total number of branches), and $\overline{N} = 10$ (total number of clients in each branch). The unbiased estimator of P is $\widehat{P}$:

$$\widehat{P} = \frac{\sum_{i \in S} A_i/(m/M)}{M\overline{N}} = \frac{1}{m\overline{N}} \sum_{i \in S} A_i = \frac{1}{m} \sum_{i \in S} \left(\frac{A_i}{\overline{N}}\right),$$

where m is the number of branches selected ($m = 40$). Since $m\overline{N}$ is the total size of the sample of clients, $\widehat{P}$ is the simple proportion of clients in the sample having a loan. We have

$$\widehat{P} = \frac{185}{40 \times 10} \approx 46.2\%.$$

3. We estimate without bias the variance with

$$\widehat{\mathrm{var}}(\widehat{P}) = \left(1 - \frac{m}{M}\right) \frac{1}{m} s_P^2,$$

where

$$s_P^2 = \frac{1}{m-1} \sum_{i \in S} (P_i - \widehat{P})^2,$$

with

$$P_i = \frac{A_i}{\overline{N}}.$$

Indeed, $\widehat{P}$ is the simple mean, on the sample of branches, of the proportions P_i of clients having a loan from branch i. Furthermore,

$$s_P^2 = \frac{\sum_{i \in S} P_i^2}{m-1} - \frac{m}{m-1} \widehat{P}^2 = \frac{1}{(m-1)\,\overline{N}^2} \left(\sum_{i \in S} A_i^2 - \frac{\left(\sum_{i \in S} A_i\right)^2}{m}\right).$$

We have

$$s_P^2 = \frac{1}{39 \times 100} \left(1\,263 - \frac{185^2}{40}\right) = 0.1045,$$

$$\widehat{\mathrm{var}}(\widehat{P}) = \left(1 - \frac{40}{3980}\right) \times \frac{1}{40} \times 0.1045 \approx 25.85 \times 10^{-4},$$

and $\widehat{\sigma} = 5.1\,\%$. We therefore have, 95 times out of 100, the estimated interval:

$$P \in [46.2\,\% - 10.2\,\%;\ 46.2\,\% + 10.2\,\%].$$

This mediocre result is due to the very small size of the sample of branches. We note that with $m = 40$ branches, the simplifying hypothesis of a normal distribution for $\widehat{P}$ can be questioned, and anyway is considered to be quite poor.

4. It is a matter of estimating without bias the variance that we would get with $n = 400$ clients selected using simple random sampling. We know that

$$\mathop{\mathrm{var}}_{\mathrm{SRS}}(\widehat{P}) = \left(1 - \frac{n}{N}\right) \frac{S_y^2}{n},$$

where, when N is large, $S_y^2 = P(1-P)$, with $P =$ the proportion of clients having obtained a loan (this is indeed the parameter from 1.).

The difficulty consists of estimating S_y^2 with the cluster sampling design. The 'trick', at least in theory, consists of using $\widehat{P}(1 - \widehat{P})$ mechanically without having calculated the expectation. Indeed, such an expression would only estimate S_y^2 without bias if the design was simple random, but this is not the case. However, according to Question 3,

$$\mathrm{E}\,\widehat{\mathrm{var}}(\widehat{P}) = \mathrm{var}(\widehat{P}) = \mathrm{E}(\widehat{P}^2) - (\mathrm{E}\widehat{P})^2.$$

In fact, according to 1., we have $\mathrm{E}\widehat{P} = P$, and thus

$$P^2 = \mathrm{E}\widehat{P}^2 - \mathrm{E}\,\widehat{\mathrm{var}}(\widehat{P}) = \mathrm{E}[\widehat{P}^2 - \widehat{\mathrm{var}}(\widehat{P})].$$

Finally,

$$S_y^2 = P - P^2 = \mathrm{E}\widehat{P} - \mathrm{E}[\widehat{P}^2 - \widehat{\mathrm{var}}(\widehat{P})] = \mathrm{E}[\widehat{P}(1 - \widehat{P}) + \widehat{\mathrm{var}}(\widehat{P})].$$

We therefore have:

$$\begin{aligned}
\mathrm{DEFF} &= \frac{\widehat{\mathrm{var}}(\widehat{P})}{\left(1 - \frac{n}{N}\right) \frac{1}{n} [\widehat{P}(1 - \widehat{P}) + \widehat{\mathrm{var}}(\widehat{P})]} \\
&= \frac{25.85 \times 10^{-4}}{\left(1 - \frac{400}{39\,800}\right) \times \frac{1}{400} \times [0.462 \times 0.538 + 0.003]} \\
&\approx 4.2.
\end{aligned}$$

We note that, numerically, the bias of $\widehat{P}(1-\widehat{P})$ is very slight. The previous sample of 40 branches thus multiplied the standard deviation by $\sqrt{4.2} \approx 2$. We are 'two times worse' than if we use a simple random sample, but in contrast the process is less expensive.

5. We have:

$$\mathrm{DEFF} = 1 + \rho(\overline{N} - 1),$$

as all the clusters are of the same size $\overline{N}$. Thus

$$\rho = \frac{4.2 - 1}{10 - 1} \approx 0.35.$$

It is a rather strong value, which expresses the 'intra-class' homogeneity of the clusters. We can simplify the situation by considering there to be two categories of branches: those which easily grant a loan (A_i close to $\overline{N}$), and those which are rather hesitant to make them (A_i close to 0).

6. With this new sampling design, we have:

$$\widehat{\mathrm{var}}_2(\widehat{P}) = \underset{\mathrm{SRS}}{\widehat{\mathrm{var}}}(\widehat{P}) \times (1 + \rho(\overline{n} - 1)),$$

where $\underset{\mathrm{SRS}}{\widehat{\mathrm{var}}}(\widehat{P})$ is the variance estimator in a simple random sample of size 400 and $\overline{n}$ is the size (constant) of the sample selected within each branch ($\overline{n} = 5$). According to 4., we have:

$$\underset{\mathrm{SRS}}{\widehat{\mathrm{var}}}(\widehat{P}) = \frac{\widehat{\mathrm{var}}(\widehat{P})}{\mathrm{DEFF}} \approx 6.15 \times 10^{-4}.$$

Therefore

$$\widehat{\mathrm{var}}_2(\widehat{P}) \approx 6.15 \times 10^{-4}(1 + 0.35 \times (5 - 1)) = 14.8 \times 10^{-4},$$

and $\widehat{\sigma}_2 = 3.8\,\%$. In comparison with cluster sampling, the length of the confidence interval is reduced by a factor of 1.3.

7. With simple random sampling, the probability that there are two clients interviewed from the same branch is extremely small. If we denote X as the number (random) of clients selected in a given branch i, the distribution of X is approximately Poisson with parameter

$$\left(400 \times \frac{10}{39\,800} \right).$$

Therefore

$$\Pr(X = 0) + \Pr(X = 1) \approx e^{-0.1}(1 + 0.1) = 0.995.$$

It is thus almost certain that there is at most one client interviewed per branch. This justifies that we send one letter per client, without previously trying to group them by branch to save money.

- With a simple sampling of clients, n is small compared to the population size $N = 39\,800$. We therefore neglect the finite population correction. We have

$$\overline{C} = C_0 + C_1 n + C_2 n,$$

and

$$\underset{\mathrm{SRS}}{\mathrm{var}}(\widehat{P}) = \frac{S_y^2}{n},$$

where $S_y^2 = P(1 - P)$. We determine $\overline{C}$ according to:

$$\overline{C} = C_0 + 400(C_1 + C_2).$$

- With a two-stage sampling, and denoting m as the number of branches selected and $\bar{n}$ as the number of clients selected from each branch, we get

$$\overline{C} = C_0 + C_1 m\bar{n} + C_2 m,$$

  and

$$\text{var}(\widehat{P}) = \frac{S_y^2}{m\bar{n}} [1 + \rho(\bar{n} - 1)].$$

Neglecting the sampling rate:

$$\frac{\text{var}(\widehat{P})}{\underset{\text{SRS}}{\text{var}}(\widehat{P})} = \frac{400}{m\bar{n}}(1 + \rho(\bar{n} - 1)) = \frac{1}{\bar{n}} \frac{C_1\bar{n} + C_2}{C_1 + C_2} (1 + \rho(\bar{n} - 1)).$$

The simple sampling is interesting if and only if:

$$\frac{\text{var}(\widehat{P})}{\underset{\text{SRS}}{\text{var}}(\widehat{P})} \geq 1,$$

which implies that

$$(\rho C_1)\,\bar{n}^2 + (\rho(C_2 - C_1) - C_2)\,\bar{n} + (1 - \rho)\,C_2 \geq 0,$$

we calculate

$$\Delta = [(C_1 + C_2)\,\rho - C_2]^2 \geq 0.$$

- Case 1: If

$$\rho = \frac{C_2}{C_1 + C_2},$$

  then

$$\frac{\text{var}(\widehat{P})}{\underset{\text{SRS}}{\text{var}}(\widehat{P})} \geq 1 \quad \Leftrightarrow \quad \frac{C_1 C_2}{C_1 + C_2} (\bar{n} - 1)^2 \geq 0,$$

  which is always true. Simple random sampling always carries this.
- Case 2:

$$\rho \neq \frac{C_2}{C_1 + C_2}.$$

The two distinct roots are 1 and $[(1 - \rho)/\rho] \times C_2/C_1 = 1.86 \times C_2/C_1$. Therefore

$$\frac{\text{var}(\widehat{P})}{\underset{\text{SRS}}{\text{var}}(\widehat{P})} \geq 1 \quad \Leftrightarrow \quad \bar{n} \text{ outside of the roots.}$$

- Case 2.a:

$$\frac{C_2}{C_1} \leq \frac{1}{1.86} \approx 0.54.$$

Simple random sampling always carries this. In the extreme case $C_2 = 0$, we very well see that two-stage sampling does not save anything, but reduces the accuracy due to ρ: it would be of no avail to use!

- Case 2.b:

$$\frac{C_2}{C_1} > 0.54.$$

The two-stage sampling is at least as advantageous as simple random sampling under the condition of selecting the suitable $\bar{n}$ in the interval, of course: if C_2/C_1 is slightly larger than 0.54, we have $\bar{n} = 1$ and therefore it is indeed a simple random sample, but if C_2/C_1 is large (for example $C_2/C_1 = 10$, and in all severity it is sufficient to have $C_2/C_1 > 2/1.86 = 1.075$), then it is worthwhile to use a genuine two-stage sampling (with $\bar{n} \geq 2$). This result is intuitively explained by the importance of the unit cost C_2: it becomes interesting to save money by limiting the number of letters sent but on the other hand to read many more records (C_1 small). Despite the cluster effect ρ, the overall 'sample size' effect eventually gets the better of this.

Exercise 5.9 *Clusters of households and number of men*

This exercise consists of a summary of cluster sampling and sampling with unequal probabilities. We consider a population of individuals of size $N = 62\ 000$. This population is made up of $M = 15\ 000$ households. We denote:

$$N_i = \text{size of household } i \text{ (number of individuals)}$$
$$A_i = \text{number of men in household } i.$$

The data from the sample required for the calculations are shown in Table 5.3.

1. First, we conduct a simple random sampling of $m = 30$ households among M (sample S), and we survey all the individuals from each of the m households selected.
   a) What do we call this type of sampling?
   b) What are the selection probabilities of the households, and what are the selection probabilities of the individuals?
   c) We denote A as the total number (unknown) of men in the population.
      i. Give an unbiased estimator $\widehat{A}$ of A and perform the numerical application.
      ii. What is the expression for its true variance?
      iii. What is the unbiased estimator of this variance (numerical application)?
      iv. What can we say about the pertinence of the total on the model of $\widehat{A}$ when we try to estimate the total number of households M? (We denote the estimator by $\widehat{M}$.) What about if we now want to estimate the total size of the population N? (We denote the estimator by $\widehat{N}$.)

Table 5.3. Sample of households: Exercise 5.9

Household identifier	N_i	A_i
1	5	1
2	6	3
3	3	1
4	3	1
5	2	1
6	3	1
7	3	1
8	3	1
9	4	2
10	4	3
...	...	...
25	2	1
26	4	3
27	3	1
28	4	2
29	2	1
30	4	2

$$\sum_{i \in S} N_i = 104, \qquad \sum_{i \in S} A_i = 53, \qquad \sum_{i \in S} N_i^2 = 404, \qquad \sum_{i \in S} A_i^2 = 117,$$

$$\sum_{i \in S} A_i N_i = 206, \qquad \sum_{i \in S} \left(\frac{A_i}{N_i}\right)^2 = 8.5, \qquad \sum_{i \in S} \frac{A_i}{N_i} = 14.9.$$

d) For this question, we are trying to estimate the total A using a 'ratio by size' expression.

   i. Give a ratio estimator $\widehat{A}$ of A that lets us perfectly and properly estimate the total size N and perform the numerical application.

   ii. Can we explain *a priori*, that is without calculating, the interest of such an estimator?

   iii. What is the expression of its true variance?

   iv. What estimator for this variance can we use? Is it biased?

   v. Perform the numerical application and conclude.

e) The goal of this question is to estimate the design effect, denoted DEFF, when we use $\widehat{A}$.

   i. How would we estimate A by assuming 'as if' the individuals in the selected households had been selected by simple random sampling directly from the population of size N, and what would be the accuracy obtained under these conditions (numerical application)?

   ii. Comparing this accuracy with the one obtained in 1.(c)*iii.*, give the DEFF obtained as part of Question 1.(c).

iii. Why could we have, intuitively, predicted the situation of DEFF in relation to the value 1?

2. Second, we decide to select m households proportionately to their size N_i. We consider that this sampling, performed in reality without replacement, can be likened to be a sampling design with replacement.

   a) Under what general conditions, when the sampling designs bring into consideration unequal probabilities, can we assimilate sampling with and without replacement?

   b) Give, as a function of N_i, the selection probability p_i of household i at the time of each primary drawing. What about the inclusion probability?

   c) Give an unbiased estimator $\widetilde{A}$ of A, and give a 95% confidence interval estimated for A. Numerical application and conclusion in comparison to the results of 1.(c) and 1.(d).

   d) What can we say about the pertinence of an estimator of the total developed on the model of $\widetilde{A}$ when we try to estimate the total number of households M (estimator $\widetilde{M}$)? What if we now want to estimate the total size N of the population (estimator $\widetilde{N}$)?

Solution

1. a) It is cluster sampling, with each household forming a cluster. The clusters are selected through simple random sampling.

   b) The inclusion probability of a household is given by:

   $$\pi_{\text{household}} = \frac{m}{M} = \frac{30}{15\ 000} = \frac{1}{500}.$$

   The inclusion probability of an individual is the inclusion probability of a household.

   c) i. The classical Horvitz-Thompson estimator is:

   $$\widehat{A} = \sum_{i \in S} \frac{A_i}{m/M} = 500 \times 53 = 26\ 500,$$

   where S is the sample of *households*.

   ii. Since S comes from a simple random sample, we have:

   $$\text{var}(\widehat{A}) = M^2 \left(1 - \frac{m}{M}\right) \frac{S_A^2}{m},$$

   where

   $$S_A^2 = \frac{1}{M-1} \sum_{i=1}^{M} (A_i - \overline{A})^2.$$

iii. Furthermore,

$$\widehat{\text{var}}(\widehat{A}) = M^2 \left(1 - \frac{m}{M}\right) \frac{s_A^2}{m},$$

where

$$s_A^2 = \frac{1}{m-1} \sum_{i \in S} \left(A_i - \frac{\widehat{A}}{M}\right)^2,$$

which again becomes

$$s_A^2 = \frac{1}{m-1} \left[\sum_{i \in S} A_i^2 - \frac{m}{M^2} (\widehat{A})^2\right].$$

The calculation gives

$$\widehat{\text{var}}(\widehat{A}) = 6\ 043\ 966.$$

Therefore,

$$\widehat{\sigma}_{\widehat{A}} = \sqrt{\widehat{\text{var}}(\widehat{A})} = 2\ 458.$$

iv. • To estimate the total number of households, we use

$$\widehat{M} = \sum_{i \in S} \frac{1}{m/M} = M.$$

This equality, true for whatever sample selected, expresses the fact that the variance of $\widehat{M}$ is *null*. We therefore perfectly and properly estimate M (this is a total like any other).

• To estimate the total number of individuals, we use

$$\widehat{N} = \sum_{i \in S} \frac{N_i}{m/M} \neq N.$$

This time, unlike for M, we do not perfectly estimate N.

d)   i. We are going to set

$$\widehat{\widehat{A}} = N \frac{\sum_{i \in S} A_i}{\sum_{i \in S} N_i} = N \frac{\widehat{A}}{\widehat{N}}.$$

In fact, $\sum_{i \in S} A_i$ estimates without bias mA/M, and $\sum_{i \in S} N_i$ estimates mN/M; therefore $\widehat{\widehat{A}}$ *perfectly* estimates N (let $A_i = N_i$, and notice then that $\widehat{\widehat{A}} = N$ for any S)

$$\widehat{\widehat{A}} = 62\ 000 \times \frac{53}{104} = 31\ 596 \quad (\neq \widehat{A}).$$

ii. *A priori*, A_i must be 'quite' proportional to N_i (we can logically think that the number of men increases more or less in proportion to the size of the household) and this would have to come back to the estimator quite precisely, as we are very well under the conditions for using a ratio (see Chapter 6).

iii. We have:

$$\widehat{\widehat{A}} = M \left[\overline{N} \frac{\widehat{\widehat{A}}}{\widehat{\widehat{N}}} \right],$$

with

$$\overline{N} = \frac{N}{M}, \quad \widehat{\widehat{N}} = \frac{\sum_{i \in S} N_i}{m}, \quad \text{and} \quad \widehat{\widehat{A}} = \frac{\sum_{i \in S} A_i}{m}.$$

This rewriting allows for the expression of $\widehat{\widehat{A}}$ under the classical form of a ratio and to immediately select the variance (approximately):

$$\mathrm{var}(\widehat{\widehat{A}}) \approx M^2 \left(1 - \frac{m}{M} \right) \frac{1}{m} S_U^2,$$

where S_U^2 indicates the population variance of the residuals U_i:

$$U_i = A_i - \overline{A} N_i,$$

where $\overline{A} = A/N$ is the true proportion of men in the population. Finally, this is

$$S_U^2 = \frac{1}{M-1} \sum_{i=1}^{M} (A_i - \overline{A} N_i)^2.$$

iv. The estimator for the variance is

$$\widehat{\mathrm{var}}(\widehat{\widehat{A}}) = M^2 \left(1 - \frac{m}{M} \right) \frac{1}{m} s_{\widehat{U}}^2,$$

where

$$s_{\widehat{U}}^2 = \frac{1}{m-1} \sum_{i \in S} \left(A_i - \frac{\widehat{\widehat{A}}}{N} N_i \right)^2.$$

The estimator $\widehat{\mathrm{var}}(\widehat{\widehat{A}})$ is biased, with a bias of $1/m$.

v. By expanding the square of $s_{\widehat{U}}^2$, we obtain:

$$s_{\widehat{U}}^2 = s_A^2 + \left(\frac{\widehat{\widehat{A}}}{N} \right)^2 s_N^2 - 2 \frac{\widehat{\widehat{A}}}{N} s_{AN},$$

where s_{AN} indicates the covariance in the sample between A_i and N_i

$$s^2_{\widehat{U}} = 0.806 + \left(\frac{31\ 596}{62\ 000}\right)^2 \times 1.5 - 2 \times \frac{31\ 596}{62\ 000} \times 0.759 \approx 0.422.$$

Therefore,

$$\widehat{\sigma}_{\widehat{A}} = 1\ 779.$$

Conclusion: Since $1\ 779 < 2\ 458$, the ratio $\widehat{\widehat{A}}$ appears to be preferable to $\widehat{A}$.

e)   i. We 'forget' for a moment the cluster aspect, and we consider that the $\sum_{i \in S} N_i$ individuals selected could have been selected by simple random sampling. We would then use the estimator:

$$A^* = N \times (\% \text{ of men selected}) = N \frac{\sum_{i \in S} A_i}{\sum_{i \in S} N_i}.$$

Numerically, $A^* = \widehat{\widehat{A}}$, but the estimated variance is calculated very differently:

$$\widehat{\text{var}}_{\text{SRS}}(A^*) = N^2 \left(1 - \frac{n}{N}\right) \frac{s^2}{n - 1},$$

with $s^2 = \widehat{P}(1 - \widehat{P})$, where

$$\widehat{P} = \frac{\sum_{i \in S} A_i}{n} = \% \text{ of men in the sample.}$$

Numerical application:

$$\widehat{\text{var}}_{\text{SRS}}(A^*) = (62\ 000)^2 \left[1 - \frac{104}{62\ 000}\right] \frac{0.51 \times 0.49}{103} = 9326313,$$

since $\widehat{P} = 53/104 = 0.51$, and therefore

$$\widehat{\sigma}_{\text{SRS}}(A^*) = \sqrt{\widehat{\text{var}}_{\text{SRS}}(A^*)} = 3\ 053.9.$$

ii. We estimate the design effect by:

$$\widehat{\text{DEFF}} = \frac{\widehat{\text{var}}(\widehat{A})}{\widehat{\text{var}}_{\text{SRS}}(A^*)} = \frac{6\ 043\ 966}{9\ 326\ 313} \approx 0.65.$$

Once again, we draw attention to the traditional difficulty encountered with each calculation of the *design effect*: the estimator of the variance used in the denominator does not estimate without bias the variance that we would obtain with simple sampling. In fact, the expression s^2 from (e) *i.* does not estimate the population variance S^2 without bias because the sampling that had taken place was not simple random (it is not even of fixed size, which greatly complicates things). We therefore only get an order of magnitude of DEFF.

iii. DEFF< 1: This time, cluster sampling was better than simple random sampling (once is not enough). We could have figured this by examining the individual data because from a gender point of view, the households seem rather *heterogeneous* from within ('negative' cluster effect: to extremely simplify, a household is composed of 50% men and 50% women).

2. a) The likening is possible under two conditions: sample size m is 'small' with respect to M *and* the sizes N_i are not much dispersed. These two conditions are in practice realised.

 b) For all i, $p_i = N_i/N$. The inclusion probability of household i is (see Exercise 5.5) $1 - (1 - p_i)^n \neq np_i$, but even so is extremely close to np_i since here $p_i \ll 1$.

 c) An unbiased estimator is

$$\widetilde{A} = \frac{1}{m} \sum_{i=1}^{m} \frac{A_i}{p_i}.$$

(Note that i = the sequence number of the drawing here, with an abuse of notation.)

$$\widehat{\mathrm{var}}(\widetilde{A}) = \frac{1}{m(m-1)} \sum_{i=1}^{m} \left(\frac{A_i}{p_i} - \widetilde{A} \right)^2$$

$$= \frac{1}{m(m-1)} \left[N^2 \sum_{i=1}^{m} \left(\frac{A_i}{N_i} \right)^2 - m\widetilde{A}^2 \right]$$

$$= \frac{1}{m(m-1)} \left[N^2 \sum_{i \in S} \left(\frac{A_i}{N_i} \right)^2 - m\widetilde{A}^2 \right].$$

This last equality expresses the comparison of sampling with and without replacement.

$$\widetilde{A} = \frac{1}{30} \times 62\,000 \times 14.9 = 30\,793 \ (\text{thus } \widetilde{A} \neq \widehat{A} \text{ and } \widetilde{A} \neq \widehat{\widehat{A}}),$$

$$\widehat{\mathrm{var}}(\widetilde{A}) = \frac{1}{30 \times 29} \left[(62\,000)^2 \times 8.5 - 30 \times (30\,793)^2 \right] = 4\,859\,460,$$

and

$$\widehat{\sigma}_{\widetilde{A}} = 2\,204.$$

Recall that

$$\widehat{\sigma}_{\widehat{A}} = 1\,779, \quad \text{and} \quad \widehat{\sigma}_{\widehat{\widehat{A}}} = 2\,458.$$

We apparently get (because it is only an *estimation*) a worse accuracy with the unequal probabilities than with the ratio according to size, but better than cluster sampling with the classical estimator. The 95% confidence interval for A is:

$$A \in [30\ 793 \pm 4\ 408].$$

Since the size m is not very large, we could hesitate to construct such a confidence interval but due to the likening made to sampling with replacement, we can depend on the central limit theorem, of which we know that it becomes usable with a few dozen units. The conditions are a bit 'tight' to consider that $\widetilde{A}$ follows a normal distribution, but must be sufficient to give an acceptable order of magnitude of uncertainty on A.

d) Since we have

$$\widetilde{M} = \frac{1'}{m} \sum_{i=1}^{m} \frac{1}{p_i} \neq M \quad (\text{var}(\widetilde{M}) > 0),$$

and

$$\widetilde{N} = \frac{1}{m} \sum_{i=1}^{m} \frac{N_i}{\left(\frac{N_i}{N}\right)} = N \quad \Rightarrow \quad \text{var}(\widetilde{N}) = 0,$$

we notice that the total size of the population is perfectly estimated, but not the total number of households. This is exactly the opposite of the Horvitz-Thompson estimator studied in Question 1.

Exercise 5.10 *Variance of systematic sampling*

In a list of N individuals, we are interested in a variable y. The individuals are identified by their order on the list, so their order goes from 1 (for the first) to N (for the last). We use systematic sampling with interval h to select n individuals from the list. We assume that: $h = N/n \in \mathbb{N}$.

1. Show that everything happens as if we selected a unique cluster of individuals from a population pre-divided into clusters. We will specify what the clusters are, what their size is, and how many there are in the population.
2. We henceforth use the following notation:

$$y_{i,k} = \text{value of } y \text{ for the } k\text{th record counted in cluster number } i.$$

We denote $\overline{Y}_i$ as the mean of the $y_{i,k}$ calculated from all the individuals of cluster number i.

a) What is the unbiased estimator $\widehat{\overline{Y}}$ of the true mean $\overline{Y}$? We will show that $\widehat{\overline{Y}}$ is effectively unbiased.

b) What is the expression of its true variance, as a function of $\overline{Y}_i$, $\overline{Y}$ and h?

c) How do we estimate this variance without bias?

3. a) Considering the natural splitting of the population into h clusters, write the analysis equation for the variance by noting:

$$S_i^2 = \frac{1}{n-1} \sum_{k=1}^{n} (y_{i,k} - \overline{Y}_i)^2.$$

b) Show that if N is large, and if we denote:

$$S_W^2 = \frac{1}{h(n-1)} \sum_{i=1}^{h} \sum_{k=1}^{n} (y_{i,k} - \overline{Y}_i)^2,$$

then we have:

$$\mathrm{var}(\widehat{\overline{Y}}) \approx S_y^2 - \frac{h(n-1)}{N} S_W^2.$$

c) Show that systematic sampling is more precise than simple random sampling if and only if: $S_y^2 < S_W^2$ by considering N as very large with respect to n.

d) In order for this condition to be satisfied, it is necessary to ensure that S_W^2 is 'large'. How does this affect $y_{i,k}$? How do we proceed in order that, in practice, this is indeed the case?

Solution

1. The configuration of the list and the different systematic samples conceivable are the following:

cluster 1: $\{1, \ 1+h, \ 1+2h, \ 1+3h, \ \ldots, \ 1+(n-1)\,h\}$,
cluster 2: $\{2, \ 2+h, \ 2+2h, \ 2+3h, \ \ldots, \ 2+(n-1)\,h\}$,
cluster h: $\{h, \ h+h, \ h+2h, \ h+3h, \ \ldots, \ \underbrace{h+(n-1)\,h}_{=nh=N}\}$.

There are thus h clusters possible in the population, each having a size n. One lone cluster is sampled.

2. a)

$$\widehat{\overline{Y}} = \frac{1}{N} \sum_{k \in S} \frac{y_k}{\frac{m}{M}},$$

where
- m = number of clusters selected = 1,
- M = number of clusters in the population = h,
- k = identifier of the individual (S is the sample of individuals).

In fact, the inclusion probability for all individuals is equal to the selection probability for the cluster in which it is contained, being m/M. We thus have

$$\widehat{\overline{Y}} = \frac{h}{N} \sum_{k \in S} y_k = \frac{1}{n} \sum_{k \in S} y_k.$$

Therefore, if cluster (unique) i is selected, we have:

$$\widehat{\overline{Y}} = \frac{1}{n} \sum_{k=1}^{n} y_{i,k} = \overline{Y}_i.$$

Demonstration of the unbiasedness of $\widehat{\overline{Y}}$: by the definition of the expected value:

$$\mathrm{E}(\widehat{\overline{Y}}) = \frac{1}{h} \sum_{i=1}^{h} \overline{Y}_i = \frac{1}{h} \sum_{i=1}^{h} \sum_{k=1}^{n} \frac{y_{i,k}}{n} = \frac{\sum_{i=1}^{h} \sum_{k=1}^{n} y_{i,k}}{N} = \overline{Y},$$

as there are h clusters in total and only one selected per simple random sampling ($\overline{Y}_i$ occurs with a probability $1/h$). Therefore, $\widehat{\overline{Y}}$ is unbiased.

b) The variance is

$$\mathrm{var}(\widehat{\overline{Y}}) = \mathrm{E}(\widehat{\overline{Y}} - \overline{Y})^2 = \sum_{i=1}^{h} \frac{1}{h} (\overline{Y}_i - \overline{Y})^2,$$

by definition of an expected value since the distribution is discrete.

c) *Trick question:* We cannot estimate this variance without bias, as we select only one cluster, and this fact prohibits the *unbiased* estimation of any population variance. In literature, we nevertheless find variance estimators for this type of sampling design but they are biased (see Wolter, 1985): under certain conditions, responding to behaviour patterns, the bias is weak, and that justifies the use of such estimators.

3. a) Recall the general expression for the decomposition of variance (classical notation):

$$S_y^2 \approx \sum_{i=1}^{H} \frac{N_i}{N} (\overline{Y}_i - \overline{Y})^2 + \sum_{i=1}^{H} \frac{N_i - 1}{N} S_i^2,$$

for any division into H sub-populations indexed by i. Here, a sub-population consists of one cluster; we have $H = h$ and $N_i = n$

$$S_y^2 \approx \sum_{i=1}^{h} \frac{n}{n \times h} (\overline{Y}_i - \overline{Y})^2 + \sum_{i=1}^{h} \frac{n - 1}{n \times h} S_i^2,$$

that is,

$$S_y^2 \approx \frac{n - 1}{n \times h} \sum_{i=1}^{h} S_i^2 + \frac{1}{h} \sum_{i=1}^{h} (\overline{Y}_i - \overline{Y})^2.$$

b) We can write

$$\text{var}(\widehat{\overline{Y}}) \approx S_y^2 - \frac{n-1}{N} \sum_{i=1}^{h} S_i^2$$

$$\approx S_y^2 - \frac{n-1}{N} \sum_{i=1}^{h} \frac{1}{n-1} \sum_{k=1}^{n} (y_{i,k} - \overline{Y}_i)^2$$

$$\approx S_y^2 - \frac{1}{N} \sum_{i=1}^{h} \sum_{k=1}^{n} (y_{i,k} - \overline{Y}_i)^2$$

$$\approx S_y^2 - \frac{1}{N} \left[h(n-1) S_W^2 \right].$$

c) Systematic sampling is more precise than simple random sampling if and only if:

$$\text{var}(\widehat{\overline{Y}}) < (1-f) \frac{S_y^2}{n}$$

$$\Leftrightarrow S_y^2 - \frac{h(n-1)}{N} S_W^2 < \left(\frac{1}{n} - \frac{1}{N} \right) S_y^2$$

$$\Leftrightarrow S_y^2 \left[1 - \frac{1}{n} + \frac{1}{N} \right] < \frac{h(n-1)}{N} S_W^2 = \left(1 - \frac{1}{n} \right) S_W^2$$

$$\Leftrightarrow S_y^2 < \frac{1 - \frac{1}{n}}{1 - \frac{1}{n} + \frac{1}{N}} S_W^2 \approx S_W^2 \left(\text{by hypothesis, } \frac{1}{n} \gg \frac{1}{N} \right)$$

$$\Leftrightarrow \quad S_y^2 < S_W^2.$$

d) We want S_W^2 to be 'large': $y_{i,k}$ must be very dispersed around their mean $\overline{Y}_i$, and this must happen for *each cluster*. In practice, a method of 'assuring' this is to *sort* the list according to an auxiliary variable x that is well correlated to y.

Exercise 5.11 *Comparison of two designs with two stages*

A population U with N individuals is divided into M primary units U_i ($i = 1, ..., M$) of size N_i. We are interested in the total Y of a variable taking the values $y_{i,k}$ ($k \in U_i$), and we denote

$$Y_i = \sum_{k \in U_i} y_{i,k}, \quad \overline{Y}_i = \frac{Y_i}{N_i}, \quad Y = \sum_{i=1}^{M} Y_i,$$

$$S_{2,i}^2 = \frac{1}{N_i - 1} \sum_{k \in U_i} \left(y_{i,k} - \overline{Y}_i \right)^2,$$

$$S_T^2 = \frac{1}{M-1} \sum_{i=1}^{M} \left(Y_i - \frac{Y}{M} \right)^2,$$

$$S_N^2 = \frac{1}{M-1} \sum_{i=1}^{M} \left(N_i - \frac{N}{M} \right)^2.$$

1. a) We select using simple random sampling (without replacement) m primary units, forming a sample S. Calculate the expected value and the variance of the estimator

$$\widehat{N}(S) = \sum_{i \in S} N_i.$$

   b) In each primary unit of S, we select (by simple random sampling without replacement) a sample of secondary units at a rate f_2. This rate is independent of S (strategy A). Calculate f_2 so that the final sample size has an expected value $\bar{n}$ fixed in advance (we assume that $f_2 \times N_i$ is an integer).

   c) What unbiased estimator $\widehat{Y}$ of Y, as a linear function of $\widehat{Y}_i$, do we propose? What is its variance? What does it become if $S_{2,i}^2$ is a constant (denoted S_2^2) for all i?

   d) For m sufficiently large, give a 95% confidence interval for the total size n of the final sample.

2. We now examine another two-stage sampling design (strategy B). The sample of primary units is selected as previously done, but in each primary unit selected in the first stage, we select using simple random sampling without replacement a sample of size $n_i = f_2 N_i$, with

$$f_2 = f_2(S) = \frac{\bar{n}}{\widehat{N}(S)}.$$

   a) Show that the sample is of fixed size (to be determined), and that for all i of S, the estimator $\widehat{Y}_i = N_i \widehat{\overline{Y}}_i$ estimates Y_i without bias. Show that $\widehat{Y}$ defined from 1.c. is always unbiased.

   b) Calculate the variance of $\widehat{Y}$ assuming that $S_{2,i}^2 = S_2^2$ for all i.

3. Compare the two strategies A and B, under the conditions that we specified. Can we say that one is indisputably better than the other?

Solution

1. a) Since

$$\widehat{N}(S) = \sum_{i \in S} N_i,$$

$$E\left(\widehat{N}(S)\right) = \sum_{i=1}^{M} N_i \frac{m}{M} = \frac{mN}{M}.$$

Furthermore, since $\widehat{N}(S)M/m$ is the unbiased estimator of $N = \sum_{i=1}^{M} N_i$,

$$\text{var}\left(\widehat{N}(S)\frac{M}{m}\right) = M^2 \frac{M-m}{Mm} S_N^2,$$

where S_N^2 is the population variance of N_i. It follows that

$$\text{var}\left(\widehat{N}(S)\right) = m^2 \frac{M-m}{Mm} S_N^2.$$

b) We fix $f_2 = n_i/N_i$ before selecting S. The n_i are therefore not random. The total sample size $n(S)$ is random, indeed

$$n(S) = \sum_{i \in S} n_i = \sum_{i \in S} N_i f_2 = f_2 \widehat{N}(S).$$

Therefore, if we fix in advance $\bar{n}$, the expected value of $n(S)$

$$\bar{n} = \text{E}(n(S)) = \frac{f_2 mN}{M},$$

and we therefore set

$$f_2 = \frac{M\bar{n}}{mN},$$

which is effectively independent of S.

c) Since $Y = \sum_{i=1}^{M} Y_i$, the estimator of the total is given by

$$\widehat{Y} = \sum_{i \in S} \frac{M}{m} \widehat{Y}_i = \sum_{i \in S} \frac{M}{m} N_i \widehat{\overline{Y}}_i,$$

where $\widehat{Y}_i$ estimates Y_i without bias, and $\widehat{\overline{Y}}_i$ is the mean of n_i secondary units sampled from i. We are in the 'classical' setting where n_i only depends on i since f_2 is independent of S. In this case, the variance, well-known, is

$$\text{var}(\widehat{Y}) = M^2 \frac{M-m}{Mm} S_T^2 + \frac{M}{m} \sum_{i=1}^{M} N_i^2 \frac{N_i - n_i}{N_i n_i} S_{2,i}^2$$

$$= M^2 \frac{M-m}{Mm} S_T^2 + \frac{M}{m}(1 - f_2) \sum_{i=1}^{M} N_i^2 \frac{m S_{2,i}^2}{M\bar{n} N_i} N$$

$$= M^2 \frac{M-m}{Mm} S_T^2 + \frac{N}{\bar{n}}(1 - f_2) \sum_{i=1}^{M} N_i S_{2,i}^2.$$

Furthermore, if $S_{2,i}^2 = S_2^2$, then

$$\text{var}(\widehat{Y}) = M^2 \frac{M-m}{Mm} S_T^2 + \frac{N^2}{\bar{n}}(1-f_2)S_2^2.$$

d) We have

$$n = n(S) = \sum_{i \in S} n_i = f_2 \widehat{N}(S).$$

Therefore,

$$\text{var}(n) = (f_2)^2 \text{var}(\widehat{N}(S)) = \left(\frac{\bar{n}}{\overline{N}}\right)^2 \left(1 - \frac{m}{M}\right) \frac{S_N^2}{m},$$

where $\overline{N} = N/M$ is the mean size of the M primary units. If m is large, then n approximately follows a normal distribution and, (about) 95 times out of 100, we have:

$$n \in \left[\bar{n} \pm 2\frac{\bar{n}}{\overline{N}}S_N\sqrt{\frac{1}{m} - \frac{1}{M}}\right].$$

2. In strategy B, the number of units selected in each primary unit becomes random

$$n_i(S) = f_2 N_i = \frac{\bar{n}N_i}{\widehat{N}(S)}.$$

The sampling of the second stage thus depends on what passes through the first stage. The invariance property of n_i is therefore not satisfied, and it is thus not a classical two-stage design. It is therefore necessary to recalculate the expected value and the variance of the total estimator, while being attentive to the fact that $n_i(S)$ is random.

a) The total size of the sample is

$$\sum_{i \in S} n_i(S) = \sum_{i \in S} \frac{\bar{n}N_i}{\widehat{N}(S)} = \frac{\bar{n}}{\widehat{N}(S)} \sum_{i \in S} N_i = \bar{n},$$

and is thus not random. We still estimate Y by

$$\widehat{Y} = \sum_{i \in S} \frac{M}{m}\widehat{Y}_i,$$

where $\widehat{Y}_i = N_i \overline{\widehat{Y}}_i$. Finally, $\widehat{Y}$ is absolutely the same estimator as in strategy A, but its distribution is not the same.

$$\text{E}\left(\widehat{Y}\right) = \text{E}\left(\sum_{i \in S} \frac{M}{m}\widehat{Y}_i\right) = \text{E}_S\text{E}\left(\sum_{i \in S} \frac{M}{m}\widehat{Y}_i | S\right)$$

$$= \text{E}_S\left(\sum_{i \in S} \frac{M}{m}\text{E}\left(\widehat{Y}_i | S\right)\right),$$

where E_S indicates the expected value with respect to the sampling distribution of S. Now, conditionally on S, the size n_i is fixed, and we therefore perform 'standard' simple random sampling in U_i: $\widehat{Y}_i$ then estimates Y_i without bias. Finally:

$$E\left(\widehat{Y}\right) = E_S\left(\sum_{i \in S}\frac{M}{m}Y_i\right) = \sum_{i=1}^{M}Y_i = Y.$$

b) The variance obtained by the classical decomposition is

$$\mathrm{var}(\widehat{Y}) = \mathrm{var}\,E\left(\widehat{Y}|S\right) + E\,\mathrm{var}\left(\widehat{Y}|S\right).$$

Indeed

$$\mathrm{var}\,E\left(\widehat{Y}|S\right) = \mathrm{var}\left(\sum_{i \in S}\frac{M}{m}Y_i\right) = M^2\frac{M-m}{Mm}S_T^2.$$

Furthermore,

$$E\,\mathrm{var}\left(\widehat{Y}|S\right) = E\left\{\sum_{i \in S}\frac{M^2}{m^2}\mathrm{var}\left(\widehat{Y}_i|S\right)\right\}.$$

There is no covariance term, as conditionally on S, the drawings within the U_i are independent from one another. Therefore

$$E\,\mathrm{var}\left(\widehat{Y}|S\right) = E\left\{\sum_{i \in S}\frac{M^2}{m^2}N_i^2\frac{N_i-n_i}{N_in_i}S_{2,i}^2\right\}$$

$$= E\left\{\sum_{i \in S}\frac{M^2}{m^2}\frac{N_i^2}{n_i}S_{2,i}^2\right\} - \sum_{i=1}^{M}\frac{M}{m}N_iS_{2,i}^2$$

$$= E\left\{\sum_{i \in S}\frac{M^2}{m^2}\frac{N_i^2\widehat{N}(S)}{\bar{n}N_i}S_{2,i}^2\right\} - \sum_{i=1}^{M}\frac{M}{m}N_iS_{2,i}^2$$

$$= E\left\{\frac{M^2}{m^2}\sum_{i \in S}\frac{N_i\widehat{N}(S)}{\bar{n}}S_{2,i}^2\right\} - \sum_{i=1}^{M}\frac{M}{m}N_iS_{2,i}^2.$$

A priori, we can no longer simplify this expression in the general case. On the other hand, if $S_{2,i}^2 = S_2^2$, we have

$$E\left[\operatorname{var}\left(\widehat{Y}|S\right)\right]$$

$$= \frac{M^2}{m^2\bar{n}}S_2^2 E\left\{\widehat{N}(S)^2\right\} - \sum_{i=1}^{M}\frac{M}{m}N_i S_2^2$$

$$= \frac{M^2}{m^2\bar{n}}S_2^2\left[\operatorname{var}\left\{\widehat{N}(S)\right\} + \left\{E\widehat{N}(S)\right\}^2\right] - \frac{M}{m}NS_2^2$$

$$= \frac{M^2}{m^2\bar{n}}S_2^2\left[m\frac{M-m}{M}S_N^2 + \left(\frac{mN}{M}\right)^2\right] - \frac{M}{m}NS_2^2 \text{ (See 1.a)}$$

$$= \frac{M(M-m)}{m\bar{n}}S_2^2 S_N^2 + \frac{N^2}{\bar{n}}S_2^2 - \frac{M}{m}NS_2^2$$

$$= \frac{M(M-m)}{m\bar{n}}S_2^2 S_N^2 + \frac{N^2}{\bar{n}}\left(1 - \frac{M\bar{n}}{mN}\right)S_2^2.$$

We finally get

$$\operatorname{var}\left(\widehat{Y}\right) = M^2\frac{M-m}{Mm}S_T^2 + \frac{M(M-m)}{m\bar{n}}S_2^2 S_N^2 + \frac{N^2}{\bar{n}}\left(1 - \frac{M\bar{n}}{mN}\right)S_2^2.$$

3. It is necessary to compare designs that are comparable; that is, designs having identical costs. In the present case, the cost is conditional on the sample size. It is therefore necessary to ensure that the expected value of the total sample size from strategy A (denoted $\bar{n}$) is equal to the total fixed size from strategy B (also denoted $\bar{n}$). Thus, $\bar{n}$ represents the same value for the two strategies. We can set $f_2 = M\bar{n}/mN$, as in strategy A, and compare:

$$\operatorname{var}(\text{strategy A}) = M^2\frac{M-m}{Mm}S_T^2 + \frac{N^2}{\bar{n}}(1 - f_2)S_2^2,$$

and

$$\operatorname{var}(\text{strategy B}) = M^2\frac{M-m}{Mm}S_T^2 + \frac{N^2}{\bar{n}}(1 - f_2)S_2^2 + \frac{M(M-m)}{m\bar{n}}S_2^2 S_N^2.$$

Strategy A, with a variable sample size, is therefore unquestionably the best, unless the M primary units are of the same size N_i (in which case $S_N^2 = 0$, and the two strategies are identical).

Exercise 5.12 *Cluster effect and variable sizes*

In a cluster sample with simple random sampling of m clusters among M, we know that if the clusters are of identical size, the variance of the estimator $\widehat{\overline{Y}}$ for the mean $\overline{Y}$ is:

$$\text{var}(\widehat{\overline{Y}}) = \frac{1-f}{m\overline{N}} \, S_y^2[1 + \rho(\overline{N} - 1)],$$

where S_y^2 is the population variance of y_k in the population, ρ is the cluster effect, and $\overline{N}$ is the common size of the clusters. The object of the exercise is to expand this expression in the case of clusters with variable sizes, with a (reasonable) hypothesis of a technical nature that is specified later on.

1. We denote $\overline{Y}$ as the true mean of $y_{i,k}$ (i is the cluster identifier, k is the identifier of the individual in the cluster), N as the total population size, Y_i as the true total (unknown) in cluster i and $\overline{\overline{Y}}$ as the mean of Y_i among the M clusters. Recall the expression of the unbiased estimator $\widehat{\overline{Y}}$ and the expression of its variance. Link $\overline{\overline{Y}}$ to $\overline{Y}$.

2. Express the population variance of the totals Y_i as a function of $y_{i,k}$, N_i (size of cluster i), Y_i, $\overline{Y}$ and $\overline{N}$.

3. Use the previous expression to derive the population variance S_y^2 of variables $y_{i,k}$, the cluster effect ρ, the population variance S_N^2 of the sizes N_i and the covariance S_{NY} of N_i and Y_i.

4. By considering M to be large, give an expression approaching $\text{var}(\widehat{\overline{Y}})$ as a function of the quantities previously defined.

5. We define the variable U_i as follows:

$$Y_i = \overline{Y} N_i + U_i \qquad \text{for all } i = 1,\, 2,\, \ldots,\, M$$

and we make the technical hypothesis (reasonable) that N_i and U_i are uncorrelated. Show that in that case $S_{NY} \approx \overline{Y} S_N^2$.

6. Show that, under this hypothesis, we have:

$$\text{var}(\widehat{\overline{Y}}) \approx \left[1 + \rho(\overline{N} - 1) + \overline{N} \left(\frac{\text{CV}_N}{\text{CV}_Y} \right)^2 \right] V_{\text{SRS}},$$

where V_{SRS} is the true variance obtained for a simple random sample of size $m\overline{N}$ (to be determined), and CV_N and CV_Y are the true coefficients of variation respectively for N_i and $y_{i,k}$.

7. Conclude, in particular by considering the reasonable orders of magnitude for the parameters involved in the variance expression.

Solution

1. We define

$$\overline{Y} = \frac{1}{N} \sum_{i=1}^{M} Y_i = \frac{1}{M\overline{N}} \sum_{i=1}^{M} Y_i,$$

which implies that

$$\widehat{\overline{Y}} = \frac{1}{Nm} \sum_{i \in S} Y_i.$$

Therefore

$$\operatorname{var}(\widehat{\overline{Y}}) = \frac{1}{N^2} \left(1 - \frac{m}{M}\right) \frac{S_T^2}{m},$$

where

$$S_T^2 = \frac{1}{M-1} \sum_{i=1}^{M} (Y_i - \overline{Y})^2.$$

We have as well: $\overline{\overline{Y}} = \overline{Y}/N$.

2. We have:

$$Y_i - \overline{\overline{Y}} = \sum_{k \in U_i} y_{i,k} - N_i \overline{Y} + N_i \overline{Y} - \overline{\overline{Y}} = \sum_{k \in U_i} (y_{i,k} - \overline{Y}) + \overline{Y}(N_i - \overline{N}).$$

Therefore,

$$\sum_{i=1}^{M} (Y_i - \overline{\overline{Y}})^2 = \sum_{i=1}^{M} \sum_{k \in U_i} (y_{i,k} - \overline{Y})^2 + \sum_{i=1}^{M} \sum_{k \in U_i} \sum_{\substack{\ell \in U_i \\ \ell \neq k}} (y_{i,k} - \overline{Y})(y_{i,\ell} - \overline{Y})$$

$$+ \overline{Y}^2 \sum_{i=1}^{M} (N_i - \overline{N})^2 + 2\overline{Y} \sum_{i=1}^{M} (N_i - \overline{N})(Y_i - N_i \overline{Y}).$$

3. Recall that

$$S_y^2 = \frac{1}{N-1} \sum_{i=1}^{M} \sum_{k \in U_i} (y_{i,k} - \overline{Y})^2,$$

and, by definition,

$$\rho = \frac{\sum_{i=1}^{M} \sum_{k=1}^{N_i} \sum_{\substack{\ell=1 \\ \ell \neq k}}^{N_i} (y_{i,k} - \overline{Y})(y_{i,\ell} - \overline{Y})}{\sum_{i=1}^{M} \sum_{k \in U_i} (y_{i,k} - \overline{Y})^2} \times \frac{1}{N-1}.$$

The term $1/(\overline{N} - 1)$ can seem strange at first glance, but it appears naturally when we divide each term of the ratio by the number of terms that it contains 'on average' (being $M\overline{N}(\overline{N}-1)$ for the numerator and $M\overline{N}$ for the denominator). This normalisation ensures a certain 'stability' in ρ. That is,

$$S_N^2 = \frac{1}{M} \sum_{i=1}^{M} (N_i - \overline{N})^2,$$

and

$$S_{NY} = \frac{1}{M} \sum_{i=1}^{M} (N_i - \overline{N})(Y_i - \overline{\overline{Y}}).$$

We consider M to be sufficiently large to assume $M - 1$ and M to be similar. We have:

$$\sum_{i=1}^{M}(Y_i - \overline{\overline{Y}})^2 = (M\overline{N} - 1)\, S_y^2 + (\overline{N} - 1)\, \rho(M\overline{N} - 1)\, S_y^2 + \overline{Y}^2 M S_N^2$$

$$+2\overline{Y}\sum_{i=1}^{M}(N_i - \overline{N})\,(Y_i - \overline{Y} + \overline{Y} - N_i\overline{Y})$$

$$= (M\overline{N} - 1)\, S_y^2(1 + \rho(\overline{N} - 1)) - M\overline{Y}^2 S_N^2 + 2M\overline{Y}S_{NY}.$$

4. When M is large, we have the approximation $M\overline{N} - 1 \approx M\overline{N} - \overline{N}$. Therefore,

$$\mathrm{var}(\widehat{\overline{Y}}) \approx \frac{1-f}{m\overline{N}}\left[S_y^2(1 + \rho(\overline{N} - 1)) + \frac{\overline{Y}}{\overline{N}}\,(2S_{NY} - \overline{Y}S_N^2)\right].$$

5. The formula proposed for Y_i simply expresses an approximately linear relation between the total Y_i and the size N_i. This relation is natural: although we do not at all need to have U_i small, it is likely that in practice this is often the case. We see that

$$\sum_{i=1}^{M}U_i = 0.$$

Furthermore: $Y_i - \overline{\overline{Y}} = \overline{Y}(N_i - \overline{N}) + U_i$. Therefore

$$MS_{NY} = \overline{Y}\sum_{i=1}^{M}(N_i - \overline{N})^2 + \sum_{i=1}^{M}(N_i - \overline{N})\,(U_i - \overline{U}) = \overline{Y}MS_N^2,$$

by the hypothesis of non-correlation between U_i and N_i.

Of course, this technical hypothesis is never exactly realised in practice, but the covariance must be small in a good number of cases. Indeed, there is no reason *a priori* for the residual $Y_i - \overline{Y}N_i$ to be linked in a particular way to the size N_i. We deduce

$$S_{NY} \approx \overline{Y}S_N^2.$$

6. The variance is

$$\mathrm{var}(\widehat{\overline{Y}}) \approx \frac{1-f}{m\overline{N}}\left[S_y^2(1 + \rho(\overline{N} - 1)) + \overline{Y}^2\frac{S_N^2}{\overline{N}}\right]. \tag{5.1}$$

If we let

$$\mathrm{CV}_N = \frac{S_N}{\overline{N}}, \qquad \text{and} \qquad \mathrm{CV}_Y = \frac{S_y}{\overline{Y}},$$

we get the desired expression by noticing that:

$$\underset{\text{SRS}}{V} = \frac{1-f}{m\overline{N}} \, S_y^2.$$

We can verify very rapidly the validity of Expression (5.1) in the particular case where $y_{i,k}$ is constant (thus equal to $\overline{Y}$), as then $S_y^2 = 0$, $Y_i = \overline{Y}N_i$ and it is instantly verified that

$$\widehat{\overline{Y}} = \frac{1}{\overline{N}}\overline{Y}\frac{1}{m}\sum_{i \in S} N_i.$$

Hence

$$\mathrm{var}(\widehat{\overline{Y}}) = \frac{\overline{Y}^2}{\overline{N}^2}\frac{1-f}{m}S_N^2 = \frac{1-f}{m\overline{N}}\left[\overline{Y}^2\frac{S_N^2}{\overline{N}}\right].$$

We observe that $m\overline{N}$ is the expected value for the size of the sample (size random):

$$\widehat{\text{Size}} = \sum_{i \in S} N_i = \sum_{i=1}^{M} N_i I_i.$$

Therefore

$$\mathrm{E}(\widehat{\text{Size}}) = \left(\sum_{i=1}^{M} N_i\right)\frac{m}{M} = m\overline{N}.$$

The expression $\underset{\text{SRS}}{V}$ was thus obtained from a sample of size comparable to the one from cluster sampling.

7. We see that there is deterioration in the quality in comparison with simple random sampling of the same size (on average), when:
   - the cluster effect is large,
   - the mean cluster size $\overline{N}$ is large,
   - the sizes N_i are varied,
   - the population variance S_y^2 is small.

The first three conclusions are well-known: heterogeneous clusters are required, of small size and of similar size. The fourth conclusion is more mechanical and expresses the great effectiveness of simple sampling when the $y_{i,k}$ are not too dispersed.

In concrete terms, take the example of the labour force survey used by INSEE: this one is built upon housing units of size 20 on average, and consisting of between 16 and 24. Using a uniform division of N_i in this interval, we will have:

$$S_N^2 = \frac{(24-16)^2}{12} \approx 5.3.$$

Therefore,

$$\mathrm{CV}_N = \frac{\sqrt{5.3}}{20} \approx 0.12.$$

We commonly find coefficients ρ of order from 5% to 10%. For a variable y having for example a coefficient of variation of 30%, and such that $\rho = 10\%$, this gives:

$$\frac{\text{var}(\widehat{\overline{Y}})}{\underset{\text{SRS}}{V}} = 1 + 0.1 \times (20 - 1) + 20 \times \left(\frac{0.12}{0.30}\right)^2 = 1 + 1.9 + 3.2 = 6.1 \approx (2.5)^2.$$

The loss in accuracy is high, and the variance of sizes N_i contributes to this appreciably more than the cluster effect ρ (even if, at first glance, the variance of the cluster sizes seems quite modest).

Exercise 5.13 *Variance and list order*

In this exercise, we are interested in the estimation of variance when we draw a systematic sample with equal probabilities, of total size n. We have a frame of N individuals, sorted in the order given. We denote i as the list number of the first individual selected and g as the sampling interval ($g = N/n$, which we consider to be an integer). Finally, we denote as $y_{i,j}$ the value of the variable of interest y for the $(j + 1)$-th individual selected ($j = 0, 1, 2, \ldots, n - 1$) when the first individual selected has list number i ($i = 1, 2, \ldots, g$).

1. What precisely is the list number of the individual corresponding to the value $y_{i,j}$? (This list number is included between 1 and N.)
2. If we denote $Y_i = \sum_{j=0}^{n-1} y_{i,j}$, give the unbiased estimator $\widehat{\overline{Y}}$ of the true mean $\overline{Y}$ and then give the expression of its (true) variance.
3. Explain why we cannot estimate this variance without bias.
4. We are henceforth going to assume that the $y_{i,j}$ are generated by a stochastic model functioning as follows for all i, j:

$$y_{i,j} = \alpha + z_{i,j},$$

where the $z_{i,j}$ are real random variables with null expected value and variance σ^2 and are uncorrelated among themselves. We denote $\mathcal{E}()$ and $\mathcal{V}()$ as the expected value and the variance associated with the distribution of z.

a) Intuitively, when do we make this type of hypothesis?
b) Calculate, under this model, the expected value of the true variance coming from 2., that is, $\mathcal{E}(\text{var}(\widehat{\overline{Y}}))$.
c) We venture to use, as a variance estimator, the expression:

$$v_1 = \left(1 - \frac{n}{N}\right) \frac{s_y^2}{n},$$

where s_y^2 is the corrected variance in the sample. Calculate the expected value under the model of v_1, being $\mathcal{E}(v_1)$.

d) Make a conclusion.
5. This time, we are interested in cases where we can reasonably make the hypothesis:
$$y_{i,j} = \alpha(i + jg) + \beta + z_{i,j},$$
still with the same hypotheses on the random variables $z_{i,j}$.
   a) In what case will we use this model?
   b) Calculate $\mathcal{E}(\text{var}(\widehat{\overline{Y}}))$ under the model.
   c) Calculate $\mathcal{E}(v_1)$ under the model.
   d) Make a conclusion.
6. With systematic sampling having been done, let us pretend 'as if' it resulted in a stratified sampling design with the simple random sampling of two individuals in each stratum, for each of the $n/2$ strata (assume n is an even number to simplify matters) and are put together by dividing the frame into blocks of $(2g)$ consecutive individuals (the systematic sampling is not 'as far off' as that for this stratified design: this last design simply gives a little more freedom in the choice of the sample, but the principle of a systematic scan of the complete frame remains more or less respected).
   a) What variance estimator v_2 should we use?
   b) With the model from 4., what would $\mathcal{E}(v_2)$ be?
   c) With the model from 5., what would $\mathcal{E}(v_2)$ be?
   d) Make a conclusion.

Solution

1. In the sorted list, the list number of the individual which takes $y_{i,j}$ for the value of y is:
$$i + jg.$$
Indeed, the first individual selected corresponds to $j = 0$ (by definition) and it indeed has the list number $i = i+0g$. The second individual selected has a list number that is larger by 'STEP' (thus by g), that is $i+g = i+1g$. The $(j + 1)$-th individual selected has a list number that is larger by jg in comparison to i, that is $i + jg$.
2. We are faced with cluster sampling, where each cluster of the population is composed of a first individual with a list number between 1 and g and the set of $(n - 1)$ individuals that are deduced from the succession of consecutive steps of length g for the entire length of the sampling frame:

   cluster 1 : list numbers 1, $1 + g$, $1 + 2g$, ..., $1 + (n - 1)\,g$,

   cluster 2 : list numbers 2, $2 + g$, $2 + 2g$, ..., $2 + (n - 1)\,g$,

   cluster g : list numbers g, $g + g$, $g + 2g$, ..., $g + (n - 1)\,g = N$.

The unbiased estimators and their variances are obtained as a consequence, by noticing that they are clusters of fixed size n. In fact, if i is initially selected:

$$\widehat{\overline{Y}} = \frac{1}{n} \sum_{j=0}^{n-1} y_{i,j} = \frac{Y_i}{n}.$$

The variance is directly obtained if we notice that $\widehat{\overline{Y}}$ takes the value Y_i/n with probability $1/g$, where i covers $1, 2, \ldots, g$:

$$\text{var}(\widehat{\overline{Y}}) = \sum_{i=1}^{g} \frac{1}{g} \left(\frac{Y_i}{n} - \frac{1}{g} \sum_{i=1}^{g} \frac{Y_i}{n} \right)^2,$$

by the definition of the variance, or again:

$$\text{var}(\widehat{\overline{Y}}) = \frac{1}{g} \sum_{i=1}^{g} \left(\frac{Y_i}{n} - \overline{Y} \right)^2 = \frac{1}{gn^2} \sum_{i=1}^{g} Y_i^2 - \overline{Y}^2.$$

3. We are in the context of cluster sampling with a single cluster selected. Surveys for which the sample size is 1 never allow for the unbiased estimation of population variances (in other words, by construction, there must be at least two units to estimate a population variance). We therefore do not have any hope of being able to estimate the variance without bias.

4. a) The model presents the individual values $y_{i,j}$ of the individuals from the frame as the realisations of uncorrelated random variables (denoted in the same way as the deterministic variables in order to simplify) with the same mean and the same variance:

$$\mathcal{E}(y_{i,j}) = \alpha \quad \text{and} \quad \mathcal{V}(y_{i,j}) = \sigma^2,$$

where $\mathcal{E}$ and $\mathcal{V}$ represent respectively the expected value and the variance in relation to the distribution of the model. Intuitively, this is realistic when there is no particular 'structure' in the sampling frame, or no apparent order: this can be because the variable y itself is not explained by any known characteristic of the individuals (variable with 'lawless' appearance) or because the frame has been mixed and that the individuals themselves appear in a random order. In one word, the over-simplicity of the model is synonymous with absolute 'chaos' at the mechanism level determining the individual values y.

b) We have, according to 2.,

$$\mathcal{E}\left[\text{var}(\widehat{\overline{Y}})\right] = \frac{1}{gn^2} \sum_{i=1}^{g} \mathcal{E}(Y_i^2) - \mathcal{E}(\overline{Y})^2.$$

We use:

$$\mathcal{E}(Y_i^2) = \mathcal{V}(Y_i) + [\mathcal{E}(Y_i)]^2,$$

and

$$\mathcal{E}(\overline{Y})^2 = \mathcal{V}(\overline{Y}) + [\mathcal{E}(\overline{Y})]^2,$$

$$\mathcal{V}(Y_i) = \sum_{j=0}^{n-1} \mathcal{V}(y_{i,j}) = n\sigma^2,$$

$$\mathcal{E}(Y_i) = \sum_{j=0}^{n-1} \mathcal{E}(y_{i,j}) = n\alpha,$$

$$\mathcal{V}(\overline{Y}) = \frac{1}{N^2} \sum_{i=1}^{g} \sum_{j=0}^{n-1} \mathcal{V}(y_{i,j}) = \frac{\sigma^2}{N} \quad (\text{recall}: ng = N),$$

$$\mathcal{E}(\overline{Y}) = \frac{1}{N} \sum_{i=1}^{g} \sum_{j=0}^{n-1} \mathcal{E}(y_{i,j}) = \alpha.$$

Therefore,

$$\mathcal{E}(\mathrm{var}(\widehat{\overline{Y}})) = \frac{1}{n^2 g} \sum_{i=1}^{g} (n\sigma^2 + n^2\alpha^2) - \left(\frac{\sigma^2}{N} + \alpha^2\right) - \left(1 - \frac{n}{N}\right)\frac{\sigma^2}{n}.$$

This expression has a well-known appearance.

c)

$$\mathcal{E}(v_1) = \left(1 - \frac{n}{N}\right)\frac{\mathcal{E}(s_y^2)}{n}.$$

In fact, if i is the list number of the ith individual selected,

$$s_y^2 = \frac{1}{n-1} \sum_{j=0}^{n-1} (y_{i,j} - \widehat{\overline{Y}})^2 = \frac{1}{n-1} \sum_{j=0}^{n-1} y_{i,j}^2 - \frac{n}{n-1}\widehat{\overline{Y}}^2.$$

Thus

$$\mathcal{E}(s_y^2) = \frac{1}{n-1} \sum_{j=0}^{n-1} \mathcal{E}(y_{i,j}^2) - \frac{n}{n-1}\mathcal{E}(\widehat{\overline{Y}}^2),$$

with

$$\widehat{\overline{Y}} = \frac{Y_i}{n}.$$

We easily verify (still from the variances, as in b.):

$$\mathcal{E}(y_{i,j}^2) = \alpha^2 + \sigma^2,$$

and $\mathcal{E}(\widehat{\overline{Y}}^2) = \sigma^2/n + \alpha^2$. Therefore, $\mathcal{E}(s_y^2) = \sigma^2$. Finally,

$$\mathcal{E}(v_1) = \left(1 - \frac{n}{N}\right)\frac{\sigma^2}{n}.$$

d) The estimator v_1 is biased for the true variance $\mathrm{var}(\widehat{\overline{Y}})$ if the inference is based on the sampling design (see 3.). On the other hand, if the inference is based on the model, we have:

$$\mathcal{E}[v_1 - \mathrm{var}(\widehat{\overline{Y}})] = 0.$$

We can therefore say that, if we take into account the risk on y expressed by the model, v_1 estimates $\mathrm{var}(\widehat{\overline{Y}})$ without bias. This property justifies the use of v_1 to estimate the accuracy of a systematic sampling when the frame is apparently 'in any order'.

5. a) $y_{i,j}$ has an expected value that clearly increases with the list number of the individual. Such a model reflects a linear tendency. This is the classical situation obtained after sorting the frame according to an auxiliary variable correlated to y. It is also in this case that we can benefit from gains in accuracy linked to systematic sampling: this is indeed because there is such a tendency that systematic sampling is used.

b)

$$\mathcal{E}(\mathrm{var}(\widehat{\overline{Y}})) = \frac{1}{gn^2} \sum_{i=1}^{g} \mathcal{E}(Y_i^2) - \mathcal{E}(\overline{Y})^2.$$

We have:

$$\mathcal{V}(Y_i) = n\sigma^2,$$

$$\mathcal{E}(Y_i) = \sum_{j=0}^{n-1} [\alpha(i + jg) + \beta] = n\left(\alpha i + \beta + \alpha g \frac{n-1}{2}\right),$$

$$\mathcal{V}(\overline{Y}) = \frac{\sigma^2}{N},$$

$$\mathcal{E}(\overline{Y}) = \sum_{i=1}^{g} \sum_{j=0}^{n-1} \frac{1}{N}(\alpha(i + jg) + \beta)$$

$$= \frac{n}{N} g\left(\alpha \frac{g+1}{2} + \beta + \alpha g \frac{n-1}{2}\right).$$

Hence

$$\mathcal{E}(\mathrm{var}(\widehat{\overline{Y}})) = \frac{1}{gn^2} \sum_{i=1}^{g} \left(n\sigma^2 + n^2\left(\alpha i + \beta + \alpha g \frac{n-1}{2}\right)^2\right) - \frac{\sigma^2}{N}$$

$$- \frac{n^2 g^2}{N^2}\left(\alpha \frac{g+1}{2} + \beta + \alpha g \frac{n-1}{2}\right)^2.$$

After a calculation that is long but not technically difficult, by using the equality:

$$\sum_{i=1}^{g} i^2 = \frac{g}{6} \left(2g^2 + 3g + 1\right),$$

and remembering that $ng = N$, we find:

$$\mathcal{E}(\mathrm{var}(\widehat{\overline{Y}})) = \left(1 - \frac{n}{N}\right) \frac{\sigma^2}{n} + \alpha^2 \frac{g^2 - 1}{12}.$$

c) By reusing 4-(c), we have:

$$\mathcal{E}(s_y^2) = \frac{1}{n-1} \sum_{j=0}^{n-1} \mathcal{E}(y_{i,j}^2) - \frac{n}{n-1} \mathcal{E}(\widehat{\overline{Y}}^2).$$

With the model from 5., we get:

$$\mathcal{E}(y_{i,j}^2) = \sigma^2 + [\alpha(i + jg) + \beta]^2,$$

and therefore

$$\mathcal{E}(\widehat{\overline{Y}}^2) = \frac{\sigma^2}{n} + \left[\frac{1}{n} \sum_{j=0}^{n-1} (\alpha(i + jg) + \beta)\right]^2$$

$$= \frac{\sigma^2}{n} + \frac{1}{n^2} n^2 \left(\alpha i + \beta + \alpha g \, \frac{n-1}{2}\right)^2.$$

The calculation (long) leads to:

$$\mathcal{E}(v_1) = \left(1 - \frac{n}{N}\right) \frac{\sigma^2}{n} + \left(1 - \frac{n}{N}\right) \left(\frac{\alpha^2 g^2}{12}\right) (n+1).$$

We notice that this expression does not depend on i.

d) If the model is true (see a for the practical conditions), we have:

$$\mathcal{E}(v_1) = \mathcal{E}(\mathrm{var}(\widehat{\overline{Y}})) + \frac{\alpha^2}{12} \left[(1 - f)(n+1) g^2 - g^2 + 1\right].$$

Under the most frequent conditions, we have $1 - f \approx 1$ and n 'large'. Hence:

$$\mathcal{E}(v_1) \approx \mathcal{E}(\mathrm{var}(\widehat{\overline{Y}})) + n \, \frac{\alpha^2 g^2}{12} = \mathcal{E}(\mathrm{var}(\widehat{\overline{Y}})) + \frac{N^2}{n} \frac{\alpha^2}{12}.$$

On average, v_1 and $\mathrm{var}(\widehat{\overline{Y}})$ differ by $(N^2\alpha^2)/(12n)$. Unless α is truly very small, close to null (in which case we find the model from 4.), this factor is very large and positive. In this case, v_1 is going to considerably overestimate $\mathrm{var}(\widehat{\overline{Y}})$. The previous calculations therefore justify a well-known principle, which states that the classical variance estimator for a simple random sample is a very bad estimator of the true variance in the case of systematic sampling for a sorted list, all the more so as the population is large and the sample small.

6. a) In a stratified sampling with H strata, we have (classical notation):

$$\widehat{\text{var}}(\widehat{Y}) = \sum_{h=1}^{H} \left(\frac{N_h}{N}\right)^2 \left(1 - \frac{n_h}{N_h}\right) \frac{s_{yh}^2}{n_h}.$$

In the context of the proposed comparison, we have $n_h = 2$ and $N_h = N/H$ with $H = n/2$, that is $N_h/N = 2/n$. If the individual with list number i is selected at the start of the list, then, in stratum h, the sample consists of two individuals with list numbers j and $(j+1)$; we easily verify then that $j = 2h - 2$, where:

$$s_{yh}^2 = \frac{1}{2-1}\left[\left(y_{i,j} - \frac{y_{i,j} + y_{i,j+1}}{2}\right)^2 + \left(y_{i,j+1} - \frac{y_{i,j} + y_{i,j+1}}{2}\right)^2\right]$$

$$= \frac{1}{2}\left(y_{i,j} - y_{i,j+1}\right)^2.$$

Hence:

$$v_2 = \sum_{j \in J} \left(\frac{2}{n}\right)^2 \left(1 - \frac{2}{2\frac{N}{n}}\right) \frac{1}{2}\frac{1}{2}\left(y_{i,j} - y_{i,j+1}\right)^2,$$

where J is the set of integer pairs included between 0 and $(n-2)$. There are $n/2$ integers in J. Finally:

$$v_2 = \left(1 - \frac{n}{N}\right)\frac{1}{n}\delta^2,$$

where

$$\delta^2 = \frac{1}{n}\sum_{j \in J}(y_{i,j} - y_{i,j+1})^2$$

$$= \frac{1}{n}\left[(y_{i,0} - y_{i,1})^2 + (y_{i,2} - y_{i,3})^2 + (y_{i,4} - y_{i,5})^2 + ...\right].$$

b) It is necessary to calculate $\mathcal{E}(\delta^2)$ with the model from 4.:

$$\mathcal{E}(\delta^2) = \frac{1}{n}\sum_{j \in J}\mathcal{E}(y_{i,j} - y_{i,j+1})^2.$$

Now:

$$\mathcal{E}(y_{i,j} - y_{i,j+1})^2 = V(y_{i,j} - y_{i,j+1}) + [\mathcal{E}y_{i,j} - \mathcal{E}y_{i,j+1}]^2$$
$$= 2\sigma^2 + (\alpha - \alpha)^2 = 2\sigma^2.$$

Therefore

$$\mathcal{E}(\delta^2) = \frac{1}{n}\frac{n}{2}2\sigma^2 = \sigma^2, \quad \text{and} \quad \mathcal{E}(v_2) = \left(1 - \frac{n}{N}\right)\frac{\sigma^2}{n}.$$

c) With the model from 5., $\mathcal{E}(\delta^2)$ takes another value:

$$\mathcal{E}(y_{i,j} - y_{i,j+1})^2 = 2\sigma^2 + [(\alpha(i+jg) + \beta) - (\alpha(i+jg+g) + \beta)]^2$$
$$= 2\sigma^2 + \alpha^2 g^2.$$

Thus,

$$\mathcal{E}(\delta^2) = \frac{1}{n}\frac{n}{2}(2\sigma^2 + \alpha^2 g^2) = \sigma^2 + \frac{\alpha^2 g^2}{2},$$

and

$$\mathcal{E}(v_2) = \left(1 - \frac{n}{N}\right)\frac{\sigma^2}{n} + \left(1 - \frac{n}{N}\right)\frac{\alpha^2 g^2}{2n}.$$

d) • With the model from 4., $\mathcal{E}[v_2 - \mathrm{var}(\widehat{\overline{Y}})] = 0$. Therefore, v_2 appears to be unbiased under the model, for the same reason as v_1. It would remain to see which of v_1 or v_2 is the 'best' estimator of $\mathrm{var}(\widehat{\overline{Y}})$ (notion of 'best' to be defined).

• With the model from 5.,

$$\mathcal{E}[v_2 - \mathrm{var}(\widehat{\overline{Y}})] = \alpha^2 \left[\left(1 - \frac{n}{N}\right)\frac{g^2}{2n} - \frac{g^2 - 1}{12}\right].$$

In the most frequent conditions, we have $1 - f \approx 1 (\Rightarrow g \gg 1)$, thus:

$$\mathcal{E}[v_2 - \mathrm{var}(\widehat{\overline{Y}})] \approx \frac{\alpha^2 g^2}{2}\left(\frac{1}{n} - \frac{1}{6}\right).$$

The fact that the bias is null for $\alpha = 0$ is not surprising, since in these conditions the models from 4. and 5. merge. We get a bias that is nearly null for $n = 6$ which raises our curiosity. If n is 'large' (more than 6), we have:

$$\mathcal{E}[v_2 - \mathrm{var}(\widehat{\overline{Y}})] \approx -\frac{\alpha^2}{12}\left(\frac{N}{n}\right)^2.$$

v_2 has rather a tendency to underestimate $\mathrm{var}(\widehat{\overline{Y}})$: this can be a very large underestimation if the STEP is very large (the approximation of systematic sampling by stratified sampling then becomes strongly questionable).

We retain that with this model of linear tendency, the absolute error introduced by v_2 is much smaller than with v_1 (see 5-d): in absolute value, the error is indeed n times smaller with v_2 than with v_1.

In summary, we distinguished here two cases:
• A sampling frame is presented in any order, and we have two variance estimators v_1 and v_2, concurrent but unbiased in the sense of the risk linked to the model.

- The frame is sorted in order to present a linear tendency. Then v_1 overestimates the true variance (in the sense of the risk of the model) while v_2 underestimates it. A combined estimator $v = v_1/(n + 1) + (n/n + 1)v_2$ is in that case unbiased.

Calibration with an Auxiliary Variable

6.1 Calibration with a qualitative variable

We assume that the sizes N_h, where $h = 1, ..., H$, of H types of a qualitative variable are known in the population. The qualitative variable specifies H parts U_h, where $h = 1, ..., H$, called post-strata in the population. If the sample S is selected in accordance with a simple design without replacement, then the size of the sample intersecting post-strata h, being $n_h = \#(U_h \cap S)$ has a hypergeometric distribution. If we denote Y_h as the true total of a variable y over U_h, we can construct the post-stratified estimator of the total

$$\widehat{Y}_{\text{post}} = \sum_{\substack{h=1 \\ n_h > 0}}^{H} N_h \widehat{\overline{Y}}_h,$$

where $\widehat{\overline{Y}}_h = \widehat{Y}_h / \widehat{N}_h$. With a simple design without replacement,

$$\widehat{\overline{Y}}_h = \frac{1}{n_h} \sum_{k \in U_h \cap S} y_k.$$

With a simple design without replacement, the post-stratified estimator is unbiased as soon as we keep to the conditions of n_h non-null for all h, and it is all the more precise since the auxiliary variable is 'linked' to the variable of interest. If n is 'large enough', the variance of $\widehat{Y}_{\text{post}}$ is approximately, for the simple design without replacement:

$\text{var}(\widehat{Y}_{\text{post}})$

$$\approx N^2 \left[\left(1 - \frac{n}{N}\right) \frac{1}{n} \left(\sum_{h=1}^{H} \frac{N_h}{N} S_{yh}^2 \right) + \left(1 - \frac{n}{N}\right) \frac{1}{n^2} \left(\sum_{h=1}^{H} \left(1 - \frac{N_h}{N}\right) S_{yh}^2 \right) \right],$$

and is estimated by

$$\widehat{\mathrm{var}}(\widehat{Y}_{\mathrm{post}})$$
$$= N^2 \left[\left(1 - \frac{n}{N}\right) \frac{1}{n} \left(\sum_{h=1}^{H} \frac{N_h}{N} s_{yh}^2\right) + \left(1 - \frac{n}{N}\right) \frac{1}{n^2} \left(\sum_{h=1}^{H} \left(1 - \frac{N_h}{N}\right) s_{yh}^2\right)\right],$$

where

$$S_{yh}^2 = \frac{1}{N_h - 1} \sum_{k \in U_h} (y_k - \overline{Y}_h)^2,$$

and

$$s_{yh}^2 = \frac{1}{n_h - 1} \sum_{k \in U_h \cap S} (y_k - \widehat{\overline{Y}}_h)^2.$$

6.2 Calibration with a quantitative variable

If the total X of a quantitative variable x is known, we can use this information to construct a more precise estimator. If $\widehat{X}_\pi$ and $\widehat{Y}_\pi$ designate respectively the Horvitz-Thompson estimators of the totals of variables x and y, then we can construct

- the difference estimator:
$$\widehat{Y}_D = \widehat{Y}_\pi + X - \widehat{X}_\pi,$$

- the ratio estimator:
$$\widehat{Y}_R = \widehat{Y}_\pi \frac{X}{\widehat{X}_\pi},$$

- the regression estimator:
$$\widehat{Y}_{\mathrm{reg}} = \widehat{Y}_\pi + (X - \widehat{X}_\pi)\hat{b},$$

where $\hat{b}$ is an estimator of the affine regression coefficient of y over x:

$$b = \frac{S_{xy}}{S_x^2},$$

and

$$S_{xy} = \frac{1}{N-1} \sum_{k \in U} (x_k - \overline{X})(y_k - \overline{Y}).$$

We can choose, to estimate b:

$$\hat{b} = \frac{\displaystyle\sum_{k \in S} \frac{1}{\pi_k} \left(x_k - \frac{\widehat{X}_\pi}{\widehat{N}_\pi}\right)\left(y_k - \frac{\widehat{Y}_\pi}{\widehat{N}_\pi}\right)}{\displaystyle\sum_{k \in S} \frac{1}{\pi_k} \left(x_k - \frac{\widehat{X}_\pi}{\widehat{N}_\pi}\right)^2}.$$

All of these estimators satisfy a fundamental property of calibration, as they estimate with null variance the total X (we are speaking about estimators calibrated on x):

$$\widehat{X}_D = \widehat{X}_R = \widehat{X}_{\text{reg}} = X.$$

We can show that:

- $\text{var}(\widehat{Y}_D) = \text{var}\left(\sum_{k \in S} \frac{y_k - x_k}{\pi_k}\right),$

- $\text{var}(\widehat{Y}_R) \approx \text{var}\left(\sum_{k \in S} \frac{y_k - \dfrac{Y}{X}x_k}{\pi_k}\right)$ (n 'large enough'),

- $\text{var}(\widehat{Y}_{\text{reg}}) \approx \text{var}\left(\sum_{k \in S} \frac{(y_k - \overline{Y}) - b(x_k - \overline{X})}{\pi_k}\right)$ (n 'large enough'),

which comes back to using the general expressions of Chapter 3 with new individual variables. Thus, with simple random sampling, we estimate these variances with:

$$N^2 \left(1 - \frac{n}{N}\right) \frac{1}{n} \frac{1}{n-1} \sum_{k \in S} (y_k - \alpha - \beta x_k)^2,$$

by holding:

- $\alpha = 0, \beta = 1$ with $\widehat{Y}_D$;

- $\alpha = 0, \beta = \dfrac{\widehat{\overline{Y}}}{\widehat{\overline{X}}}$ with $\widehat{Y}_R$;

- $\alpha = \widehat{\overline{Y}} - \widehat{b}\,\widehat{\overline{X}}, \beta = \widehat{b} = \dfrac{\displaystyle\sum_{k \in S}(x_k - \widehat{\overline{X}})(y_k - \widehat{\overline{Y}})}{\displaystyle\sum_{k \in S}(x_k - \widehat{\overline{X}})^2}$ with $\widehat{Y}_{\text{reg}}$.

EXERCISES

Exercise 6.1 *Ratio*

In a population of 10 000 businesses, we want to estimate the average sales $\overline{Y}$. For that, we sample $n = 100$ businesses using simple random sampling. Furthermore, we have at our disposal the auxiliary information 'number of employees', denoted by x, for each business. The data coming from the sample are:

- $\overline{X} = 50$ employees (true mean for x_k),
- $\widehat{\overline{Y}} = 5.2 \times 10^6$ Euros (average sales in the sample),

- $\overline{\widehat{X}} = 45$ employees (sample mean),
- $s_y^2 = 25 \times 10^{10}$ (corrected sample variance of y_k),
- $s_x^2 = 15$ (corrected sample variance of x_k),
- $\widehat{\rho} = 0.80$ (linear correlation coefficient between x and y calculated in the sample).

1. What is the ratio estimator? (We denote this as $\widehat{\overline{Y}}_R$.) Is this estimator biased?
2. Recall the 'true' variance formula for this estimator.
3. Calculate an estimate of the true variance. Is the variance estimator used biased?
4. Give a 95% confidence interval for $\overline{Y}$.

Solution

1. By definition:

$$\widehat{\overline{Y}}_R = \overline{\widehat{X}} \frac{\widehat{\overline{Y}}}{\overline{\widehat{X}}} = 50 \times \frac{5.2 \times 10^6}{45} \approx 5.8 \times 10^6 \text{ Euros.}$$

We have $\widehat{\overline{Y}}_R > \widehat{\overline{Y}}$ because the sample contains businesses that are on average too small (in terms of employees), and thus with sales that are a little bit too small. *A priori,* the estimator is biased: the $1/n$ term appearing in the bias is null when

$$\frac{S_x}{\overline{X}} = \rho \frac{S_y}{\overline{Y}}.$$

None of the terms of this equality can be estimated without bias, but a calculation of magnitudes (bias $1/n$) compares:

$$\frac{s_x}{\overline{\widehat{X}}} \approx 0.086 \quad \text{and} \quad \widehat{\rho} \frac{s_y}{\overline{\widehat{Y}}} \approx 0.077.$$

Numerically, they are close values, which lets us think that the bias must be very small.

2. For n 'large', we have:

$$\text{var}(\widehat{\overline{Y}}_R) \approx \frac{1-f}{n} S_u^2 = \frac{1-f}{n} [S_y^2 + R^2 S_x^2 - 2RS_{xy}].$$

S_u^2 is the population variance of u_i, where $u_i = y_i - Rx_i$ with $R = \overline{Y}/\overline{X}$.

3. We have

$$\widehat{\text{var}}(\widehat{\overline{Y}}_R) = \frac{1-f}{n} [s_y^2 + \widehat{R}^2 s_x^2 - 2\widehat{R}s_{xy}].$$

Indeed

$$s_{xy} = \widehat{\rho}s_x s_y = 0.8 \times \sqrt{25 \times 10^{10}} \times \sqrt{15} \approx 1\,549\,193,$$

and $f \approx 0$. Therefore

$$\widehat{\text{var}}(\overline{\widehat{Y}}_R)$$
$$= \frac{1}{100} \left[25 \times 10^{10} + \left(\frac{5.2 \times 10^6}{45}\right)^2 \times 15 - 2 \times \frac{5.2 \times 10^6}{45} \times 1\,549\,293 \right]$$
$$= 0.0923 \times 10^{10} \approx (0.03 \times 10^6)^2.$$

This variance estimator is biased (because it is not written as a linear combination of estimators that are themselves unbiased).

4. Since n is large, $\overline{\widehat{Y}}_R$ is going to approximately follow a normal distribution. The estimated confidence interval (at 95%) is:

$$\overline{Y} = 5.8 \times 10^6 \pm 0.06 \times 10^6.$$

Due to the estimates ($\widehat{\text{var}}$ biased, passing on to the root) and hypotheses (normal distribution), the real probability of covering $\overline{Y}$ is not 95% but a percentage that is close to it.

Exercise 6.2 *Post-stratification*

Consider an agricultural region consisting of $N = 2010$ farms. We draw a simple random sample of farms of size $n = 100$. We possess information on the total surface area cultivated for each farm. In particular, we know that there are 1 580 farms of less than 160 hectares (post-stratum 1) and 430 farms of more than 160 hectares (post-stratum 2). We try to estimate the mean surface area of cereals cultivated $\overline{Y}$. Using simple random sampling without replacement (having denoted with the indices 1 and 2 the two post-strata thus defined), we have:

$$n_1 = 70, \quad n_2 = 30, \quad \overline{\widehat{Y}}_1 = 19.40, \quad \overline{\widehat{Y}}_2 = 51.63, \quad s_{y1}^2 = 312, \quad s_{y2}^2 = 922.$$

1. What is the post-stratified estimator $\overline{\widehat{Y}}_{\text{post}}$? Is it different than the simple mean $\overline{\widehat{Y}}$?
2. What is the distribution of n_1? What is its expected value? What is its variance?
3. Give the unbiased estimator of the variance $\widehat{\text{var}}(\overline{\widehat{Y}}_{\text{post}})$ and a 95% confidence interval.

Solution

1. The post-stratified estimator is

$$\widehat{\overline{Y}}_{\text{post}} = \frac{N_1}{N} \widehat{\overline{Y}}_1 + \frac{N_2}{N} \widehat{\overline{Y}}_2,$$

where $\widehat{\overline{Y}}_h$ is the simple mean in post-stratum h $(h = 1, 2)$.

- We know that the post-stratified estimator is unbiased (by assuming that $Pr(n_h = 0) \approx 0$):

$$\text{E}(\widehat{\overline{Y}}_{\text{post}}) = \overline{Y}.$$

- *Numerical application:*
  The post-stratified estimator is:

$$\widehat{\overline{Y}}_{\text{post}} = \frac{1\,580}{2\,010} \times 19.40 + \frac{430}{2\,010} \times 51.63 = 26.30 \text{ hectares.}$$

Furthermore, the simple mean is:

$$\widehat{\overline{Y}} = \frac{n_1}{n} \widehat{\overline{Y}}_1 + \frac{n_2}{n} \widehat{\overline{Y}}_2 = \frac{70}{100} \times 19.40 + \frac{30}{100} \times 51.63$$

$$= 29.07 \text{ hectares} \neq \widehat{\overline{Y}}_{\text{post}}.$$

The adjustment is interpreted as a reweighting method: we go from initial weights equal to $1/n$ (which we find in $\widehat{\overline{Y}}$) to adjusted weights equal to $N_h/(Nn_h)$ for an individual in post-stratum h (which we find in $\widehat{\overline{Y}}_{\text{post}}$).

2. Because the sampling is simple random and without replacement,

$$n_1 \sim \text{hypergeometric } (N, n, P),$$

with

$$P = \frac{N_1}{N}.$$

Therefore,

$$\text{E}(n_1) = nP = n\frac{N_1}{N},$$

$$\text{var}\left(\frac{n_1}{n}\right) = (1 - f) \frac{PQ}{n} \frac{N}{N - 1},$$

where

$$Q = 1 - P = \frac{N_2}{N} \text{ and } f = \frac{n}{N}.$$

This last expression comes from the estimation theory of proportions in the case of simple random sampling. We get:

$$\text{var}(n_1) = n^2 \left(1 - \frac{n}{N}\right) \frac{N}{N - 1} \frac{PQ}{n} = n\frac{N - n}{N - 1} PQ.$$

3. As $n = 100$, we can use the slightly biased estimator as follows:

$$\widehat{\text{var}}(\widehat{\overline{Y}}_{\text{post}}) = \frac{1-f}{n} \sum_{h=1}^{H} \frac{N_h}{N} s_{yh}^2 + \frac{1-f}{n^2} \sum_{h=1}^{H} \left(1 - \frac{N_h}{N}\right) s_{yh}^2.$$

Numerical application:

$$\widehat{\text{var}}(\widehat{\overline{Y}}_{\text{post}}) = \frac{1 - \frac{100}{2\,010}}{100} \left[\frac{1\,580}{2\,010} \times 312 + \frac{430}{2\,010} \times 922\right]$$
$$+ \frac{1 - \frac{100}{2\,010}}{(100)^2} \left[\left(1 - \frac{1\,580}{2\,010}\right) 312 + \left(1 - \frac{430}{2\,010}\right) 922\right]$$
$$\approx 4.205 + 0.075$$
$$\approx 4.28.$$

We notice that the first term of $\widehat{\text{var}}(\widehat{\overline{Y}}_{\text{post}})$ is numerically predominant and that it could have been sufficient for the calculation. Here, $n = 100$, which is 'sufficiently large' to approach the distribution of $\widehat{\overline{Y}}_{\text{post}}$ through a normal distribution. We can therefore construct a confidence interval:

$$\overline{Y} \in [26.30 \pm 1.96 \times \sqrt{4.28}] = [22.25\,;\ 30.35] \quad 95 \text{ times out of } 100.$$

Exercise 6.3 *Ratio and accuracy*

We are placed in the context of Exercise 6.2 but we now exploit the auxiliary variable x (measuring the total surface area cultivated) to construct a ratio estimator. We are given:

$$\overline{X} = 118.32 \text{ hectares}, \quad \widehat{\overline{X}} = 131.25 \text{ hectares}, \quad \widehat{\overline{Y}} = 29.07 \text{ hectares},$$

and

$$s_x^2 = 9\,173, \quad s_y^2 = 708, \quad \widehat{\rho} = 0.57.$$

where $\widehat{\rho}$ is the estimator of the 'true' unknown linear correlation coefficient ρ.

1. Recall the expression of ρ.
2. How do we define $\widehat{\rho}$? Is the estimator $\widehat{\rho}$ biased?
3. Show that the ratio estimator of $\overline{Y}$ appears to be preferable to the simple mean $\widehat{\overline{Y}}$ if and only if:

$$\widehat{\rho} > \frac{1}{2} \frac{\widehat{\text{CV}}(x)}{\widehat{\text{CV}}(y)},$$

where the $\widehat{\text{CV}}$ estimate the coefficients of variation. Do the numerical application.

4. Calculate $\widehat{\overline{Y}}_R$, the ratio estimator of $\overline{Y}$.
5. Estimate its accuracy, and give a 95% confidence interval.

Solution

1. The correlation coefficient is:

$$\rho = \frac{\sum_{k \in U}(x_k - \overline{X})\,(y_k - \overline{Y})}{\sqrt{\sum_{k \in U}(x_k - \overline{X})^2}\,\sqrt{\sum_{k \in U}(y_k - \overline{Y})^2}}.$$

2. With the classical notations, we have:

$$\widehat{\rho} = \frac{\sum_{k \in S}(x_k - \widehat{\overline{X}})(y_k - \widehat{\overline{Y}})}{\sqrt{\sum_{k \in S}(x_k - \widehat{\overline{X}})^2}\sqrt{\sum_{k \in S}(y_k - \widehat{\overline{Y}})^2}}.$$

Obviously $E(\widehat{\rho}) \neq \rho$, since $\widehat{\rho}$ is a complex ratio. The denominator of $\widehat{\rho}$ is for that matter not even an unbiased estimator of the denominator of ρ, since it is a product of square roots.

3. The ratio estimator is

$$\widehat{\overline{Y}}_R = \overline{X}\frac{\widehat{\overline{Y}}}{\widehat{\overline{X}}},$$

and its estimated variance

$$\widehat{\mathrm{var}}(\widehat{\overline{Y}}_R) = \frac{1-f}{n}\,s^2_{y - \widehat{R}x} = \frac{1-f}{n}\,[s^2_y + \widehat{R}^2 s^2_x - 2\widehat{R}\,s_{xy}],$$

where $\widehat{R} = \widehat{\overline{Y}}/\widehat{\overline{X}}$.

- Note: $s^2_{y - \widehat{R}x} = s^2_{\widehat{u}}$ is the sample variance of the estimated residuals, which is:

$$\widehat{u}_k = y_k - \widehat{R}x_k.$$

Furthermore,

$$\widehat{\mathrm{var}}(\widehat{\overline{Y}}) = \frac{1-f}{n}\,s^2_y.$$

Therefore,

$$\widehat{\mathrm{var}}(\widehat{\overline{Y}}_R) < \widehat{\mathrm{var}}(\widehat{\overline{Y}}) \Leftrightarrow s^2_y + \widehat{R}^2 s^2_x - 2\widehat{R}s_{xy} < s^2_y$$

$$\Leftrightarrow \widehat{R}s^2_x < 2s_{xy} = 2\widehat{\rho}s_x s_y$$

$$\Leftrightarrow \widehat{R}s_x < 2\widehat{\rho}s_y$$

$$\Leftrightarrow \frac{\widehat{\overline{Y}}}{\widehat{\overline{X}}}\,s_x < 2\widehat{\rho}s_y$$

$$\Leftrightarrow \widehat{\rho} > \frac{1}{2}\frac{\widehat{\mathrm{CV}}(x)}{\widehat{\mathrm{CV}}(y)}.$$

Attention: This comparison is made, in practice, on variance estimates and not on true values of variance. It is therefore not totally 'assured'.

- *Numerical application:*

$$
\begin{cases}
\widehat{\rho} = 0.57 \\[2mm]
\widehat{CV}(y) = \dfrac{s_y}{\overline{\widehat{Y}}} = \dfrac{\sqrt{708}}{29.07} \approx 91.5\% \\[3mm]
\widehat{CV}(x) = \dfrac{s_x}{\overline{\widehat{X}}} = \dfrac{\sqrt{9\,173}}{131.25} \approx 73\%,
\end{cases}
$$

which gives

$$
\frac{1}{2}\,\frac{\widehat{CV}(x)}{\widehat{CV}(y)} \approx 0.40 < 0.57 = \widehat{\rho}.
$$

The estimator $\overline{\widehat{Y}}_R$ effectively appears to be better than $\overline{\widehat{Y}}$.

4. The ratio estimator is

$$
\overline{\widehat{Y}}_R = \overline{X}\frac{\overline{\widehat{Y}}}{\overline{\widehat{X}}} = 118.32 \times \frac{29.07}{131.25} = 26.21 \text{ hectares.}
$$

5. Seeing as n is 'sufficiently large', we estimate the variance of the estimator by

$$
\begin{aligned}
\widehat{\mathrm{var}}(\overline{\widehat{Y}}_R) &= \frac{1-f}{n}\,[s_y^2 + \widehat{R}^2 s_x^2 - 2\widehat{R}\widehat{\rho}s_x s_y] \\[2mm]
&= \frac{1 - \frac{100}{2010}}{100}\left[708 + \left(\frac{29.07}{131.25}\right)^2 9173 \right. \\[2mm]
&\qquad\left. -2 \times \left(\frac{29.07}{131.25}\right)0.57 \times \sqrt{708 \times 9173}\right] \\[2mm]
&\approx 4.90.
\end{aligned}
$$

The sample size (here $n = 100$) is sufficient in order that we liken the distribution of $\overline{\widehat{Y}}_R$ to a normal distribution. Therefore, the estimated confidence interval is

$$
\overline{Y} \in [26.21 \pm 1.96\ \sqrt{4.90}] = [21.88\,;\ 30.54] \text{ 95 times out of 100.}
$$

Note: With this data, the accuracy of the ratio estimator is a little worse than that for the post-stratified estimator (see Exercise 6.2). It is not necessary to select this in general, as we cannot say that the ratio estimator is systematically preferable to the post-stratified estimator.

Exercise 6.4 *Comparison of estimators*

We propose to estimate the mean $\overline{Y}$ of a characteristic y by way of a sample selected according to a simple random design without replacement of size 1000 in a population of size 1 000 000. We know the mean $\overline{X} = 15$ of an auxiliary characteristic x. We have the following results:

$$s_y^2 = 20, s_x^2 = 25, s_{xy} = 15, \widehat{\overline{X}} = 14, \widehat{\overline{Y}} = 10.$$

1. Estimate $\overline{Y}$ by way of the Horvitz-Thompson, difference, ratio and regression estimators. Estimate the variances of these estimators.
2. Which estimator should we choose to estimate $\overline{Y}$?

Solution

1. a) The Horvitz-Thompson estimator is $\widehat{\overline{Y}}_\pi = \widehat{\overline{Y}} = 10$ and the estimator of its variance is given by

$$\widehat{\mathrm{var}}\left(\widehat{\overline{Y}}_\pi\right) = \frac{N - n}{Nn}s_y^2 = \frac{1\ 000\ 000 - 1000}{1\ 000\ 000 \times 1000} \times 20 \approx 0.020.$$

b) The difference estimator is given by

$$\widehat{\overline{Y}}_D = \widehat{\overline{Y}} + \overline{X} - \widehat{\overline{X}} = 10 + 15 - 14 = 11.$$

Its estimated variance is

$$\widehat{\mathrm{var}}\left(\widehat{\overline{Y}}_D\right) = \frac{N - n}{Nn}\left\{s_y^2 - 2s_{xy} + s_x^2\right\}$$
$$= \frac{1\ 000\ 000 - 1000}{1\ 000\ 000 \times 1000} \times \{20 - 2 \times 15 + 25\}$$
$$\approx 0.015.$$

c) The ratio estimator is given by

$$\widehat{\overline{Y}}_R = \frac{\widehat{\overline{Y}}\ \overline{X}}{\widehat{\overline{X}}} = \frac{10 \times 15}{14} = 10.71.$$

Its variance is comparable to its mean square error (MSE), given the large sample size (the true variance varies by $1/n$, the square of the bias by $1/n^2$):

$$\widehat{\mathrm{MSE}}\left(\widehat{\overline{Y}}_R\right) \approx \frac{N - n}{Nn}\left\{s_y^2 - 2\frac{\widehat{\overline{Y}}}{\widehat{\overline{X}}}s_{xy} + \frac{\widehat{\overline{Y}}^2}{\widehat{\overline{X}}^2}s_x^2\right\}$$
$$= \frac{1\ 000\ 000 - 1000}{1\ 000\ 000 \times 1000} \times \left\{20 - 2 \times \frac{10}{14} \times 15 + \frac{10^2}{14^2} \times 25\right\}$$
$$\approx 0.0113.$$

d) The regression estimator is given by

$$\widehat{\overline{Y}}_{\text{reg}} = \widehat{\overline{Y}} + \frac{s_{xy}}{s_x^2}(\overline{X} - \widehat{\overline{X}}) = 10 + \frac{15}{25}(15 - 14) = 10.6.$$

Its estimated variance is approximately equal to its estimated MSE:

$$\widehat{\text{MSE}}\left(\widehat{\overline{Y}}_{\text{reg}}\right) \approx \frac{N-n}{Nn}s_y^2(1 - \hat{\rho}^2),$$

where $\hat{\rho}^2 = s_{xy}^2/s_x^2 s_y^2 = 15^2/20 \times 25 = 0.45$ represents the square of the linear correlation coefficient between x and y in the sample

$$\widehat{\text{MSE}}\left(\widehat{\overline{Y}}_{\text{reg}}\right) \approx 0.0110.$$

2. The smallest variance estimated is that for the regression estimator, which is expected given the large sample size. Nevertheless, the relationship between x and y is strongly linear: the regression line passes close to the origin, so that the ratio estimator appears (almost) as effective as the regression estimator.

Exercise 6.5 *Foot size*

The director of a business that makes shoes wants to estimate the average length of right feet of adult men in a city. Let y be the characteristic 'length of right foot' (in centimetres) and x be the height of the individual (in centimetres). The director knows moreover from the results of a census that the average height of adult men in this city is 168 cm. To estimate the foot length, the director draws a simple random sample without replacement of 100 adult men. The results are the following:

$$\widehat{\overline{X}} = 169, \widehat{\overline{Y}} = 24, s_{xy} = 15, s_x = 10, s_y = 2.$$

Knowing that 400 000 adult men live in this city,

1. Calculate the Horvitz-Thompson estimator, the ratio estimator, the difference estimator and the regression estimator.
2. Estimate the variances of these four estimators.
3. Which estimator would you recommend to the director?
4. Express the literal difference between the estimated variance of the ratio estimator and the estimated variance of the regression estimator, as a function of $\overline{X}$, $\widehat{\overline{Y}}$ and the slope $\hat{b}$ of the regression of y on x in the sample. Comment on this.

Solution

1. The estimator $\widehat{\overline{Y}}$ is the Horvitz-Thompson estimator, being

$$\widehat{\overline{Y}} = 24 \text{ cm.}$$

Furthermore,

$$\widehat{\overline{Y}}_R = \frac{\widehat{\overline{Y}}\,\overline{X}}{\widehat{\overline{X}}} = \frac{24 \times 168}{169} = 23.86 \text{ cm,}$$

$$\widehat{\overline{Y}}_D = \widehat{\overline{Y}} + \overline{X} - \widehat{\overline{X}} = 24 + 168 - 169 = 23 \text{ cm,}$$

$$\widehat{\overline{Y}}_{\text{reg}} = \widehat{\overline{Y}} + \frac{s_{xy}}{s_x^2}(\overline{X} - \widehat{\overline{X}}) = 24 + \frac{15}{10^2}(168 - 169) = 23.85 \text{ cm.}$$

2. The variance estimators are:

$$\widehat{\text{var}}\left(\widehat{\overline{Y}}\right) = \frac{N-n}{Nn}s_y^2 = \frac{400000 - 100}{400000 \times 100}2^2 = 0.0399,$$

$$\widehat{\text{var}}\left(\widehat{\overline{Y}}_R\right) = \frac{N-n}{Nn}\left(s_y^2 - 2\frac{\widehat{\overline{Y}}}{\widehat{\overline{X}}}s_{xy} + \frac{\widehat{\overline{Y}}^2}{\widehat{\overline{X}}^2}s_x^2\right)$$

$$= \frac{400000 - 100}{400000 \times 100}\left(2^2 - 2\frac{24}{169} \times 15 + \frac{24^2}{169^2} \times 10^2\right)$$

$$= 0.0176.$$

We verify that

$$\widehat{\text{var}}\left(\widehat{\overline{Y}}_R\right) < \widehat{\text{var}}\left(\widehat{\overline{Y}}\right) \Leftrightarrow \frac{\widehat{\overline{Y}}}{\widehat{\overline{X}}} < 2\frac{s_{xy}}{s_x^2}.$$

Here,

$$\frac{\widehat{\overline{Y}}}{\widehat{\overline{X}}} = \frac{24}{169} = 0.142 < 2 \times \frac{15}{100} = 0.3,$$

$$\widehat{\text{var}}\left(\widehat{\overline{Y}}_D\right) = \frac{N-n}{Nn}\left(s_y^2 - 2s_{xy} + s_x^2\right)$$

$$= \frac{400000 - 100}{400000 \times 100}\left(2^2 - 2 \times 15 + 10^2\right)$$

$$= 0.7398,$$

$$\widehat{\text{var}}\left(\widehat{\overline{Y}}_{\text{reg}}\right) = \frac{N-n}{Nn}s_y^2(1 - \widehat{\rho}^2), \text{ avec } \widehat{\rho}^2 = \frac{s_{xy}^2}{s_x^2 s_y^2} = 0.5625$$

$$= \frac{400000 - 100}{400000 \times 100} \times 2^2 \times (1 - 0.5625)$$

$$= 0.0175.$$

3. We recommend the ratio estimator which has a variance distinctly smaller than the two other estimators and is identical to that of the regression estimator, but is simpler to use compared to the latter.

4. The variance estimators are

$$\widehat{\mathrm{var}}\left(\widehat{\overline{Y}}_R\right) = \frac{N-n}{Nn}(s_y^2 + \widehat{R}^2 s_x^2 - 2\widehat{R}s_{xy}), \text{ with } \widehat{R} = \frac{\widehat{\overline{Y}}}{\widehat{\overline{X}}},$$

and

$$\widehat{\mathrm{var}}\left(\widehat{\overline{Y}}_{\mathrm{reg}}\right) \approx \frac{N-n}{Nn}(1 - \widehat{\rho}^2)s_y^2 \text{ with } \widehat{\rho}^2 = \frac{s_{xy}^2}{s_x^2 s_y^2}.$$

Thus:

$$D = \left(\frac{N-n}{Nn}\right)^{-1}\left(\widehat{\mathrm{var}}(\widehat{\overline{Y}}_R) - \widehat{\mathrm{var}}(\widehat{\overline{Y}}_{\mathrm{reg}})\right) = s_y^2 + \widehat{R}^2 s_x^2 - 2\widehat{R}s_{xy} - s_y^2 + \frac{s_{xy}^2}{s_x^2}.$$

If we denote $\widehat{b} = s_{xy}/s_x^2$, the slope of the regression line of y on x in the sample,

$$D = \widehat{R}^2 s_x^2 - 2\widehat{R}\widehat{b}s_x^2 + \widehat{b}^2 s_x^2 = s_x^2(\widehat{R} - \widehat{b})^2.$$

Therefore,

$$\widehat{\mathrm{var}}(\widehat{\overline{Y}}_R) - \widehat{\mathrm{var}}(\widehat{\overline{Y}}_{\mathrm{reg}}) = \left(\frac{N-n}{nN}s_x^2\right)(\widehat{R} - \widehat{b})^2 = \left(\frac{\widehat{\overline{Y}}}{\widehat{\overline{X}}} - \frac{s_{xy}}{s_x^2}\right)^2 \widehat{\mathrm{var}}(\widehat{\overline{X}}_\pi) \geq 0.$$

The difference between the accuracies depends on the difference between the slopes of the regression lines, going through the origin or not. In the previous numerical example, we have:

$$\frac{\widehat{\overline{Y}}}{\widehat{\overline{X}}} = 0.142 \quad \text{and} \quad \frac{s_{xy}}{s_x^2} = \frac{15}{100} = 0.150.$$

The gap between the two slopes is very small: from this fact, the regression estimator hardly provides anything more than the ratio estimator.

Exercise 6.6 *Cavities and post-stratification*

Two dentists conduct a survey on the condition of teeth of 200 children in a village. The first dentist selects using simple random sampling 20 children among the 200, and counts the data in the sample according to the number of teeth with cavities. The results are presented in Table 6.1. The second dentist examines the 200 children but with the sole goal of determining who has no cavities. He notices that 50 children are in this category.

Table 6.1. Teeth with cavities: Exercise 6.6

Number of teeth with cavities	0 1 2 3 4 5 6 7 8
Number of children	8 4 2 2 1 2 0 0 1

1. Estimate the mean number of teeth with cavities per child in the village using only the results of the first dentist. What is the accuracy of the unbiased estimator obtained? Estimate this accuracy and the associated confidence interval.
2. Propose another estimator for the mean number of teeth with cavities per child using the results of the two dentists. Calculate the new estimate, and estimate the gain in efficiency obtained.
3. Find a reason showing whether or not post-stratification is appropriate: it can end up *in fine* in comparing the survey unit cost α of the first dentist with the survey unit cost β of the second dentist.

Solution

1. Since it is a simple random sample, if we denote y_k as the number of teeth with cavities for child k, we use

$$\widehat{\overline{Y}} = \frac{1}{n} \sum_{k \in S} y_k$$

$$= \frac{1}{20}(0 \times 8 + 1 \times 4 + 2 \times 2 + 3 \times 2 + 4 \times 1 + 5 \times 2 + +8 \times 1)$$

$$= \frac{36}{20} \approx 1.8.$$

We have

$$\text{var}\left(\widehat{\overline{Y}}\right) = \frac{N-n}{Nn} S_y^2, \text{ and } \widehat{\text{var}}\left(\widehat{\overline{Y}}\right) = \frac{N-n}{Nn} s_y^2,$$

$$s_y^2 = \frac{1}{n-1} \sum_{k \in S} \left(y_k - \widehat{\overline{Y}}\right)^2$$

$$= \frac{1}{n-1} \sum_{k \in S} y_k^2 - \frac{n}{n-1} \left(\widehat{\overline{Y}}\right)^2$$

$$= \frac{1}{19}(0 + 1 \times 4 + 4 \times 2 + 9 \times 2 + 16 + 25 \times 2 + 64) - \frac{20}{19}1.8^2$$

$$= 5.0105,$$

$$\widehat{\text{var}}\left(\widehat{\overline{Y}}\right) = \frac{N-n}{Nn} s_y^2 = \left(1 - \frac{20}{200}\right) \frac{s_y^2}{20} = 0.2255,$$

$$\overline{Y} \in \left[\widehat{\overline{Y}} \pm 1.96 \sqrt{\widehat{\text{var}}\left(\widehat{\overline{Y}}\right)} \right] = [0.87; 2.73].$$

The accuracy is mediocre, but the sample size is very small. We are under the limits of the utilisation conditions of the normal distribution for $\overline{Y}$: it is highly likely that the true probability of covering the interval that we have just calculated is noticeably different from 0.95.

2. We can post-stratify: post-stratum 1 contains the children who have no cavities (size N_1) and post-stratum 2 contains the children who have at least one cavity (size N_2).

$$\widehat{\overline{Y}}_{\text{post}} = \frac{1}{N} \left(N_1 \widehat{\overline{Y}}_1 + N_2 \widehat{\overline{Y}}_2 \right) = \frac{N_2}{N} \widehat{\overline{Y}}_2.$$

We see that the post-stratified estimator is equal to the ratio estimator constructed from the auxiliary variable which is 1 for post-stratum 2 and 0 for post-stratum 1. Since we have $N_1 = 50, N_2 = 150, \overline{Y}_2 = 36/12 = 3$, we get

$$\widehat{\overline{Y}}_{\text{post}} = 3 \times \frac{150}{200} = 2.25.$$

The variance is

$$\text{var}(\widehat{\overline{Y}}_{\text{post}}) \approx \frac{N-n}{nN^2} \sum_{h=1}^{H} N_h S_{yh}^2 + \frac{(N-n)}{n^2 N} \sum_{h=1}^{H} \frac{N - N_h}{N} S_{yh}^2,$$

and is estimated by

$$\widehat{\text{var}}(\widehat{\overline{Y}}_{\text{post}}) = \frac{N-n}{nN^2} \sum_{h=1}^{H} N_h s_{yh}^2 + \frac{(N-n)}{n^2 N} \sum_{h=1}^{H} \frac{N - N_h}{N} s_{yh}^2.$$

As $S_{y1}^2 = s_{y1}^2 = 0$, we have

$$\widehat{\text{var}}(\widehat{\overline{Y}}_{\text{post}}) = \frac{N-n}{nN^2} N_2 s_{y2}^2 + \frac{(N-n)}{n^2 N} \frac{N - N_2}{N} s_{y2}^2.$$

After a few calculations, we obtain

$$s_{y2}^2 = 4.7273,$$

$$\frac{N-n}{N^2 n} N_2 = 0.03375,$$

$$\frac{N-n}{n^2 N} \frac{N - N_2}{N} = 0.00056,$$

thus $\widehat{\text{var}}(\widehat{\overline{Y}}_{\text{post}}) = 0.1622$ and therefore, with a 95% probability,

$$\overline{Y} \in \left[\widehat{\overline{Y}}_{\text{post}} \pm 1.96\sqrt{\widehat{\text{var}}\left(\widehat{\overline{Y}}_{\text{post}}\right)}\right] = [2.25 \pm 0.79] = [1.46; 3.04].$$

The gain due to the post-stratification, measurable by:

$$\frac{\widehat{\text{var}}\left(\widehat{\overline{Y}}_{\text{post}}\right)}{\widehat{\text{var}}\left(\widehat{\overline{Y}}\right)} = \frac{0.1622}{0.2255} = 72\% = (0.85)^2$$

is thus not very large since the length of the confidence interval is reduced by 15%: the question henceforth consists in knowing if the cost related to the contribution of the second dentist is or is not made up for by the reduction by 15% of the length of the confidence interval of $\overline{Y}$.

3. If we neglect the numerical value of the second term of the variance of $\widehat{\overline{Y}}_{\text{post}}$ (in 2., it is 60 times smaller than the first term), and if we hold on to the small sample sizes compared to N, the standard deviation of $\widehat{\overline{Y}}_{\text{post}}$ varies by the inverse of $\sqrt{n}$, where n is the sample size examined by the first dentist:

$$\sigma_p(n) \approx \frac{\delta}{\sqrt{n}} \quad (\delta \text{ is a complex expression}).$$

The total cost of the process is $n\alpha + 200\beta$ (since the second dentist must examine the 200 children). If we choose not to post-stratify, the first dentist interviews 20 children and the accuracy obtained is (still n negligible compared to N):

$$\sigma(\widehat{\overline{Y}}) = \frac{\delta'}{\sqrt{20}}.$$

Since under these conditions δ and δ' do not (nearly) depend on n, we can also write:

$$\frac{\delta}{\sqrt{20}} = \frac{\delta'}{\sqrt{20}} \times 0.85, \text{ according to 2.}$$

In this case, the cost is $20\,\alpha$.

To make comparisons between the two methods (and to thus determine which is the most worthwhile) we are going to think with constant accuracy, being for example the accuracy obtained with simple random sampling of 20 children without post-stratification, which serves as a reference situation. To attain this accuracy, it would be necessary with post-stratification that the first dentist examines n_0 children, by setting:

$$\sigma_p(n_0) = \frac{\delta}{\sqrt{n_0}} = \frac{\delta'}{\sqrt{n_0}} \times 0.85 = \frac{\delta'}{\sqrt{20}},$$

thus:

$$n_0 = 20 \times (0.85)^2 \approx 14.45, \text{ rounded to 14 or 15.}$$

With this sample size, the cost is $n_0\alpha + 200\beta$, while the cost without post-stratification is 20α. The accuracy being fixed, post-stratification is therefore worthwhile if and only if:

$$n_0\alpha + 200\beta < 20\alpha \quad \text{with} \quad n_0 = 14 \quad \text{or} \quad n_0 = 15.$$

Thus,

$$\alpha > 33\beta \quad (n_0 = 14) \quad \text{or} \quad \alpha > 40\beta \quad (n_0 = 15).$$

In the first approximation, we can therefore conclude that post-stratification is most likely worthwhile if the hourly rate of the second dentist is at least 40 times less than that of the first dentist (who works more, since he must count the cavities).

Exercise 6.7 *Votes and difference estimation*

A television channel enters into a contract with a survey institute for the next election. This institute is in charge of providing, on election night, a first estimate at 8 o'clock (the definitive results not being known until two hours later). The methodology put into place can be described in the following way. The population considered is that of polling stations. We denote N as the number of polling stations (the statistical unit is therefore not the individual voting but the polling station). The objective is to estimate at a national level the percentage of votes for a political party A. We consider that the polling stations are comprised in a manner of grouping the same number of voters. We select, according to a certain method, a sample (denoted S) of polling stations. At 7:50, we have available the exact percentage of votes y_k obtained by party A at polling station k (for each k in S). Furthermore, we have available auxiliary information defined by:

- x_k, the percentage obtained by A at polling station k at the time of the previous election,
- X, the percentage obtained by A at the national level at the time of the previous election.

Preliminary: How do we simply write the desired percentage as a function of y_k, $k = 1, ..., N$?

1. a) We assume that we select a single polling station, denoted k. We propose to estimate $\overline{Y}$ by:

$$y_k + (\overline{X} - x_k).$$

Under what condition (in terms of political behaviour) does this estimator seem to have to be better than the 'naive' estimator y_k? Justify in an intuitive manner.

b) We select polling stations according to simple random sampling of size n. We propose to estimate $\overline{Y}$ with:

$$\widehat{\overline{Y}}_D = \frac{1}{n} \sum_{k \in S} \{y_k + (\overline{X} - x_k)\} = \widehat{\overline{Y}} + \left(\overline{X} - \widehat{\overline{X}}\right).$$

- Show that $\widehat{\overline{Y}}_D$ estimates $\overline{Y}$ without bias.
- Calculate the variance of $\widehat{\overline{Y}}_D$ (we will put it under the form

$$\left(1 - \frac{n}{N}\right)\frac{D}{n},$$

where D is a corrected population variance).
- Give a simple condition, necessary and sufficient, with respect to the slope of the regression line of y_k on x_k so that $\widehat{\overline{Y}}_D$ is better than the Horvitz-Thompson estimator $\widehat{\overline{Y}}_\pi$.

2. We wish to improve the estimator $\widehat{\overline{Y}}_D$ by acting on the sampling phase. We propose to create strata of polling stations according to the prevailing political party of these polling stations at the time of the previous election. For example: Stratum 1 = 'extreme right', Stratum 2 = 'right', Stratum 3 = 'left', Stratum 4 = 'extreme left'.

a) Does this stratification appear to you to be of good judgement?

b) We draw a stratified sample of fixed size n with a simple random design without replacement of fixed size in each stratum. We denote n_h as the sample size in stratum U_h, $h = 1, ..., H$, and N_h as the size of U_h.

- Propose an unbiased estimator $\widehat{\overline{Y}}_N$ of $\overline{Y}$ using the auxiliary information x_k (at the estimation stage). We suppose that the percentage obtained by A in stratum U_h at the time of the previous election is known (we denote this $\overline{X}_h$).
- Express its variance.

c) We are trying to get a constant accuracy (in terms of variance) for the estimators of $\overline{Y}_h$, for all $h = 1, ..., H$.

- Explain the functional relationship g_h such that $n_h = g_h(e)$ where e indicates the fixed constant accuracy.
- Deduce that there exists a sole allocation guaranteeing the equality of variances of the estimators of $\overline{Y}_h$ for a fixed size n of the final sample.

d) Give the optimal allocation assuring the best accuracy for $\widehat{\overline{Y}}_N$ (without any calculation).

Solution

Preliminary

We are in the situation where all the polling stations have the same number of voters: the national percentage desired is then (see the theory of cluster sampling with clusters of constant size)

$$\overline{Y} = \frac{1}{N} \sum_{k \in U} y_k.$$

It is therefore a question of estimating a simple mean $\overline{Y}$.

1. a) If $\overline{X} - x_k$ is positive, that signifies that at the time of the last election polling station k underestimated the percentage of voters at the national level. Under a hypothesis of stability in the political voting structure at the time ($x_k < \overline{X}$ therefore $y_k < \overline{Y}$), then intuitively $y_k + (\overline{X} - x_k)$ would have to be better than the naive estimator y_k which probably underestimates $\overline{Y}$, since we add a positive corrective term. For example, if $\overline{X} = 25\%$ and $x_k = 22\%$, station k is '3 points' below the national average. If we consider that the gap remains at the time, we are naturally adding three points to y_k as compensation, to get nearer to $\overline{Y}$.

   b) The estimator $\widehat{\overline{Y}}_D$ is what we call a difference estimator

   $$E(\widehat{\overline{Y}}_D) = E\left(\frac{1}{n} \sum_{k \in S} y_k\right) + \overline{X} - E\left(\frac{1}{n} \sum_{k \in S} x_k\right) = \overline{Y} + \overline{X} - \overline{X} = \overline{Y}.$$

   If we denote $z_k = y_k - x_k$, we have

   $$\widehat{\overline{Y}}_D = \widehat{\overline{Z}} + \overline{X},$$

   where

   $$\widehat{\overline{Z}} = \frac{1}{n} \sum_{k \in S} z_k.$$

   Therefore

   $$\text{var}\left(\widehat{\overline{Y}}_D\right) = \text{var}\left(\widehat{\overline{Z}}\right) = \frac{N-n}{Nn} S_z^2,$$

   where

   $$S_z^2 = \frac{1}{N-1} \sum_{k \in U} \left(z_k - \overline{Z}\right)^2,$$

   and

   $$\overline{Z} = \frac{1}{N} \sum_{k \in U} z_k.$$

   In order for $\text{var}\left(\widehat{\overline{Y}}_D\right)$ to be null, it is necessary and sufficient that $S_z^2 = 0$, which is obtained when z_k is constant, that is to say $z_k = C$. It is therefore necessary and sufficient that

$$y_k = x_k + C.$$

This model indeed corresponds to the intuitive idea from 1.(a) pushed to its extreme: it expresses the perfect stability of the political structure at the time, which is to say that the interest in party A can develop, but in the same way at all the polling stations. In order that

$$\text{var}\left(\widehat{\overline{Y}}_D\right) \le \text{var}\left(\widehat{\overline{Y}}_\pi\right),$$

it is necessary and sufficient that $S_z^2 \le S_y^2$. Indeed

$$S_z^2 = S_x^2 + S_y^2 - 2S_{xy}.$$

It is necessary and sufficient that $S_x^2 + S_y^2 - 2S_{xy} \le S_y^2$, which is to say that

$$\frac{S_{xy}}{S_x^2} \ge \frac{1}{2}.$$

It is therefore necessary and sufficient that the slope of the affine regression line of y_k on x_k in the population is larger than $1/2$. This threshold quantifies and specifies what can be a 'certain' political stability at the time.

2. a) This stratification must be good to estimate the voting proportions for the parties situated in a 'left-right' dimension. It is however not very relevant for a party that is not situated in this dimension. Also, in this case, a better solution would be to stratify into classes according to the x_k which are known at the time of the previous election, where x_k is the percentage of votes relative to the party in question.

   b) We are going to set:

$$\widehat{\overline{Y}}_N = \sum_{h=1}^H \frac{N_h}{N} \widehat{\overline{Y}}_{Dh},$$

with

$$\widehat{\overline{Y}}_{Dh} = \frac{1}{n_h} \sum_{k \in S_h} (y_k + \overline{X}_h - x_k),$$

where S_h indicates the sample obtained in stratum h. We have

$$\text{E}\left(\widehat{\overline{Y}}_{Dh}\right) = \overline{Y}_h \text{ see 1.(b), thus } \text{E}\left(\widehat{\overline{Y}}_N\right) = \sum_{h=1}^H \frac{N_h}{N} \overline{Y}_h = \overline{Y},$$

and

$$\text{var}\left(\widehat{\overline{Y}}_N\right) = \sum_{h=1}^H \frac{N_h^2}{N^2} \frac{N_h - n_h}{N_h n_h} S_{zh}^2,$$

where S_{zh}^2 is the corrected population variance of z_k in stratum U_h (see 1.(b) for the definition of z_k).

c) To get a constant accuracy e in each stratum, it is necessary and sufficient that

$$\frac{N_h - n_h}{N_h n_h} S_{zh}^2 = e,$$

for all h, which is equivalent to

$$n_h = \frac{N_h S_{zh}^2}{N_h e + S_{zh}^2} = g_h(e).$$

We verify that $g_h(e)$ is decreasing and continues on $\mathbb{R}^+$,

$$g_h(0) = N_h, g_h(+\infty) = 0.$$

The sample size can be written

$$n = \sum_{h=1}^{H} g_h(e) = g(e).$$

The function $g(e)$ is also decreasing and continues on $\mathbb{R}^+$,

$$g(0) = N, g(+\infty) = 0.$$

For n fixed, there thus exists e^* unique such that $n = g(e^*)$, which allows to set, for all h, $n_h = g_h(e^*)$. Thus, there is existence and uniqueness of n_h assuring a given 'local' variance e^* subject to the fixed sample size n.

d) It is sufficient to remark that $\widehat{\overline{Y}}_N = \overline{X} + \widehat{\overline{Z}}_{\text{strat}}$, where

$$\widehat{\overline{Z}}_{\text{strat}} = \sum_{h=1}^{H} \frac{N_h}{N} \widehat{\overline{Z}}_h \text{ with } \widehat{\overline{Z}}_h = \frac{1}{n_h} \sum_{k \in S_h} z_k.$$

To the additive constant near $\overline{X}$, it is the classical stratified estimator obtained with the variable z_k. Thus

$$\min_{n_1,\dots,n_H} \text{var}(\widehat{\overline{Y}}_N) = \min_{n_1,\dots,n_H} \text{var}(\widehat{\overline{Z}}_{\text{strat}}),$$

subject to

$$\sum_{h=1}^{H} n_h = n.$$

The optimal allocation is the famous Neyman optimal allocation:

$$n_h = \frac{n S_{zh} N_h}{\sum_{\ell=1}^{H} S_{z\ell} N_\ell} \text{ (if } n_h \le N_h).$$

This is the allocation optimising the quality of a global estimator $\widehat{\overline{Y}}_N$: it differs from the allocation of (c) which completely had another objective.

Exercise 6.8 *Combination of ratios*

In a population of size N, we have available two quantitative auxiliary variables x_1 and x_2. We are interested in estimating a total

$$Y = \sum_{k \in U} y_k.$$

We denote $X_i, i = 1, 2$, the true known total of the information x_i. A simple random sample without replacement is performed in the population. Throughout the exercise, we consider that the sample size is 'large'.

Preliminary question:

Show that the sampling covariance between two simple means $\widehat{\overline{Y}}$ and $\widehat{\overline{X}}_i$ is:

$$\mathrm{cov}\left(\widehat{\overline{X}}_i, \widehat{\overline{Y}}\right) = \frac{N-n}{Nn} S_{x_i y}, i = 1, 2,$$

where $S_{x_i y}$ is the corrected covariance between x_{ik} and y_k in the population.

1. Write the two ratio estimators of the total Y that we are able to form. We denote them $\widehat{Y}_{R1}$ and $\widehat{Y}_{R2}$.

2. We then construct the synthetic estimator

$$\widehat{Y}_R = \alpha \widehat{Y}_{R1} + \beta \widehat{Y}_{R2}.$$

   What reasonable relationship are we trying to impose between α and β? Is the estimator $\widehat{Y}_R$ biased?

3. Calculate $\mathrm{var}(\widehat{Y}_R)$ as a function of

$$\alpha, \beta, \mathrm{var}(\widehat{Y}_{R1}), \mathrm{var}(\widehat{Y}_{R2}) \text{ and } \mathrm{cov}(\widehat{Y}_{R1}, \widehat{Y}_{R2}),$$

   this last term representing the covariance between $\widehat{Y}_{R1}$ and $\widehat{Y}_{R2}$. Deduce an optimal value for α, denoted α_{opti}, then the optimal variance of $\widehat{Y}_R$, denoted $\mathrm{var}_{\text{opti}}(\widehat{Y}_R)$.

4. Using a technique of limited development, express $\mathrm{cov}(\widehat{Y}_{R1}, \widehat{Y}_{R2})$ as a function of the following quantities:

$$n, N, S_y^2, S_{x_1 y}, S_{x_2 y}, S_{x_1 x_2}, R_1 \text{ and } R_2, \text{ where } R_i = \frac{Y}{X_i}, i = 1, 2.$$

5. How do we estimate $\mathrm{var}(\widehat{Y}_{R1})$, $\mathrm{var}(\widehat{Y}_{R2})$, and $\mathrm{cov}(\widehat{Y}_{R1}, \widehat{Y}_{R2})$? Deduce an estimator $\widehat{\alpha}_{\text{opti}}$, of α_{opti}, and an optimal variance estimator $\mathrm{var}_{\text{opti}}(\widehat{Y}_R)$.

6. *Numerical application:*

   We want to estimate the mean population of 195 large cities in 1999 (denoted $\overline{Y} = Y/N$). Furthermore, from censuses, we know the mean population $\overline{X}_1$ in 1990 and the mean population $\overline{X}_2$ in 1980. The simple random sample of cities is of size 50. We have, in millions of residents:

$$\overline{X}_1 = 1482, \overline{X}_2 = 1420, \widehat{\overline{Y}} = 1896, \widehat{\overline{X}}_1 = 1693, \widehat{\overline{X}}_2 = 1643,$$

$$s_y^2 = (2088)^2, s_{x_1}^2 = (1932)^2, s_{x_2}^2 = (1931)^2,$$

$$s_{x_1y} = (1996)^2, s_{x_2y} = (1978)^2, s_{x_1x_2} = (1926)^2.$$

Calculate the two ratio estimators of $\overline{Y}$, denoted $\widehat{\overline{Y}}_{R1}$ and $\widehat{\overline{Y}}_{R2}$, and the estimator $\widehat{\overline{Y}}_{\text{opti}}$ obtained with $\widehat{\alpha}_{\text{opti}}$. For each of these estimators, give an estimate of the accuracy. Make a conclusion.

Solution
Preliminary question:
We propose two methods of resolution.
Method 1
Since

$$\text{var}(\widehat{\overline{X}} + \widehat{\overline{Y}}) = \text{var}(\widehat{\overline{X}}) + \text{var}(\widehat{\overline{Y}}) + 2\text{cov}(\widehat{\overline{X}}, \widehat{\overline{Y}}),$$

we have

$$\text{cov}(\widehat{\overline{X}}, \widehat{\overline{Y}}) = \frac{1}{2}\left[\text{var}(\widehat{\overline{X}} + \widehat{\overline{Y}}) - \text{var}(\widehat{\overline{X}}) - \text{var}(\widehat{\overline{Y}})\right] = \frac{N-n}{2Nn}\left[S_z^2 - S_x^2 - S_y^2\right],$$

where $z_k = x_k + y_k$. Therefore:

$$S_z^2 = \frac{1}{N-1}\sum_{k \in U}\left(x_k + y_k - \overline{X} - \overline{Y}\right)^2 = S_x^2 + S_y^2 + 2S_{xy},$$

where

$$S_{xy} = \frac{1}{N-1}\sum_{k \in U}\left(x_k - \overline{X}\right)\left(y_k - \overline{Y}\right).$$

We finally obtain:

$$\text{cov}(\widehat{\overline{X}}, \widehat{\overline{Y}}) = \frac{N-n}{2Nn}\left[S_x^2 + S_y^2 + 2S_{xy} - S_x^2 - S_y^2\right] = \frac{N-n}{Nn}S_{xy}.$$

Method 2
Denoting I_k as the random variable for the presence of unit k in the sample, we have:

$$\text{cov}(\widehat{\overline{X}}, \widehat{\overline{Y}}) = \text{cov}\left(\frac{1}{n}\sum_{k \in S}x_k, \frac{1}{n}\sum_{\ell \in S}y_\ell\right) = \frac{1}{n^2}\sum_{k \in U}\sum_{\ell \in U}x_k y_\ell \text{cov}\left(I_k, I_\ell\right).$$

Indeed

$$\text{cov}\left(I_k, I_\ell\right) = \begin{cases} \pi_{k\ell} - \pi_k\pi_\ell = \dfrac{n(n-1)}{N(N-1)} - \dfrac{n^2}{N^2} = -\dfrac{n(N-n)}{N^2(N-1)}, & \text{if } k \neq \ell \\ \pi_k(1 - \pi_k) = \dfrac{n}{N}\left(\dfrac{N-n}{N}\right), & \text{if } k = \ell. \end{cases}$$

Therefore

$$\text{cov}(I_k, I_\ell) = \frac{N-n}{N}\frac{n}{N} \times \begin{cases} -\dfrac{1}{N-1} & \text{if } k \neq \ell \\ 1 & \text{if } k = \ell. \end{cases}$$

We therefore obtain:

$$\text{cov}(\widehat{\overline{X}}, \widehat{\overline{Y}}) = \frac{1}{n^2}\sum_{k \in U} x_k y_k \text{cov}(I_k, I_k) + \frac{1}{n^2}\sum_{k \in U}\sum_{\ell \in U, \ell \neq k} x_k y_\ell \text{cov}(I_k, I_\ell)$$

$$= \frac{1}{n^2}\frac{N-n}{N}\frac{n}{N}\left[\sum_{k \in U} x_k y_k - \frac{1}{N-1}\sum_{k \in U}\sum_{\ell \in U, \ell \neq k} x_k y_\ell\right]$$

$$= \frac{N-n}{nN^2}\left[\sum_{k \in U} x_k y_k - \frac{1}{N-1}\sum_{k \in U}\sum_{\ell \in U} x_k y_\ell + \frac{1}{N-1}\sum_{k \in U} x_k y_k\right]$$

$$= \frac{N-n}{nN^2}\left[\frac{N}{N-1}\sum_{k \in U} x_k y_k - \frac{N^2}{N-1}\overline{X}\,\overline{Y}\right]$$

$$= \frac{N-n}{n(N-1)}\left[\frac{1}{N}\sum_{k \in U} x_k y_k - \overline{X}\,\overline{Y}\right]$$

$$= \frac{N-n}{nN}S_{xy}.$$

1. Successively using the two variables x_1 and x_2, we have

$$\widehat{Y}_{R1} = X_1\frac{\widehat{\overline{Y}}}{\widehat{\overline{X}}_1} \quad \text{and} \quad \widehat{Y}_{R2} = X_2\frac{\widehat{\overline{Y}}}{\widehat{\overline{X}}_2}.$$

2. The expected value of the estimator is

$$\text{E}\left(\widehat{Y}_R\right) = \alpha\text{E}\left(\widehat{Y}_{R1}\right) + \beta\text{E}\left(\widehat{Y}_{R2}\right)$$

$$= \alpha\left[Y + O\left(\frac{1}{n}\right)\right] + \beta\left[Y + O\left(\frac{1}{n}\right)\right]$$

$$= (\alpha + \beta)Y + O\left(\frac{1}{n}\right),$$

where the notation $O(1/n)$ represents a quantity which remains restricted when multiplied by n. We would like that $\alpha + \beta = 1$. Thus, the estimator $\widehat{Y}_R$ is biased but its bias is of $O(1/n)$, which is negligible when n is large.

3. The variance is

$$\text{var}\left(\widehat{Y}_R\right) = \alpha^2\text{var}\left(\widehat{Y}_{R1}\right) + \beta^2\text{var}\left(\widehat{Y}_{R2}\right) + 2\alpha\beta\text{cov}\left(\widehat{Y}_{R1}, \widehat{Y}_{R2}\right). \quad (6.1)$$

If we minimise $\text{var}\left(\widehat{Y}_R\right)$ by α, after having set $\beta = 1 - \alpha$, we find

$$\alpha_{\text{opti}} = \frac{\text{var}\left(\widehat{Y}_{R2}\right) - \text{cov}\left(\widehat{Y}_{R1}, \widehat{Y}_{R2}\right)}{\text{var}\left(\widehat{Y}_{R1}\right) + \text{var}\left(\widehat{Y}_{R2}\right) - 2\text{cov}\left(\widehat{Y}_{R1}, \widehat{Y}_{R2}\right)}.$$

Replacing in (6.1) α by α_{opti}, we get after a few calculations

$$\text{var}_{\text{opti}}\left(\widehat{Y}_R\right) = \frac{\text{var}\left(\widehat{Y}_{R1}\right)\text{var}\left(\widehat{Y}_{R2}\right) - \left[\text{cov}\left(\widehat{Y}_{R1}, \widehat{Y}_{R2}\right)\right]^2}{\text{var}\left(\widehat{Y}_{R1}\right) + \text{var}\left(\widehat{Y}_{R2}\right) - 2\text{cov}\left(\widehat{Y}_{R1}, \widehat{Y}_{R2}\right)}.$$

The Schwarz inequality ensures that the numerator is indeed positive.

4. Since $Y = \text{E}(\widehat{Y}_{Ri}) + O\left(\frac{1}{n}\right)$,

$$\text{cov}\left(\widehat{Y}_{R1}, \widehat{Y}_{R2}\right) = \text{E}\left\{\left(\widehat{Y}_{R1} - \text{E}(\widehat{Y}_{R1})\right)\left(\widehat{Y}_{R2} - \text{E}(\widehat{Y}_{R2})\right)\right\}$$

$$= \text{E}\left\{\left(\widehat{Y}_{R1} - Y + O\left(\frac{1}{n}\right)\right)\left(\widehat{Y}_{R2} - Y + O\left(\frac{1}{n}\right)\right)\right\}.$$

The mean square error of $\widehat{Y}_{Ri}$ being $1/n$, we can write

$$\widehat{Y}_{Ri} - Y = O\left(1/\sqrt{n}\right),$$

which yields $\widehat{Y}_{Ri} - Y$ leading to $O\left(1/n\right)$, thus

$$\text{cov}\left(\widehat{Y}_{R1}, \widehat{Y}_{R2}\right) \approx \text{E}\left\{\left(\widehat{Y}_{R1} - Y\right)\left(\widehat{Y}_{R2} - Y\right)\right\},$$

which gives

$$\text{cov}\left(\widehat{Y}_{R1}, \widehat{Y}_{R2}\right) = \text{E}\left\{X_1 X_2 \left(\frac{\widehat{\overline{Y}}}{\widehat{\overline{X}}_1} - R_1\right)\left(\frac{\widehat{\overline{Y}}}{\widehat{\overline{X}}_2} - R_2\right)\right\}$$

$$= X_1 X_2 \text{E}\left\{\left(\frac{\widehat{\overline{Y}}}{\widehat{\overline{X}}_1} - R_1\right)\left(\frac{\widehat{\overline{Y}}}{\widehat{\overline{X}}_2} - R_2\right)\right\}$$

$$= X_1 X_2 \text{E}\left\{\left(\frac{\widehat{\overline{Y}} - R_1 \widehat{\overline{X}}_1}{\widehat{\overline{X}}_1}\right)\left(\frac{\widehat{\overline{Y}} - R_2 \widehat{\overline{X}}_2}{\widehat{\overline{X}}_2}\right)\right\}.$$

By a limited development of $\widehat{\overline{X}}_i$ around its expected value $\overline{X}_i$, we get

$$\frac{\widehat{\overline{Y}} - R_i \widehat{\overline{X}}_i}{\widehat{\overline{X}}_i} \approx \frac{\widehat{\overline{Y}} - R_i \widehat{\overline{X}}_i}{\overline{X}_i}\left(1 - \frac{\widehat{\overline{X}}_i - \overline{X}_i}{\overline{X}_i}\right), i = 1, 2.$$

Only keeping the term of order $1/n$ in the limited development, we finally have

$$\text{cov}\left(\widehat{Y}_{R1}, \widehat{Y}_{R2}\right)$$

$$\approx \frac{X_1 X_2}{\overline{X}_1 \overline{X}_2} \text{E}\left[(\widehat{\overline{Y}} - R_1\widehat{\overline{X}}_1)(\widehat{\overline{Y}} - R_2\widehat{\overline{X}}_2)\left(1 - \frac{\widehat{\overline{X}}_1 - \overline{X}_1}{\overline{X}_1}\right)\left(1 - \frac{\widehat{\overline{X}}_2 - \overline{X}_2}{\overline{X}_2}\right)\right].$$

Indeed $\widehat{\overline{Y}} - R_i\widehat{\overline{X}}_i$ has a null expected value and a variance of $1/n$; it is of order of magnitude $1/\sqrt{n}$. Likewise $(\widehat{\overline{X}}_i - \overline{X}_i)/\overline{X}_i$ has a null expected value and a variance of $1/n$; it is therefore of order of magnitude $1/\sqrt{n}$ as well. Save the $1/n$ terms and reject those in $1/n^{3/2}$, leading to keep only the product $(\widehat{\overline{Y}} - R_1\widehat{\overline{X}}_1)(\widehat{\overline{Y}} - R_2\widehat{\overline{X}}_2)$. Therefore

$$\text{cov}\left(\widehat{Y}_{R1}, \widehat{Y}_{R2}\right)$$

$$\approx \frac{X_1 X_2}{\overline{X}_1 \overline{X}_2} \text{E}\left[(\widehat{\overline{Y}} - R_1\widehat{\overline{X}}_1)(\widehat{\overline{Y}} - R_2\widehat{\overline{X}}_2)\right]$$

$$= N^2 \text{cov}\left(\widehat{\overline{Y}} - R_1\widehat{\overline{X}}_1, \widehat{\overline{Y}} - R_2\widehat{\overline{X}}_2\right)$$

$$= N^2\left[\text{var}(\widehat{\overline{Y}}) - R_1\text{cov}(\widehat{\overline{X}}_1, \widehat{\overline{Y}}) - R_2\text{cov}(\widehat{\overline{X}}_2, \widehat{\overline{Y}}) + R_1 R_2\text{cov}(\widehat{\overline{X}}_1, \widehat{\overline{X}}_2)\right].$$

Since the sample is simple random, by using the preliminary question,

$$\text{var}(\widehat{\overline{Y}}) = \frac{N-n}{Nn}S_y^2, \quad \text{cov}(\widehat{\overline{X}}_1, \widehat{\overline{Y}}) = \frac{N-n}{Nn}S_{x_1 y},$$

$$\text{cov}(\widehat{\overline{X}}_2, \widehat{\overline{Y}}) = \frac{N-n}{Nn}S_{x_2 y}, \quad \text{cov}(\widehat{\overline{X}}_1, \widehat{\overline{X}}_2) = \frac{N-n}{Nn}S_{x_1 x_2},$$

we finally get

$$\text{cov}\left(\widehat{Y}_{R1}, \widehat{Y}_{R2}\right) \approx N^2\frac{N-n}{Nn}\left[S_y^2 - R_1 S_{x_1 y} - R_2 S_{x_2 y} + R_1 R_2 S_{x_1 x_2}\right].$$

5. We are going to use:

$$\widehat{\alpha}_{\text{opti}} = \frac{\widehat{\text{var}}\left(\widehat{Y}_{R2}\right) - \widehat{\text{cov}}\left(\widehat{Y}_{R1}, \widehat{Y}_{R2}\right)}{\widehat{\text{var}}\left(\widehat{Y}_{R1}\right) + \widehat{\text{var}}\left(\widehat{Y}_{R2}\right) - 2\widehat{\text{cov}}\left(\widehat{Y}_{R1}, \widehat{Y}_{R2}\right)},$$

with

$$\widehat{\text{var}}\left(\widehat{Y}_{R1}\right) = N^2\frac{N-n}{Nn}s_u^2,$$

where

$$u_k = y_k - \widehat{R}_1 x_{1k},$$

and

$$\widehat{\text{var}}\left(\widehat{Y}_{R2}\right) = N^2\frac{N-n}{Nn}s_v^2,$$

where

$$v_k = y_k - \widehat{R}_2 x_{2k},$$

with

$$\widehat{R}_1 = \frac{\widehat{\overline{Y}}}{\widehat{\overline{X}}_1} \text{ and } \widehat{R}_2 = \frac{\widehat{\overline{Y}}}{\widehat{\overline{X}}_2},$$

$$s_u^2 = s_y^2 + \widehat{R}_1^2 s_{x1}^2 - 2\widehat{R}_1 s_{x_1 y},$$

and

$$s_v^2 = s_y^2 + \widehat{R}_2^2 s_{x2}^2 - 2\widehat{R}_2 s_{x_2 y}.$$

Furthermore, we set

$$\widehat{\mathrm{cov}}\left(\widehat{Y}_{R1}, \widehat{Y}_{R2}\right) = N^2 \frac{N-n}{Nn}\left[s_y^2 - \widehat{R}_1 s_{x_1 y} - \widehat{R}_2 s_{x_2 y} + \widehat{R}_1 \widehat{R}_2 s_{x_1 x_2}\right].$$

All of these estimators are obviously biased, but the biases are very small when n is large (bias $1/n$). The optimal variance estimated is immediately obtained and without problem, each component being estimated as above.

6. The two estimators are

$$\widehat{\overline{Y}}_{R1} = 1482 \times \frac{1896}{1693} = 1660, \quad \widehat{\overline{Y}}_{R2} = 1420 \times \frac{1896}{1643} = 1639.$$

In sampling, $n = 50$ can be considered as 'large', even if we are at the limits of accepting such an assertion.

$$\widehat{\mathrm{var}}\left(\widehat{\overline{Y}}_{R1}\right)$$
$$= \frac{195-50}{195 \times 50}\left[(2088)^2 + \left(\frac{1896}{1693}\right)^2 (1932)^2 - 2 \times \frac{1896}{1693} \times (1996)^2\right] \approx 1750.$$

Therefore,

$$\sqrt{\widehat{\mathrm{var}}\left(\widehat{\overline{Y}}_{R1}\right)} \approx 41.8.$$

Furthermore,

$$\widehat{\mathrm{var}}\left(\widehat{\overline{Y}}_{R2}\right) \approx 4393,$$

which gives

$$\sqrt{\widehat{\mathrm{var}}\left(\widehat{\overline{Y}}_{R2}\right)} \approx 66.3.$$

The increase in variance obtained by going from $\widehat{\overline{Y}}_{R1}$ to $\widehat{\overline{Y}}_{R2}$ is logical since information x_2 is older than information x_1, and is therefore less correlated with y. Furthermore,

$$\widehat{\mathrm{cov}}\left(\widehat{\overline{Y}}_{R1}, \widehat{\overline{Y}}_{R2}\right) = 2632,$$

and, after a few calculations, we find $\widehat{a}_{\text{opti}} = 2$, which gives

$$\widehat{\overline{Y}}_{\text{opti}} = 2\widehat{\overline{Y}}_{R1} - \widehat{\overline{Y}}_{R2} = 1681,$$

and

$$\widehat{\text{var}}\left(\widehat{\overline{Y}}_{\text{opti}}\right) = 865.$$

We notice a net improvement in accuracy with the optimal linear combination of $\widehat{\overline{Y}}_{R1}$ and $\widehat{\overline{Y}}_{R2}$.

Exercise 6.9 *Overall ratio or combined ratio*

The goal of this exercise is to compare the performance of several sampling designs using stratification and ratios, when the sample size is large. We consider that the sample is *stratified* (H strata with simple random sampling in each stratum), and we have available an auxiliary variable x.

1. A stratified estimator of $\overline{Y}$ can be constructed on the model:

$$\widehat{\overline{Y}}_{\text{com}} = \overline{X} \frac{\sum_{h=1}^{H} \frac{N_h}{N} \widehat{\overline{Y}}_h}{\sum_{h=1}^{H} \frac{N_h}{N} \widehat{\overline{X}}_h},$$

where $\widehat{\overline{X}}_h$ and $\widehat{\overline{Y}}_h$ represent the simple means of x and y in the sample of stratum h.

a) Justify this expression (we are speaking about a *combined* ratio).
b) Using limited developments, give an approximation of the bias of order $1/n$ (we are therefore placed in the case where n is 'large'). Under what condition is this bias null?
c) Give an approximation of the mean square error and then of the variance of order $1/n$.
d) For what relationship between x_k and y_k is the estimator $\widehat{\overline{Y}}_{\text{com}}$ interesting?

2. A second estimator can be constructed from the ratio estimators considered stratum by stratum, being:

$$\widehat{\overline{Y}}_{\text{strat}} = \frac{1}{N} \sum_{h=1}^{H} X_h \frac{\widehat{\overline{Y}}_h}{\widehat{\overline{X}}_h},$$

where X_h represents the true total (known) of x in stratum h (we are speaking here about a *stratified* ratio). Go back to Questions 1.(a), 1.(b) and 1.(c) and compare, from the viewpoint of bias and then of the variance, the performances of $\widehat{\overline{Y}}_{\text{com}}$ and $\widehat{\overline{Y}}_{\text{strat}}$.

3. *Numerical application:* We return to the example from Exercise 6.2, where
we would consider a population of 2 010 farms. We stratify into two parts:
the farms where the total surface area cultivated x is less than 160 hectares
(stratum 1) and the other farms (stratum 2). The data are presented in
Table 6.2. The selected allocation is $n_1 = 70$ and $n_2 = 30$ (we are restricted
in selecting 100 farms in total).

Table 6.2. Total surface area cultivated x, and surface area cultivated in cereals y
in two strata: Exercise 6.9

Stratum	N_h	$\overline{Y}_h$	s_{yh}^2	s_{xh}^2	s_{xyh}	$\overline{X}_h$	X_h
1	1580	19.40	312	2055	494	82.56	84
2	430	51.63	922	7357	858	244.85	241.32
Total	2010	––	620	7619	1453	–	–

a) What is the property of this allocation?
b) Compare the estimated variances of the mean estimators for the fol-
lowing five concurrent sampling designs:
- Simple random sampling with the simple mean $\widehat{\overline{Y}}$;
- Simple random sampling with ratio;
- Stratified sampling, 'classical' estimator;
- Stratified sampling, combined ratio estimator;
- Stratified sampling, stratified ratio estimator.

We neglect the sampling rates. To estimate the variances of the unstratified designs, we will act as if the individuals had been selected using
simple random sampling (note that this only poses a problem of bias in
the estimators).

Solution

1. a) We know that for a stratified survey with simple random sampling in
each stratum, we have:

$$E\left[\sum_{h=1}^{H} \frac{N_h}{N} \widehat{\overline{Y}}_h\right] = \overline{Y}.$$

Therefore, $\widehat{\overline{Y}}_{\text{com}}$ naturally estimates

$$\overline{X} \frac{\sum_{h=1}^{H} \frac{N_h}{N} \overline{Y}_h}{\sum_{h=1}^{H} \frac{N_h}{N} \overline{X}_h} = \frac{\overline{Y}}{\overline{X}} \overline{X} = \overline{Y}.$$

b) We denote:

$$\widehat{\overline{Y}} = \sum_{h=1}^{H} \frac{N_h}{N} \widehat{\overline{Y}}_h, \text{ and } \widehat{\overline{X}} = \sum_{h=1}^{H} \frac{N_h}{N} \widehat{\overline{X}}_h.$$

Therefore,

$$\widehat{\overline{Y}}_{\text{com}} = \overline{X} \frac{\widehat{\overline{Y}}}{\widehat{\overline{X}}}.$$

We write

$$\widehat{\overline{X}} = \overline{X} \left[1 + \frac{\widehat{\overline{X}} - \overline{X}}{\overline{X}} \right] \quad \text{and} \quad \widehat{\overline{Y}} = \overline{Y} \left[1 + \frac{\widehat{\overline{Y}} - \overline{Y}}{\overline{Y}} \right],$$

which gives

$$\widehat{\overline{Y}}_{com}$$

$$= \overline{X} \frac{\overline{Y}}{\overline{X}} \left[1 + \frac{\widehat{\overline{Y}} - \overline{Y}}{\overline{Y}} \right] \left[1 - \frac{\widehat{\overline{X}} - \overline{X}}{\overline{X}} + \left(\frac{\widehat{\overline{X}} - \overline{X}}{\overline{X}} \right)^2 + \varepsilon_1 \right]$$

$$= \overline{Y} \left[1 - \frac{\widehat{\overline{X}} - \overline{X}}{\overline{X}} + \left(\frac{\widehat{\overline{X}} - \overline{X}}{\overline{X}} \right)^2 + \frac{\widehat{\overline{Y}} - \overline{Y}}{\overline{Y}} - \frac{(\widehat{\overline{Y}} - \overline{Y})(\widehat{\overline{X}} - \overline{X})}{\overline{X}\,\overline{Y}} + \varepsilon_2 \right],$$

where ε_2 is an expression containing an infinite number of terms originating from the limited development of $1/\widehat{\overline{X}}$. Finally,

$$\text{E}[\widehat{\overline{Y}}_{\text{com}}] \approx \overline{Y} + \frac{\overline{Y}}{\overline{X}^2} \text{E}[\widehat{\overline{X}} - \overline{X}]^2 - \frac{1}{\overline{X}} \text{E}(\widehat{\overline{Y}} - \overline{Y})(\widehat{\overline{X}} - \overline{X}).$$

Both of the expected values are manifestly $1/n$, in regards respectively to a variance and a covariance. We are convinced that all the other terms neglected here being $\text{E}(\varepsilon_2)$ are $1/n^{\alpha}$ with $\alpha > 1$. (We can even say that the 'forgotten' first-order varies by $1/n^{3/2}$.) We can neglect them as soon as n is 'large'. We have

$$\text{E}[\widehat{\overline{X}} - \overline{X}]^2 = \text{var}[\widehat{\overline{X}}] = \sum_{h=1}^{H} \left(\frac{N_h}{N} \right)^2 \text{var}(\widehat{\overline{X}}_h),$$

$$\text{E}(\widehat{\overline{X}} - \overline{X})(\widehat{\overline{Y}} - \overline{Y}) = \text{cov}(\widehat{\overline{X}}, \widehat{\overline{Y}})$$

$$= \text{cov} \left(\sum_{h=1}^{H} \frac{N_h}{N} \widehat{\overline{X}}_h, \sum_{h=1}^{H} \frac{N_h}{N} \widehat{\overline{Y}}_h \right)$$

$$= \sum_{h=1}^{H} \left(\frac{N_h}{N} \right)^2 \text{cov}(\widehat{\overline{X}}_h, \widehat{\overline{Y}}_h),$$

as the cross-covariances $h \times k$ $(h \neq k)$ are null due to the independence of drawings from one stratum to another. The covariance is

$$\text{cov}(\widehat{\overline{X}}_h, \ \widehat{\overline{Y}}_h) = \frac{1 - f_h}{n_h} \, S_{xyh},$$

where S_{xyh} indicates the true covariance between x_k and y_k in stratum h and

$$\text{var}(\widehat{\overline{X}}_h) = \frac{1 - f_h}{n_h} S_{xh}^2.$$

Conclusion: The approximate bias obtained by only keeping the largest terms $(1/n)$ is:

$$E[\widehat{\overline{Y}}_{\text{com}}] - \overline{Y}$$

$$\approx \overline{Y} \left[\frac{\text{var}(\widehat{\overline{X}})}{\overline{X}^2} - \frac{\text{cov}(\widehat{\overline{X}}, \ \widehat{\overline{Y}})}{\overline{X} \ \overline{Y}} \right]$$

$$\approx \overline{Y} \left[\frac{\sum_{h=1}^{H} \left(\frac{N_h}{N}\right)^2 \frac{S_{xh}^2}{n_h} (1 - f_h)}{\overline{X}^2} - \frac{\sum_{h=1}^{H} \left(\frac{N_h}{N}\right)^2 \frac{S_{xyh}}{n_h} (1 - f_h)}{\overline{X} \ \overline{Y}} \right]$$

$$\approx \sum_{h=1}^{H} \left(\frac{N_h}{N}\right)^2 \frac{1 - f_h}{n_h} \left[\frac{\overline{Y}}{\overline{X}^2} S_{xh}^2 - \frac{S_{xyh}}{\overline{X}} \right].$$

This bias is null if and only if, for all h,

$$\frac{S_{xyh}}{S_{xh}^2} = \frac{\overline{Y}}{\overline{X}} = \text{constant}.$$

The combined ratio is thus unbiased (or very slightly biased) if and only if the affine regression lines developed in each of the *strata* are of the *same slope*, and that this common slope is $\overline{Y}/\overline{X}$. That comes back to saying that all the regression lines have the same slope and pass through the origin in each stratum.

c) We calculate the mean square error:

$$\text{MSE}[\widehat{\overline{Y}}_{\text{com}}] = E \left(\overline{X} \frac{\widehat{\overline{Y}}}{\widehat{\overline{X}}} - \overline{Y} \right)^2 = \overline{X}^2 E \left[\frac{\widehat{\overline{Y}}}{\widehat{\overline{X}}} - R \right]^2 = \overline{X}^2 E \left[\frac{\widehat{\overline{Y}} - R\widehat{\overline{X}}}{\widehat{\overline{X}}} \right]^2,$$

where $R = \overline{Y}/\overline{X}$. By developing, we find

$$\text{MSE}[\widehat{\overline{Y}}_{\text{com}}] = E \left[\widehat{\overline{Y}} - R\widehat{\overline{X}} \right]^2 \left[1 - \frac{\widehat{\overline{X}} - \overline{X}}{\overline{X}} + \left(\frac{\widehat{\overline{X}} - \overline{X}}{\overline{X}} \right)^2 + ... \right]^2$$

$$\approx E[\widehat{\overline{Y}} - R\widehat{\overline{X}}]^2,$$

by keeping only the $1/n$ terms. Using the technique of limited development for n large, and by only keeping the $1/n$ terms, the calculation of the last line comes back to replacing $\widehat{\overline{X}}$ of the denominator with $\overline{X}$. Indeed

$$\widehat{\overline{Y}} - R\widehat{\overline{X}} = \sum_{h=1}^{H} \frac{N_h}{N} (\widehat{\overline{Y}}_h - R\widehat{\overline{X}}_h)$$

is the null expected value by definition of R. Therefore

$$E[\widehat{\overline{Y}} - R\widehat{\overline{X}}]^2 = \text{var}[\widehat{\overline{Y}} - R\widehat{\overline{X}}] = \sum_{h=1}^{H} \left(\frac{N_h}{N}\right)^2 \text{var}(\widehat{\overline{Y}}_h - R\widehat{\overline{X}}_h).$$

Indeed

$$\text{var}(\widehat{\overline{Y}}_h - R\widehat{\overline{X}}_h) = \text{var}(\widehat{\overline{Y}}_h) + R^2\text{var}(\widehat{\overline{X}}_h) - 2R\text{cov}(\widehat{\overline{Y}}_h, \widehat{\overline{X}}_h),$$

which gives

$$\text{MSE}[\widehat{\overline{Y}}_{\text{com}}] \approx \sum_{h=1}^{H} \left(\frac{N_h}{N}\right)^2 \frac{1-f_h}{n_h} [S_{yh}^2 + R^2 S_{xh}^2 - 2R S_{xyh}],$$

with

$$R = \overline{Y}/\overline{X}.$$

$$\text{var}(\widehat{\overline{Y}}_{\text{com}}) = \text{MSE}[\widehat{\overline{Y}}_{\text{com}}] - \text{Bias}^2.$$

The bias indeed depends on $1/n_h$, which is the same as MSE. As the bias is squared, $\text{MSE}[\widehat{\overline{Y}}_{\text{com}}]$ and $\text{var}(\widehat{\overline{Y}}_{\text{com}})$ have the same approximation of order $1/n$.

d) The estimator $\widehat{\overline{Y}}_{\text{com}}$ is interesting once $\text{var}(\widehat{\overline{Y}}_h - R\widehat{\overline{X}}_h)$ is *small*, which is as soon as the *population variance* of the variable

$$y_k - Rx_k$$

is small in each stratum as well, which is when

$$y_k - Rx_k \approx C_h,$$

for all k of stratum h where C_h is a constant only depending on stratum h. The favourable situation (from the point of view of the variance) is therefore presented when

$$y_k = C_h + Rx_k \quad \text{for all individuals } k \text{ of stratum } h.$$

Then,

$$\overline{Y} = \sum_{h=1}^{H} \frac{N_h}{N} \overline{Y}_h = \sum_{h=1}^{H} \frac{N_h}{N} C_h + R\overline{X} = \sum_{h=1}^{H} \frac{N_h}{N} C_h + \overline{Y}.$$

Thus

$$\sum_{h=1}^{H} \frac{N_h}{N} C_h = 0.$$

The 'ideal model' is then:

$$y_k = C_h + Rx_k, \quad \text{with} \quad \sum_{h=1}^{H} N_h C_h = 0.$$

In practice, we instead expect relationships where C_h is close to null, of the type: $y_k \approx Rx_k$, which is a proportionality between x and y with the same proportionality factor in all strata. Thus, it is a rather restricting 'model'.

2. a) $X_h \dfrac{\widehat{\overline{Y}}_h}{\overline{X}_h}$ is the ratio estimator of the unknown true total Y_h in stratum

h: the expression of $\widehat{\overline{Y}}_{\text{strat}}$ is therefore natural.

b) We denote

$$\widehat{Y}_{Rh} = X_h \frac{\widehat{\overline{Y}}_h}{\overline{X}_h}.$$

Going back to the expression of bias (approximate) for the ratio estimator (see 1.(b)), we deduce this as:

$$\text{bias in stratum } h : \text{E}\widehat{Y}_{Rh} - Y_h \approx Y_h \left[\frac{\text{var}(\overline{X}_h)}{\overline{X}_h^2} - \frac{\text{cov}(\overline{X}_h, \, \widehat{\overline{Y}}_h)}{\overline{X}_h \overline{Y}_h} \right].$$

The approximate bias of $\widehat{\overline{Y}}_{\text{strat}}$ is therefore:

$$\text{E}(\widehat{\overline{Y}}_{\text{strat}}) - \overline{Y} \approx \frac{1}{N} \sum_{h=1}^{H} Y_h \left(\frac{1-f_h}{n_h} \right) \left[\frac{S_{xh}^2}{\overline{X}_h^2} - \frac{S_{xyh}}{\overline{X}_h \overline{Y}_h} \right]$$

$$= \sum_{h=1}^{H} \left(\frac{N_h}{N} \right) \frac{1-f_h}{n_h} \left[\frac{\overline{Y}_h}{\overline{X}_h^2} S_{xh}^2 - \frac{S_{xyh}}{\overline{X}_h} \right].$$

It is of course not possible to compare in a rigorous manner the biases of $\widehat{\overline{Y}}_{\text{com}}$ and $\widehat{\overline{Y}}_{\text{strat}}$. We can however notice that N_h/N is larger than $(N_h/N)^2$: if the terms in square brackets are mostly of the same sign (for example, if the regression lines most often have positive y-intercepts) we can think that $\widehat{\overline{Y}}_{\text{strat}}$ is more biased than $\widehat{\overline{Y}}_{\text{com}}$ (especially if there are many strata). This bias of $\widehat{\overline{Y}}_{\text{strat}}$ is null if and only if, for all h,

$$\frac{S_{xyh}}{S_{xh}^2} = \frac{\overline{Y}_h}{\overline{X}_h},$$

which is the classical condition in the absence of bias for the ratio estimator: the slope of the affine regression line of y on x is equal to the true ratio, which comes back to saying that this line *passes through the origin,* but stratum by stratum. From this point of view, we get an appreciably less restrictive condition than that which corresponds to the uselessness of the bias of $\widehat{\overline{Y}}_{\text{com}}$.

c) We use here the approximated variance expression for the ratio estimator (n_h large), obtained in 1.(c).

$$\text{var}[\widehat{\overline{Y}}_{\text{strat}}] = \frac{1}{N^2} \sum_{h=1}^{H} \text{var}\,[\widehat{Y}_{Rh}] \approx \frac{1}{N^2} \sum_{h=1}^{H} N_h^2 \text{var}\,[\widehat{\overline{Y}}_h - R_h \widehat{\overline{X}}_h],$$

with

$$R_h = \frac{\overline{Y}_h}{\overline{X}_h}.$$

Thus

$$\text{var}(\widehat{\overline{Y}}_{\text{strat}}) \approx \sum_{h=1}^{H} \left(\frac{N_h}{N}\right)^2 \frac{1 - f_h}{n_h} [S_{yh}^2 + R_h^2\, S_{xh}^2 - 2R_h S_{xyh}].$$

$\widehat{\overline{Y}}_{\text{strat}}$ is 'good' as soon as $\widehat{Y}_{Rh}$ is 'good' from stratum to stratum, which is as soon as the relationship between x and y is sufficiently *linear* in each stratum. However, with such an estimator, it is possible that the slope of the regression line is quite variable from one stratum to another, without being penalising, *contrary* to $\widehat{\overline{Y}}_{\text{com}}$.

Conclusions: A priori, except for a slightly 'bizarre' configuration:

- $\widehat{\overline{Y}}_{\text{com}}$ is instead less biased than $\widehat{\overline{Y}}_{\text{strat}}$;
- $\widehat{\overline{Y}}_{\text{com}}$ is instead more variable than $\widehat{\overline{Y}}_{\text{strat}}$;
- $\widehat{\overline{Y}}_{\text{com}}$ requires *less* auxiliary information x than $\widehat{\overline{Y}}_{\text{strat}}$. In fact, to use $\widehat{\overline{Y}}_{\text{strat}}$, we have to know X_h for all h (but only $X = \sum_{h=1}^{H} X_h$ to use $\widehat{\overline{Y}}_{\text{com}}$).

A selection rule could therefore be of the following type:

- With *small* n_h, we use $\widehat{\overline{Y}}_{\text{com}}$ to avoid in the first place biases that are too large.
- With *large* n_h, we use $\widehat{\overline{Y}}_{\text{strat}}$, under the condition that the squared biases are 'small' compared to the variances (which is expected, since the bias and variance vary $1/n_h$).

The variance estimates are obtained by replacing the population variance parameters with their counterparts in the sample.

3. a) This allocation is (almost) the Neyman optimal allocation. Indeed, with optimal allocation n_h is proportional to $N_h S_{yh}$ (S_{yh} was 'estimated' here). Therefore,

$$\begin{cases} n_1 \text{ proportional to } 1\,580 \times \sqrt{312} \\ n_2 \text{ proportional to } 430 \times \sqrt{922} \end{cases} \quad \text{with} \quad n_1 + n_2 = 100.$$

This calculation technically leads, after rounding, to $n_1 = 68$ and $n_2 = 32$. Nevertheless, we know that the optimum is 'flat'; that is, the neighbouring allocations of the optimal allocation (such as 70 and 30) lead to variances (nearly) equal to the minimum variance.

b) In the calculations of the two estimates which follow (unstratified case), the population variances S_y^2, S_x^2 and the true covariance S_{xy} are estimated from the data which are in reality obtained from stratified sampling: as a result, we lose the property of the absence of bias, strictly speaking. In regards to sampling with slightly unequal probabilities, we can however assume that the bias is small when the sample size is 'large' (which is the case with 100 units; see Exercise 3.21).

- *Simple design and simple mean:*

  If we denote $\widehat{\overline{Y}}$ as the simple mean in the sample, then

$$\widehat{\text{var}}_{\text{SRS}}(\widehat{\overline{Y}}) \approx \frac{s_y^2}{n} = \frac{620}{100} = 6.2.$$

- *Simple design and ratio:*

  If $\widehat{\overline{X}}$ and $\widehat{\overline{Y}}$ indicate the simple means of x and y in the sample, then

$$\widehat{\text{var}}_{\text{SRS}}\left(\overline{X}\frac{\widehat{\overline{Y}}}{\widehat{\overline{X}}}\right) \approx \frac{1}{n}\,[s_y^2 + \widehat{R}^2 s_x^2 - 2\widehat{R}s_{xy}]$$

$$\approx \frac{1}{100}\,[620 + (0.2215)^2 \times 7\,619 - 2(0.2215) \times 1\,453],$$

as

$$\widehat{\overline{Y}} = \sum_{h=1}^{2} \frac{n_h}{n}\,\widehat{\overline{Y}}_h = 0.7 \times 19.4 + 0.3 \times 51.63 = 29.07,$$

and by the same calculation $\widehat{\overline{X}} = 131.25$ and thus $\widehat{R} = \frac{\widehat{\overline{Y}}}{\widehat{\overline{X}}} \approx 0.2215$. Thus

$$\widehat{\text{var}}_{\text{SRS}}\left(\overline{X}\frac{\widehat{\overline{Y}}}{\widehat{\overline{X}}}\right) \approx \frac{350.12}{100} \approx 3.50.$$

- *Stratified design with 'classical' estimator:*

$$\widehat{\text{var}}\left(\sum_{h=1}^{H} \frac{N_h}{N}\,\widehat{\overline{Y}}_h\right) = \sum_{h=1}^{H} \left(\frac{N_h}{N}\right)^2 \frac{1}{n_h}\,s_{yh}^2$$

$$= \left(\frac{1\,580}{2\,010}\right)^2 \times \frac{1}{70} \times 312 + \left(\frac{430}{2\,010}\right)^2 \times \frac{1}{30} \times 922$$

$$= 4.16.$$

- *Stratified design with combined ratio estimator.*
  Let us go back to the initial notations from 1.(b). We are going to estimate the true ratio R with $\widehat{\overline{Y}}/\widehat{\overline{X}}$, being:

$$\widehat{\widehat{R}} = \frac{\sum_{h=1}^{H} N_h \widehat{\overline{Y}}_h}{\sum_{h=1}^{H} N_h \widehat{\overline{X}}_h},$$

which is preferable in comparison to the ratio of the simple means.

Note: We could have used this expression $\widehat{\widehat{R}}$ to estimate the variance of the ratio with simple random sampling (case 2), but we kept the ratio of simple means because it is the classical approach. The calculation gives

$$\widehat{\widehat{R}} = \frac{1\ 580 \times 19.40 + 430 \times 51.63}{1\ 580 \times 82.56 + 430 \times 244.85} \approx 0.2242.$$

We have

$$\widehat{\text{var}}[\widehat{\overline{Y}}_{\text{com}}]$$

$$= \left(\frac{1\ 580}{2\ 010}\right)^2 \times \frac{1}{70} \times [312 + (0.2242)^2 \times 2\ 055 - 2 \times 0.2242 \times 494]$$

$$+ \left(\frac{430}{2\ 010}\right)^2 \times \frac{1}{30} [922 + (0.2242)^2 \times 7\ 357 - 2 \times 0.2242 \times 858]$$

$$= 3.10.$$

- *Stratified design with separate ratio estimator.*

$$\widehat{\text{var}}[\widehat{\overline{Y}}_{\text{strat}}]$$

$$= \left(\frac{1580}{2\ 010}\right)^2 \frac{1}{70} \left[312 + \left(\frac{19.40}{82.56}\right)^2 \times 2055 - 2 \times \frac{19.40}{82.56} \times 494\right]$$

$$+ \left(\frac{430}{2010}\right)^2 \frac{1}{30} \left[922 + \left(\frac{51.63}{244.85}\right)^2 \times 7357 - 2 \times \frac{51.63}{244.85} \times 858\right]$$

$$= 3.06.$$

Finally, we get the following classification:

$$\widehat{\text{var}}(\widehat{\overline{Y}}_{\text{strat}}) < \widehat{\text{var}}(\widehat{\overline{Y}}_{\text{com}}) < \underset{\text{SRS}}{\widehat{\text{var}}}\left(\overline{X}\frac{\widehat{\overline{Y}}}{\widehat{\overline{X}}}\right) < \widehat{\text{var}}\left(\sum_{h=1}^{H} \frac{N_h}{N}\widehat{\overline{Y}}_h\right) < \underset{\text{SRS}}{\widehat{\text{var}}}\left(\widehat{\overline{Y}}\right).$$

Remaining cautious with the interpretation of calculations for the two unstratified designs, we notice that the three ratio estimators seem of comparable quality and produce the smallest variances, with a small advantage for those that take into account the stratification.

Exercise 6.10 *Calibration and two phases*

A regional agricultural cooperative wishes to estimate the average surface area of wheat cultivated $\overline{Y}$ in N farms of the region. To do this, a sample S^* of n^* farms is selected and the average surface area cultivated $\widehat{\overline{X}}^*$ is computed in the sample from land registers. Afterwards, n farms are resampled from the previous sample (we get the sample S) and one calculates the average surface area cultivated $\widehat{\overline{X}}$ and the average surface area of wheat cultivated $\widehat{\overline{Y}}$ in this sample, resulting from a trip of investigators into the field. Each sample is simple random (without replacement). The cooperative chooses to use an estimator of type

$$\widehat{\overline{Y}}_c = \widehat{\overline{Y}} + c(\widehat{\overline{X}}^* - \widehat{\overline{X}}),$$

where c is a known fixed value (thus non-random).
First of all, show that S can be considered as coming from a simple random sample without replacement of size n in a population of size N (hint: think about conditioning with respect to S^*).

1. a) Justify the expression $\widehat{\overline{Y}}_c$ with consideration to the accuracy (without calculation). In particular, give the relationship that would have to exist between x_k and y_k so that $\widehat{\overline{Y}}_c$ is precise, and interpret the constant c.
   b) Justify the expression $\widehat{\overline{Y}}_c$ with consideration to the cost.
   c) What do we call the type of sampling that is applied here?
2. Show that $\widehat{\overline{Y}}_c$ estimates $\overline{Y}$ without bias (hint: think about conditioning).
3. a) Write the decomposition formula of the variance allowing to express $\mathrm{var}(\widehat{\overline{Y}}_c)$ as a function of terms necessitating successively the sampling of S^* and the sampling of S.
   b) We define the following notation:
   - $\widehat{\overline{Y}}^*$: mean surface area of wheat cultivated for S^*,
   - $u_k = y_k - cx_k$,
   - s_u^{2*}: sample variance of u_k calculated on S^*,
   - S_u^2: population variance of u_k calculated on the entire population U,
   - S_y^2: population variance of y_k calculated on the entire population U.

   From each of the previously defined terms, show that

   $$\mathrm{var}(\widehat{\overline{Y}}_c) = \frac{N - n^*}{Nn^*}S_y^2 + \frac{n^* - n}{nn^*}S_u^2. \tag{6.2}$$

4. a) What is the gain (if a gain exists) offered by this sampling design compared to a simple random sample without replacement of size n with estimator $\widehat{\overline{Y}}$?

b) From the previous result, define in practice the context in which the estimator $\widehat{\overline{Y}}_c$ is a good estimator and again, find the result from 1.(a).

5. We are placed in the favourable context defined in 4.(b) and we contribute, in this question only, some considerations of cost. We denote c_1 as the cost of listing the surface area of a farm from the land registers, and c_2 as the survey cost in the field to list the surface area of wheat cultivated for a farm. We call $\overline{C}$ the total budget that we have available.

   a) Write out the budget constraint.

   b) Find the sizes n^* and n allowing to obtain the best accuracy for $\widehat{\overline{Y}}_c$ and note that S_u^2 must be situated in a certain interval (which we will determine) so that the sample is not reduced to a classical simple random sample.

6. a) Write out the population variance S_u^2 as a function of c and the population variances and covariances of y and x. What are we going to naturally impose on c?

   b) From the expression of the accuracy for $\widehat{\overline{Y}}_c$, determine the constant c which permits to get the most precise estimator $\widehat{\overline{Y}}_c$.

   c) What difficulty (difficulties) do we have in practice to calculate this optimal constant? Under these conditions, what 'natural' estimator are we tempted to use?

Solution

Preliminary question:

We denote $\binom{N}{n}$ as the number of ways of choosing n individuals among N without replacement (this is also $N!/n!(N-n)!$). We denote $p(s)$ as the probability of selecting sample s.

$$p(s) = \sum_{\{s^* | s \subset s^*\}} p(s|s^*)p(s^*) = \frac{\#\{s^* \subset U \mid s \subset s^*\}}{\binom{n^*}{n}\binom{N}{n^*}}.$$

The sum involves all the samples s^* containing s: n individuals are fixed, and it remains to select $(n^* - n)$ of them (to form s^*) among the 'remaining' $(N - n)$ individuals in the population. Therefore, the sum contains $\binom{N-n}{n-n^*}$ identical terms.

$$p(s) = \frac{\binom{N-n}{n^*-n}}{\binom{n^*}{n}\binom{N}{n^*}} = \frac{(N-n)!n!(n^*-n)!n^*!(N-n^*)!}{(n^*-n)!(N-n^*)!n^*!N!} = \binom{N}{n}^{-1},$$

which characterises a simple random sampling without replacement of fixed size n in a population of size N.

1. a) The estimator resembles the regression estimator, but this is not the regression estimator as on the one hand, the true mean $\overline{X}$ was replaced

with an estimate $\widehat{\overline{X}}^*$ and on the other hand, c is chosen *a priori*. As $\widehat{\overline{X}}^*$ is going to estimate $\overline{X}$, through a reasoning similar to that which permits the construction of the regression estimator, we can suspect that $\widehat{\overline{Y}}_c$ leads to a better accuracy than $\widehat{\overline{Y}}$ if we have a relationship of the type

$$y_k = a + bx_k + u_k,$$

where the u_k are small and of null sum. The constant c is then an estimate *a priori* (independent of the sample) of the slope of the regression line of y_k on x_k, and therefore a value close to b. This modelling corresponds well, *a priori*, to the concrete situation coming from the statement as we can reasonably think that the surface area of wheat cultivated is a function more or less linear to the total surface area cultivated.

b) The cost of obtaining $\overline{X}$, the true mean cultivated surface area, is high *a priori*, since it is necessary to get the cadastral information for all of the existing farms. By replacing $\overline{X}$ with $\widehat{\overline{X}}^*$, we certainly lose some accuracy but we use auxiliary information at less cost (it is 'sufficient' to consult the registers for some of the farms).

c) This is a two-phase sample. At the first phase, we select S^* and at the second phase, we select S.

2. In all of the calculations that follow, it is necessary to remember that S^* is a simple random sample and that, conditionally on S^*, S is a simple random sample in S^*.

$$\mathrm{E}\left(\widehat{\overline{Y}}_c\right) = \mathrm{E}\,\mathrm{E}\left(\widehat{\overline{Y}}_c | S^*\right),$$

the first expected value agrees in comparison to the distribution $p(s^*)$, and the second in comparison to the conditional distribution $p(s|s^*)$.

$$\mathrm{E}\left(\widehat{\overline{Y}}_c\right) = \mathrm{E}\,\mathrm{E}\left(\widehat{\overline{Y}} + c(\widehat{\overline{X}}^* - \widehat{\overline{X}})|S^*\right) = \mathrm{E}\left(\widehat{\overline{Y}}^* + c(\widehat{\overline{X}}^* - \widehat{\overline{X}}^*)\right)$$

$$= \mathrm{E}\left(\widehat{\overline{Y}}^*\right) = \overline{Y}.$$

Another method consists of using the preliminary question:

$$\mathrm{E}(\widehat{\overline{Y}}_c) = \mathrm{E}_S(\widehat{\overline{Y}}) + c(\mathrm{E}_{S^*}\widehat{\overline{X}}^* - \mathrm{E}_S\widehat{\overline{X}}),$$

where E_S and E_{S^*} respectively indicate the expected value in relation to the sampling distributions $p(s)$ and $p(s^*)$. As S^* and S are simple random samples in the population U, we have directly

$$\mathrm{E}_S(\widehat{\overline{Y}}) = \overline{Y}, \quad \mathrm{E}_S(\widehat{\overline{X}}) = \overline{X}, \quad \text{and} \quad \mathrm{E}_{S^*}(\widehat{\overline{X}}^*) = \overline{X},$$

which leads to $\mathrm{E}(\widehat{\overline{Y}}_c) = \overline{Y}$.

3. a) The decomposition of the variance (respective distributions $p(s^*)$ and $p(s|s^*)$) gives:

$$\mathrm{var}\left(\widehat{\overline{Y}}_c\right) = \mathrm{var}\,\mathrm{E}\left(\widehat{\overline{Y}}_c|S^*\right) + \mathrm{E}\,\mathrm{var}\left(\widehat{\overline{Y}}_c|S^*\right)$$
$$= \mathrm{var}\left(\widehat{\overline{Y}}^*\right) + \mathrm{E}\,\mathrm{var}\left(\widehat{\overline{Y}} - c\widehat{\overline{X}}|S^*\right).$$

b) Setting $u_k = y_k - cx_k$, and

$$s_u^{*2} = \frac{1}{n^* - 1}\sum_{k\in S^*}(u_k - \widehat{\overline{U}}^*)^2,$$

where

$$\widehat{\overline{U}}^* = \frac{1}{n^*}\sum_{k\in S^*}u_k,$$

we have

$$\mathrm{var}\left(\widehat{\overline{Y}}_c\right) = \mathrm{var}\left(\widehat{\overline{Y}}^*\right) + \mathrm{E}\,\mathrm{var}\left(\widehat{\overline{U}}|S^*\right)$$
$$= \frac{N - n^*}{Nn^*}S_y^2 + \mathrm{E}\left(\frac{n^* - n}{nn^*}s_u^{*2}\right)$$
$$= \frac{N - n^*}{Nn^*}S_y^2 + \frac{n^* - n}{nn^*}S_u^2,$$

as $\mathrm{E}(s_u^{*2}) = S_u^2$.

4. a) In a 'direct' simple random design without replacement of size n, the variance of the unbiased estimator $\widehat{\overline{Y}}$ is

$$\mathrm{var}\left(\widehat{\overline{Y}}\right) = \frac{N - n}{Nn}S_y^2.$$

The gain in the two-phase design with $\widehat{\overline{Y}}_c$ is therefore

$$\mathrm{var}\left(\widehat{\overline{Y}}\right) - \mathrm{var}\left(\widehat{\overline{Y}}_c\right) = \frac{N - n}{Nn}S_y^2 - \frac{N - n^*}{Nn^*}S_y^2 - \frac{n^* - n}{nn^*}S_u^2$$
$$= \frac{n^* - n}{nn^*}\left(S_y^2 - S_u^2\right).$$

This difference can be both positive and negative according to the sign of $S_y^2 - S_u^2$.

b) We notice that $\widehat{\overline{Y}}_c$ is much better when S_u^2 is small (with fixed sample sizes, but it is here a question of budget). This is the only term that we can try to keep at a minimum since S_y^2 is set for us. To obtain S_u^2 small, it is necessary and sufficient that $y_k - cx_k$ is not very dispersed, which means that it is approximately constant. In other words, we would like that

$$y_k \approx a + cx_k, \text{ for all } k,$$

which very well returns to the idea presented in 1.(a). The gain therefore depends on the choice of c.

For the following, we are placed in the case where $\widehat{\overline{Y}}_c$ is a 'good' estimator, which indicates in the first place that it is preferable to $\overline{Y}$, and therefore that $S_y^2 - S_u^2 > 0$.

5. a) The total budget is

$$\overline{C} = c_1 n^* + c_2 n.$$

b) We are therefore trying to minimise

$$\frac{N - n^*}{N} \times \frac{S_y^2}{n^*} + \frac{n^* - n}{n^*} \frac{S_u^2}{n}, \tag{6.3}$$

subject to $\overline{C} = c_1 n^* + c_2 n$ and $n^* \geq n$. We immediately verify that the function (6.3) to minimise can be replaced with

$$\frac{1}{n^*}(S_y^2 - S_u^2) + \frac{1}{n}S_u^2.$$

If we 'forget' the inequality constraint in the first place, then by differentiating the Lagrangian linked to this last expression in relation to n^* and n, we get

$$\begin{cases} -\dfrac{S_y^2 - S_u^2}{n^{*2}} - \lambda c_1 = 0 \\ -\dfrac{S_u^2}{n^2} - \lambda c_2 = 0, \end{cases}$$

where λ is the Lagrange multiplier. By making a ratio from these two equations, we get

$$\frac{S_y^2 - S_u^2}{S_u^2}\left(\frac{n}{n^*}\right)^2 = \frac{\lambda c_1}{\lambda c_2} = \frac{c_1}{c_2}.$$

We therefore have

$$n^* = n\sqrt{\frac{S_y^2 - S_u^2}{S_u^2}\frac{c_2}{c_1}}.$$

With the cost constraint

$$c_1 n^* + c_2 n = c_1 n\sqrt{\frac{S_y^2 - S_u^2}{S_u^2}\frac{c_2}{c_1}} + c_2 n = \overline{C},$$

we obtain:

$$n = \frac{\overline{C}}{c_2 + \sqrt{c_1 c_2 \frac{S_y^2 - S_u^2}{S_u^2}}}.$$

and

$$n^* = n\sqrt{\frac{S_y^2 - S_u^2}{S_u^2}\frac{c_2}{c_1}}.$$

It remains to verify if we had reason to not count on the constraint $n^* \geq n$, that is, if

$$\frac{S_y^2 - S_u^2}{S_u^2}\frac{c_2}{c_1} \geq 1,$$

which is also written

$$S_u^2 \leq \frac{S_y^2}{1 + \frac{c_1}{c_2}} \leq S_y^2.$$

It is therefore necessary and sufficient that

$$S_u^2 \in \left[0, \frac{S_y^2}{1 + \frac{c_1}{c_2}}\right]$$

(in which case $S_y^2 - S_u^2 \geq 0$) so that we have a genuine two-phase sample. If

$$S_u^2 \geq \frac{S_y^2}{1 + \frac{c_1}{c_2}},$$

then we stumble upon the constraint, which is to say that $n^* = n$ and we are then brought back to a simple random sample without replacement of fixed size

$$n = n^* = \frac{\overline{C}}{c_2},$$

in which case $\widehat{\overline{Y}}_c = \widehat{\overline{Y}}$. This is the case as soon as $S_u^2 > S_y^2$, since then $\widehat{\overline{Y}}$ appears to be better than $\widehat{\overline{Y}}_c$.

6. a) In the population, the variance of u_k is

$$S_u^2 = S_{y-cx}^2 = S_y^2 + c^2 S_x^2 - 2cS_{xy}.$$

So that $S_u^2 < S_y^2$, that is $\mathrm{var}(\widehat{\overline{Y}}) > \mathrm{var}(\widehat{\overline{Y}}_c)$ with $c > 0$, it is necessary and sufficient that

$$c < \frac{S_{xy}}{S_x^2} \times 2.$$

b) From the variance (see 3.b), we want to minimise S_u^2 by c, or in other words to minimise

$$S_y^2 + c^2 S_x^2 - 2cS_{xy}.$$

By setting the derivative with respect to c equal to zero, we got the optimal value $\tilde{c}$

$$2\tilde{c}S_x^2 - 2S_{xy} = 0,$$

which gives

$$\tilde{c} = \frac{S_{xy}}{S_x^2},$$

which is the slope of the regression line of y on x in the population U.

c) That problem is that $\tilde{c}$ is incalculable, since we do not know the population variances. We are tempted to estimate $\tilde{c}$ by

$$\hat{c} = \frac{s_{xy}}{s_x^2},$$

the slope of the regression line of y on x in the sample. But be careful, as this is no longer the difference estimator, as $\hat{c}$ depends on the sample. It is then a regression estimator, where the estimation of the slope is done from the x_k and y_k collected, which is to say by calling only on the second phase.

Exercise 6.11 *Regression and repeated surveys*

The object of this exercise is to show how a regression estimator can improve the quality of mean estimation when we perform two surveys on the same theme y on two successive dates $t = 1$ and $t = 2$. At the same time, we can study the (delicate) question of 'optimal' replacement of a part of the sample. We are interested in the estimation of the mean of y on date $t = 2$, denoted $\overline{Y}_2$. At period $t = 1$, we select by simple random sampling a sample S_1 of size n in a very large population (f negligible). At period $t = 2$, we re-*interview* c individuals that were part of the previous sample (those coming from a simple random sample in S_1), and we select r *new* individuals using simple random sampling in a way that the total size remains equal to n (we therefore have $c + r = n$). We consider that the identifiers for the population are exactly the same for the two dates (the population does not change between the two dates), and we denote:

- y_{tk} = value of y for individual k on date t,
- $\widehat{\overline{Y}}_{tc}$ = simple mean in the *common* sample of size c, calculated on date t. This sample is denoted S_c,
- $\widehat{\overline{Y}}_{tr}$ = simple mean in the sample *to replace or replaced* of size r, calculated on date t. This sample is denoted S_{tr}.

Finally, we denote:

$$S_t^2 = \frac{1}{N-1} \sum_{k \in U} (y_{tk} - \overline{Y}_t)^2 \ (t = 1, 2).$$

1. a) Conditionally on the composition of S_1, the sample S_{2r} comes from a simple random sample in the population outside of S_1. Show that if we 'decondition' from S_1, we can again consider that S_{2r} comes from a simple random sample in the *total* population.

   b) Give the expression for the variance $\mathrm{var}(\widehat{\overline{Y}}_{2r})$.

2. As the estimator of $\overline{Y}_2$, we first propose the following 'regression' estimator:

$$\widehat{\overline{Y}}_{2,\,\mathrm{reg}} = \widehat{\overline{Y}}_{2c} + \widehat{b}(\widehat{\overline{Y}}_1 - \widehat{\overline{Y}}_{1c}),$$

   where $\widehat{\overline{Y}}_1$ is the simple mean of y_{1k} from S_1 on date $t = 1$.

   a) Justify this formula, and specify the expression of the estimator $\widehat{b}$.

   b) Calculate the approximate variance of $\widehat{\overline{Y}}_{2,\,\mathrm{reg}}$.

   Hint: since $\widehat{\overline{Y}}_1$ is random (in that, $\widehat{\overline{Y}}_{2,\,\mathrm{reg}}$ is not a 'genuine' regression estimator) we use the decomposition formula of the variance, conditioning on S_1 in the first place. We will find:

$$\mathrm{var}(\widehat{\overline{Y}}_{2,\,\mathrm{reg}}) \approx \frac{S_2^2}{n} + \left(\frac{1}{c} - \frac{1}{n}\right)(1 - \rho^2)\,S_2^2,$$

   where ρ is the linear correlation coefficient between y_{1k} and y_{2k} in the total population.

3. Still to estimate $\overline{Y}_2$, we now propose the following estimator where α is a fixed value in $[0,1]$:

$$\widehat{\overline{Y}}_2(\alpha) = \alpha\,\widehat{\overline{Y}}_{2r} + (1 - \alpha)\,\widehat{\overline{Y}}_{2,\,\mathrm{reg}}.$$

   a) If we denote the replacement rate as $x = r/n$, and if we consider that $\widehat{\overline{Y}}_{2r}$ and $\widehat{\overline{Y}}_{2,\,\mathrm{reg}}$ are uncorrelated, explain why this approximation is reasonable, and give the optimal value of α, denoted α_{opti}.

   b) Calculate $\mathrm{var}[\widehat{\overline{Y}}_2(\alpha_{\mathrm{opti}})]$, as a function of x and ρ^2.

   c) Deduce x^*, the optimum replacement rate, as well as the optimal variance $\mathrm{var}_{\mathrm{opti}}(\rho)$ obtained with this rate.

   d) What is the gain of this strategy $\widehat{\overline{Y}}_2(\alpha_{\mathrm{opti}})$ with the rate x^* compared to the 'classical' estimator $\widehat{\overline{Y}}_2$ coming from a simple random sample of size n on date $t = 2$? (In other words, what is the design effect?) Study the variation of this gain as a function of ρ, where $0 \leq \rho \leq 1$. Make a conclusion.

4. Indicate without calculation the strategy to adopt if we want to best estimate, not the mean $\overline{Y}_2$, but the difference $\overline{Y}_2 - \overline{Y}_1$, where $\overline{Y}_1$ is the true mean of y_{1k}.

Solution

On the two successive dates t_1 and t_2, the samples can be represented by Figure 6.1, with the part S_c being common in the two samples S_1 and S_2:

Fig. 6.1. Samples on two dates: Exercise 6.11

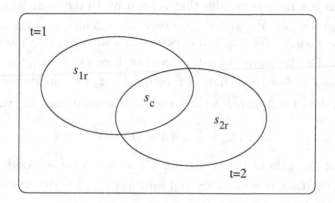

1. a) Let us look at the distributions in play. Due to the occurrence of the adopted sampling method, we have:

$$p(s_1) = \frac{1}{\binom{N}{n}},$$

$$p(s_{2r} \mid s_1) = \begin{cases} \dfrac{1}{\binom{N-n}{r}} & \text{if } s_{2r} \cap s_1 = \emptyset \\ 0 & \text{otherwise,} \end{cases}$$

therefore

$$p(s_{2r}) = \sum_{\{s_1/s_1 \cap s_{2r} = \emptyset\}} p(s_{2r} \mid s_1)p(s_1).$$

There are $\binom{N-r}{n}$ identical terms in the sum, thus:

$$p(s_{2r}) = \binom{N-r}{n} \frac{1}{\binom{N-n}{r}} \frac{1}{\binom{N}{n}} = \frac{1}{\binom{N}{r}}.$$

Everything happens 'as if' we selected r individuals from N using simple random sampling.

b) According to a), $\widehat{\overline{Y}}_{2r}$ is the mean coming from a simple random sample of size r in a population of size N and with population variance S_2^2. Therefore

$$\mathrm{var}(\widehat{\overline{Y}}_{2r}) = \frac{S_2^2}{r}.$$

2. a) The estimator proposed $\widehat{\overline{Y}}_{2,reg}$ resembles a regression estimator: $\widehat{\overline{Y}}_1$ has the role of auxiliary information (this is *not* however a true value; this is a *random* variable that is sensitive to the composition of S_1). Furthermore, $\widehat{\overline{Y}}_{1c}$ and $\widehat{\overline{Y}}_{2c}$ are very well calculated on the *same* sample, here S_c. We thus find ourselves in a situation that resembles the one for the regression estimator, but here we set $\widehat{\overline{Y}}_{2c}$ on the value taken on date $t = 1$. Indeed, if we use $\widehat{\overline{Y}}_{2,\,reg}$ to estimate the mean $\overline{Y}_1$ on date $t = 1$, and if $\widehat{b}$ takes under these conditions the value 1, we have:

$$\widehat{\overline{Y}}_{1,reg} = \widehat{\overline{Y}}_{1c} + 1 \times (\widehat{\overline{Y}}_1 - \widehat{\overline{Y}}_{1c}) = \widehat{\overline{Y}}_1,$$

and we again find the estimator $\widehat{\overline{Y}}_1$ on which 'we are calibrated'. To justify that, it is necessary and sufficient that $\widehat{b}$ is the estimated slope in the regression of y_{2k} on y_{1k}:

$$y_{2k} = a + b y_{1k} + u_k, \text{ with } \sum_{k \in U} u_k = 0 \text{ and } (a, b) \text{ minimising } \sum_{k \in U} u_k^2,$$

thus

$$\widehat{b} = \frac{\sum_{k \in S_c} (y_{2k} - \widehat{\overline{Y}}_{2c})(y_{1k} - \widehat{\overline{Y}}_{1c})}{\sum_{k \in S_c} (y_{1k} - \widehat{\overline{Y}}_{1c})^2}.$$

This formula permits as it were to be armed against the 'bizarre' compositions of S_c in comparison to S_1. If, for example, S_c plainly *overestimates* $\overline{Y}_1$, which is to say if $\widehat{\overline{Y}}_{1c} > \widehat{\overline{Y}}_1$, then, from the fact of the natural correlation between the variables y_k from dates 1 and 2, we can think that S_c continues to overestimate $\overline{Y}_2$ on date 2. We indeed verify that in that case the corrective coefficient $\widehat{b}(\widehat{\overline{Y}}_1 - \widehat{\overline{Y}}_{1c})$ is *negative* and that we have as a consequence $\widehat{\overline{Y}}_{2,\,reg} < \widehat{\overline{Y}}_{2c}$, which makes the estimation evolve 'in a good way'.

b) By the decomposition of the variance, we have:

$$\mathrm{var}[\widehat{\overline{Y}}_{2,\,reg}] = \mathrm{var}\left\{\mathrm{E}[\widehat{\overline{Y}}_{2,\,reg} \mid S_1]\right\} + \mathrm{E}\left\{\mathrm{var}[\widehat{\overline{Y}}_{2,\,reg} \mid S_1]\right\}.$$

First, we are going to condition on S_1: in this case, the risk is *not* carried on S_1, which is 'fixed'. Therefore,

$$E[\widehat{\overline{Y}}_{2,\,\text{reg}} \mid S_1]$$

$$= E\left[\widehat{\hat{b}}\,\widehat{\overline{Y}}_1 + (\widehat{\overline{Y}}_{2c} - \widehat{\hat{b}}\,\widehat{\overline{Y}}_{1c}) \mid S_1\right]$$

$$= \widehat{\overline{Y}}_1 E(\widehat{\hat{b}} \mid S_1) + E(\widehat{\overline{Y}}_{2c} \mid S_1) - E(\widehat{\hat{b}}\,\widehat{\overline{Y}}_{1c} \mid S_1)$$

$$= \widehat{\overline{Y}}_1 E(\widehat{\hat{b}} \mid S_1) + E(\widehat{\overline{Y}}_{2c} \mid S_1) - E(\widehat{\hat{b}} \mid S_1)E(\widehat{\overline{Y}}_{1c} \mid S_1) - \text{cov}(\widehat{\hat{b}}, \widehat{\overline{Y}}_{1c} \mid S_1).$$

Indeed, S_c comes from a simple random sample in S_1. That leads to:

$$E(\widehat{\overline{Y}}_{1c} \mid S_1) = \widehat{\overline{Y}}_1 \quad \text{and} \quad E(\widehat{\overline{Y}}_{2c} \mid S_1) = \widehat{\overline{Y}}_2,$$

where $\widehat{\overline{Y}}_2$ is the simple mean for y_{2k} of $S_1 = S_c \cup S_{1r}$ ($\widehat{\overline{Y}}_2$ is thus calculated on S_1, the sample selected at $t = 1$, and not on $S_c \cup S_{2r}$, which is the sample in use on date $t = 2$). Therefore

$$E[\widehat{\overline{Y}}_{2,reg} \mid S_1] = \widehat{\overline{Y}}_2 - \text{cov}(\widehat{\hat{b}}, \widehat{\overline{Y}}_{1c} \mid S_1).$$

The covariance is a very complex term, but which is of the form $\widehat{\theta}(S_1)/c$, where $\widehat{\theta}(S_1)$ is a function of S_1. Its variance compared to the risk on S_1 is negligible with respect to that of $\widehat{\overline{Y}}_2$, due to the $1/c^2$ term. Therefore

$$\text{var}\left[E(\widehat{\overline{Y}}_{2,\,\text{reg}} \mid S_1)\right] \approx \text{var}(\widehat{\overline{Y}}_2) = \frac{S_2^2}{n}.$$

Furthermore, if we set

$$\widehat{b} = \frac{\sum_{k \in S_1}(y_{2k} - \widehat{\overline{Y}}_2)(y_{1k} - \widehat{\overline{Y}}_1)}{\sum_{k \in S_1}(y_{1k} - \widehat{\overline{Y}}_1)^2},$$

$$\text{var}[\widehat{\overline{Y}}_{2,\,\text{reg}} \mid S_1] \approx \text{var}[\widehat{\overline{Y}}_{2c} - \widehat{b}\,\widehat{\overline{Y}}_{1c} \mid S_1] = \text{var}[\widehat{\overline{U}}_c \mid S_1] = \frac{s_u^2}{c}\left(1 - \frac{c}{n}\right),$$

with, for all k of S_1,

$$u_k = y_{2k} - \widehat{b}\,y_{1k} - \widehat{a},$$

the 'true' residual in the linear regression for S_1 of y_{2k} on y_{1k}, and s_u^2 is the sample variance of u_k in S_1.

Attention: $(1 - c/n)$ is not *a priori* negligible. We know that:

$$s_u^2 = (1 - \widehat{\rho}^2)\,s_{y2}^2,$$

where

- $\widehat{\rho}$ is the linear correlation coefficient between y_{1k} and y_{2k} (in S_1),
- s_{y2}^2 is the sample variance of y_{2k} in S_1 (our 'population' as there is conditioning).

Thus,

$$\text{var}[\widehat{\overline{Y}}_{2,\,\text{reg}} \mid S_1] = (1 - \widehat{\rho}^2)\left(1 - \frac{c}{n}\right)\frac{s_{y2}^2}{c},$$

with

$$E[s_{y2}^2] = S_2^2,$$

which is a classical result due to the simple random sampling of S_1. That leads to:

$$E\left[\text{var}(\widehat{\overline{Y}}_{2,\,\text{reg}} \mid S_1)\right] \approx \left(\frac{1}{c} - \frac{1}{n}\right)(1 - \rho^2)\,S_2^2.$$

Note: we replaced $\widehat{\rho}$ with ρ, calculated on the entire population of size N. Indeed, the standard deviation of $\widehat{\rho}$, which is a ratio, is of order $1/\sqrt{n}$. If n is 'large', we neglect the difference $\mid \rho^2 - \widehat{\rho}^2 \mid$ compared to ρ^2 and therefore

$$1 - \widehat{\rho}^2 = 1 - \rho^2 + \underbrace{(\rho^2 - \widehat{\rho}^2)}_{\text{negligible}} \approx 1 - \rho^2.$$

Finally, we get:

$$\text{var}[\widehat{\overline{Y}}_{2,\,\text{reg}}] \approx \frac{S_2^2}{n} + \left(\frac{1}{c} - \frac{1}{n}\right)(1 - \rho^2)\,S_2^2.$$

3. a) S_{2r} is practically independent of S_1 and S_c; the difference in independence is due to the *sole* fact that S_{2r} is selected, on date 2, in the population *outside* of S_1, but S_1 is small with respect to this population: in other words, a direct sampling of S_{2r} in the complete population (without taking count of S_1 as a result) would give 'almost surely' the same results. We therefore have

$$\text{cov}(\widehat{\overline{Y}}_{2r},\ \widehat{\overline{Y}}_{2,\,\text{reg}}) \approx 0,$$

and thus

$$\text{var}[\widehat{\overline{Y}}_2(\alpha)] \approx \alpha^2 \underbrace{\text{var}[\widehat{\overline{Y}}_{2r}]}_{\text{see 1) b)}} + (1 - \alpha)^2 \underbrace{\text{var}[\widehat{\overline{Y}}_{2,\,\text{reg}}]}_{\text{see 2) b)}}.$$

If we minimise this variance as a function of α, we very easily find:

$$\alpha_{\text{opti}} = \frac{\text{var}[\widehat{\overline{Y}}_{2,\,\text{reg}}]}{\text{var}[\widehat{\overline{Y}}_{2r}] + \text{var}[\widehat{\overline{Y}}_{2,\,\text{reg}}]} \in [0, 1].$$

b) By including the value of α_{opti} in the expression $\text{var}[\widehat{\overline{Y}}_2(\alpha)]$ and by noting that $c/n = 1 - x$, we find:

$$\text{var}\left[\widehat{\overline{Y}}_2(\alpha_{\text{opti}})\right] = \frac{S_2^2}{n}\frac{1 - \rho^2 x}{1 - \rho^2 x^2}.$$

c) With ρ 'fixed', we easily verify that $\mathrm{var}[\widehat{\overline{Y}}_2(\alpha)]$ is a convex x function and that:

$$\left[x^* \leq 1 \text{ and } \left(\frac{\partial \mathrm{var}}{\partial x} \right)_{x^*} = 0 \right] \Leftrightarrow x^* = \frac{1}{1 + \sqrt{1 - \rho^2}}.$$

Then,

$$\mathrm{var}_{\mathrm{opti}}(\rho) = \frac{S_2^2}{n} \frac{(1 + \sqrt{1 - \rho^2})}{2},$$

where

- S_2^2/n is the variance of $\widehat{\overline{Y}}_2$ in the frame of a simple random sample of size n (accuracy of reference),
- $(1 + \sqrt{1 - \rho^2})/2$ is the corrective coefficient applied to the accuracy of reference (*design effect*).

d) The design effect as a function of ρ is:

$$\mathrm{DEFF}(\rho) = \frac{1 + \sqrt{1 - \rho^2}}{2}.$$

It is a function strictly decreasing over $[0, 1]$.

Conclusion: for all $\rho \in [0, 1]$, $\mathrm{DEFF}(\rho) \leq 1$, we always improve the accuracy compared to the 'classical' estimator $\widehat{\overline{Y}}_2$, and particularly since $|\rho|$ is large, which is modelled on the 'philosophy' of regression estimation. *Whatever the value may be for ρ, we see that we get as the optimum:*

$$x^* \geq 50 \ \%.$$

It is therefore necessary to always replace at least half of the sample to reach this optimum. This result is remarkable, as intuitively it presents a certain logic: if $\rho = 1$, there is a perfect link between y_{1k} and y_{2k}. In this case corresponding to $x^* = 100\%$, it is necessary to replace at most the sample as otherwise, we collect two times in a row the *same information*! (At the limit, by choosing $c = 2$, we take only two individuals back between the two dates to be able to calculate $\widehat{b}$, and that is sufficient to ensure that $\widehat{\overline{Y}}_{2,\mathrm{reg}} = \widehat{\overline{Y}}_2$.) We are then well aware that everything happens 'as if' we had a sample of size $2n$ on date $t = 2$:

- the n individuals of S_1 corresponding to n pieces of *recollected* information, allowing for the calculation of $\widehat{\overline{Y}}_{2,\mathrm{reg}} = \widehat{\overline{Y}}_2$,
- the n individuals of S_{2r} corresponding to n pieces of *supplementary* information, allowing for the calculation of $\widehat{\overline{Y}}_{2r}$.

4. The estimation strategy of an evolution is a *panel* strategy: we renew the entire sample. This is the contrary logic ($x^* = 0$ in this case): $\widehat{\overline{Y}}_2 - \widehat{\overline{Y}}_1$ estimates $\overline{Y}_2 - \overline{Y}_1$ without bias, and $\mathrm{cov}(\widehat{\overline{Y}}_1, \widehat{\overline{Y}}_2)$ is *positive* if $\rho > 0$.

$$\text{var}(\widehat{\overline{Y}}_2 - \widehat{\overline{Y}}_1) = \text{var}(\widehat{\overline{Y}}_1) + \text{var}(\widehat{\overline{Y}}_2) - 2\text{cov}(\widehat{\overline{Y}}_1, \widehat{\overline{Y}}_2) < \text{var}(\widehat{\overline{Y}}_1) + \text{var}(\widehat{\overline{Y}}_2).$$

Therefore, we benefit from the sign of the covariance: conserving the sample over time decreases the variance of $\widehat{\overline{Y}}_2 - \widehat{\overline{Y}}_1$, which is preferable to a partial replacement, even for the total sample.

Exercise 6.12 *Bias of a ratio*

This exercise is a little atypical in spirit. It is to get here an approximation of the bias of a ratio and to propose a method to decrease this bias, without proceeding in a rigorous way but by underlining the difficulties that there would be to do this. We are placed in the case of a simple random sample of large size n. We are interested in the classical ratio constructed from the variables x and y. We denote $R = \overline{Y}/\overline{X}$ and $\widehat{R} = \widehat{\overline{Y}}/\widehat{\overline{X}}$.

1. Write $\widehat{R} - R$ under the form of a ratio utilising $\widehat{\overline{X}}, \widehat{\overline{Y}}, R, \overline{X}$ and $\Delta_X = (\widehat{\overline{X}} - \overline{X})/\overline{X}$. Specify the variance and the expected value of Δ_X.
2. Using the limited development of the function $1/(1+u)$ in the vicinity of 0, where $u \in \mathbb{R}$, rewrite $\widehat{R} - R$ in a way so that there is no longer any random variable in the denominator.
3. When n is 'large', what can we say about the random variable Δ_X?
4. We decide (arbitrarily) to hold the first two terms random from the limited development. What are the questions that we can ask ourselves if we want to practise rigorously?
5. Coming from the previous approximation that we assume to be good, express the expected value $E(\widehat{R} - R)$, by isolating the $1/n$ term. Conclude on the approximate bias of the ratio estimator when the sample size is large.
6. We consider from now on that the expected value of the ratio is written under the form:

$$E(\widehat{R}) = R + \frac{A}{n} + \frac{B}{n^{3/2}}.$$

For each individual i of the sample S, we construct the estimator $\widehat{R}^{(i)}$ on the model of $\widehat{R}$ but by removing individual i, then we are interested in the estimator:

$$\widehat{\widehat{R}} = n\widehat{R} - \frac{n-1}{n} \sum_{i \in S} \widehat{R}^{(i)}.$$

Express $E(\widehat{\widehat{R}})$ and conclude on the interest (from the point of view of bias in any case) in choosing $\widehat{\widehat{R}}$ instead of $\widehat{R}$ to estimate R.

Solution

1. As

$$\widehat{R} = \frac{\widehat{\overline{Y}}}{\widehat{\overline{X}}},$$

we have

$$\widehat{R} - R = \frac{\widehat{\overline{Y}}}{\widehat{\overline{X}}} - R = \frac{\widehat{\overline{Y}} - R\widehat{\overline{X}}}{\widehat{\overline{X}}} = \frac{\widehat{\overline{Y}} - R\widehat{\overline{X}}}{\overline{X}(1 + \Delta_X)},$$

where

$$\Delta_X = \frac{\widehat{\overline{X}} - \overline{X}}{\overline{X}}.$$

We notice that

$$\mathrm{E}(\Delta_X) = 0, \tag{6.4}$$

and

$$\mathrm{var}(\Delta_X) = \frac{1}{\overline{X}^2} \mathrm{var}(\widehat{\overline{X}}) = \frac{1}{\overline{X}^2} \frac{N-n}{N} \frac{S_x^2}{n}, \tag{6.5}$$

where

$$S_x^2 = \frac{1}{N-1} \sum_{k \in U} (x_k - \overline{X})^2.$$

2. We have, with u close to 0,

$$\frac{1}{1+u} = \sum_{j=0}^{\infty} (-u)^j,$$

therefore

$$\widehat{R} - R = \frac{\widehat{\overline{Y}} - R\widehat{\overline{X}}}{\overline{X}} \sum_{j=0}^{\infty} (-\Delta_X)^j = \frac{\widehat{\overline{Y}} - R\widehat{\overline{X}}}{\overline{X}} \left\{ 1 - \Delta_X + \sum_{j=2}^{\infty} (-\Delta_X)^j \right\}.$$

The random variable Δ_X is *a priori* close to 0, since it is of null expected value and of variance $1/n$, with n large.

3. From (6.4) and (6.5), we can write

$$\Delta_X = O_p\left(\frac{1}{\sqrt{n}}\right),$$

where $O_p(1/x)$ is a quantity which remains bounded in probability when multiplied by x, which is written

for all $\varepsilon > 0$, there exists M_ε such that $\mathrm{Pr}\left[\frac{|\Delta_X|}{\frac{1}{\sqrt{n}}} \geq M_\varepsilon\right] \leq \varepsilon.$

We can in fact show that if X is a random variable of the null expected value and of the variance $\text{var}(X) = f(n)$ where f is a given function, then $X = O_p\left(\sqrt{f(n)}\right)$.

4. We can therefore write, in neglecting the random term $\sum_{j=3}^{\infty}(-\Delta_X)^j$,

$$\widehat{R} - R = \frac{\widehat{\overline{Y}} - R\widehat{\overline{X}}}{\overline{X}}\{1 - \Delta_X + O_p\left(1/n\right)\}.$$

Indeed, $\Delta_X^2 = O_p\left(\frac{1}{n}\right)$ according to the following result:

$$O_p\left(f(n)\right) \times O_p\left(g(n)\right) = O_p\left(f(n) \times g(n)\right),$$

where $f(n)$ and $g(n)$ are any two functions. It is legitimate to do this approximation if n is large, which comes back to neglecting $1/n^{3/2}$ (which is of the order Δ_X^3) compared to $1/n$.

5. The approximate bias is

$$\text{E}(\widehat{R} - R) \approx \text{E}\left\{\frac{\widehat{\overline{Y}} - R\widehat{\overline{X}}}{\overline{X}}\left(1 - \Delta_X + O_p\left(\frac{1}{n}\right)\right)\right\}$$

$$= \frac{1}{\overline{X}}\left\{\text{E}(\widehat{\overline{Y}} - R\widehat{\overline{X}}) - \text{E}[(\widehat{\overline{Y}} - R\widehat{\overline{X}})\Delta_X] + \text{E}\left[(\widehat{\overline{Y}} - R\widehat{\overline{X}})O_p\left(\frac{1}{n}\right)\right]\right\}$$

$$= \frac{1}{\overline{X}^2}\left\{R\,\text{var}(\widehat{\overline{X}}) - \text{cov}(\widehat{\overline{X}}, \widehat{\overline{Y}}) + O_p\left(\frac{1}{n^{3/2}}\right)\right\}.$$

Indeed, $\widehat{\overline{Y}} - R\widehat{\overline{X}}$ is $O_p\left(1/\sqrt{n}\right)$, due to the null expected value and the $1/n$ variance. We select

$$(\widehat{\overline{Y}} - R\widehat{\overline{X}})O_p\left(\frac{1}{n}\right) = O_p\left(\frac{1}{n^{3/2}}\right),$$

$$\text{E}(\widehat{R} - R) \approx \frac{1}{\overline{X}^2}\frac{N-n}{Nn}\left(RS_x^2 - S_{xy}\right) + O_p\left(\frac{1}{n^{3/2}}\right),$$

where

$$S_{xy} = \frac{1}{N-1}\sum_{k \in U}(x_k - \overline{X})(y_k - \overline{Y}).$$

Finally, the expected value can be written

$$\text{E}(\widehat{R}) \approx R + \frac{A}{n} + O_p\left(\frac{1}{n^{3/2}}\right),$$

where

$$A = \frac{1}{\overline{X}^2}\frac{N-n}{N}(RS_x^2 - S_{xy}) = \left(1 - \frac{n}{N}\right)\left(\frac{S_x}{\overline{X}}\right)^2\left[\frac{\overline{Y}}{\overline{X}} - \frac{S_{xy}}{S_x^2}\right].$$

We therefore have

$$E(\widehat{R}) \approx R + \frac{A}{n} + \frac{B}{n^{3/2}}.$$

We notice that A is small in absolute value if the affine regression line for y_k on x_k in the population passes close to the origin. Furthermore, if n is large, A/n is negligible and the estimator is approximately unbiased.

6. In this question, we transform the approximation obtained in 5. into an equality. We have:

$$E(\widehat{\widehat{R}}) = nE(\widehat{R}) - \frac{n-1}{n}E\left(\sum_{i \in S} \widehat{R}^{(i)}\right).$$

Now, for every sample S collected, and for every element i selected afterwards in S, $\widehat{R}^{(i)}$ is a standard ratio constructed from a sample of size $(n-1)$ selected by simple random sampling in the complete population, being:

$$E\widehat{R}^{(i)} = R + \frac{A}{n-1} + \frac{B}{(n-1)^{3/2}} \quad \text{(approximately).}$$

Since this expected value does not depend on i, we have approximately:

$$\begin{aligned}
E\widehat{\widehat{R}} &= n\left(R + \frac{A}{n} + \frac{B}{n^{3/2}}\right) - \frac{n-1}{n}\left[n\left(R + \frac{A}{n-1} + \frac{B}{(n-1)^{3/2}}\right)\right] \\
&= R + \frac{B}{\sqrt{n}} - \frac{B}{\sqrt{n-1}} \\
&= R - \frac{B}{\sqrt{n(n-1)}\left(\sqrt{n} + \sqrt{n-1}\right)}.
\end{aligned}$$

The bias of $\widehat{\widehat{R}}$ is manifestly of order of magnitude $1/(n^{3/2})$; for n large, the bias of $\widehat{\widehat{R}}$ is approximately less than that for $\widehat{R}$ (which is of $1/n$). This technique for reducing the bias is known under the name *jackknife*.

7

Calibration with Several Auxiliary Variables

7.1 Calibration estimation

The totals of p auxiliary variables $x_1, ..., x_p$ are assumed to be known for the population U. Let us consider the vector $\mathbf{x_k} = (x_{k1}, ..., x_{kj}, ..., x_{kp})'$ of values taken by the p auxiliary variables on unit k. The total

$$\mathbf{X} = \sum_{k \in U} \mathbf{x_k}$$

is assumed to be known. The objective is always to estimate the total

$$Y = \sum_{k \in U} y_k,$$

using the information given by $\mathbf{X}$. Furthermore, we denote

$$\widehat{Y}_\pi = \sum_{k \in S} \frac{y_k}{\pi_k}, \quad \text{and} \quad \widehat{\mathbf{X}}_\pi = \sum_{k \in S} \frac{\mathbf{x_k}}{\pi_k},$$

the Horvitz-Thompson estimators of Y and $\mathbf{X}$. The general idea of calibration methods (see on this topic Deville and Särndal, 1992) consists of defining weights $w_k, k \in S$, which benefit from a calibration property, or in other words which are such that

$$\sum_{k \in S} w_k \mathbf{x_k} = \sum_{k \in U} \mathbf{x_k}. \tag{7.1}$$

To obtain such weights, we minimise a pseudo-distance $G_k(.,.)$ between w_k and $d_k = 1/\pi_k$,

$$\min_{w_k} \sum_{k \in S} \frac{G_k(w_k, d_k)}{q_k},$$

under the constraints of calibration given in (7.1). The weights $q_k, k \in S$, form a set of strictly positive known coefficients. The function $G_k(.,.)$ is assumed to be strictly convex, positive and such that $G_k(d_k, d_k) = 0$. The weights w_k are then defined by

$$w_k = d_k F_k(\lambda' \mathbf{x}_k),$$

where $d_k F_k(.)$ is the reciprocal of the function $g_k(., d_k)/q_k$, with

$$g_k(w_k, d_k) = \frac{\partial G_k(w_k, d_k)}{\partial w_k},$$

and λ is the Lagrange multiplier following from the constraints. The vector λ is obtained by solving the calibration equations:

$$\sum_{k \in S} d_k F_k(\lambda' \mathbf{x}_k) \mathbf{x}_k = \sum_{k \in U} \mathbf{x}_k.$$

7.2 Generalised regression estimation

If the function $G_k(.,.)$ is chi-square,

$$G_k(w_k, d_k) = \frac{(w_k - d_k)^2}{d_k},$$

then the calibrated estimator is equal to the generalised regression estimator which is

$$\widehat{Y}_{\text{reg}} = \widehat{Y}_\pi + (\mathbf{X} - \widehat{\mathbf{X}}_\pi)' \widehat{\mathbf{b}},$$

where

$$\widehat{\mathbf{b}} = \left(\sum_{k \in S} \frac{\mathbf{x}_k \mathbf{x}_k' q_k}{\pi_k} \right)^{-1} \sum_{k \in S} \frac{\mathbf{x}_k y_k q_k}{\pi_k}.$$

7.3 Marginal calibration

A particularly important case is obtained when the auxiliary variables are the indicator variables of the strata, and the function $G_k(w_k, d_k) = w_k \log(w_k/d_k)$. We can show that we then obtain weights equivalent to those given by the calibration algorithm on the margins (also known under the name *raking ratio*). In the case where the sample leads to a table of real values estimated $\widehat{N}_{ij}, i = 1, \ldots, I$, and $j = 1 \ldots, J$, and the true marginals $N_{i.}, i = 1, \ldots, I$, and $N_{.j}, j = 1, \ldots, J$, of this table are known in the population, the equivalent calibration method consists of adjusting the estimated table successively by row and by column. The algorithm is thus the following. We initialise by having:

$$N_{ij}^{(0)} = \widehat{N}_{ij}, \text{ for all } i = 1, \ldots I, j = 1, \ldots J.$$

Next, we successively adjust the rows and columns. For $t = 1, 2, 3, \ldots$

$$N_{ij}^{(2t-1)} = N_{ij}^{(2t-2)} \frac{N_{i.}}{\sum_j N_{ij}^{(2t-2)}}, \text{ for all } i = 1, \ldots I, j = 1, \ldots J,$$

$$N_{ij}^{(2t)} = N_{ij}^{(2t-1)} \frac{N_{.j}}{\sum_i N_{ij}^{(2t-1)}}, \text{ for all } i = 1, \ldots I, j = 1, \ldots J.$$

The algorithm rapidly converges if the table $\widehat{N}_{ij}$ is not composed of null values.

EXERCISES

Exercise 7.1 *Adjustment of a table on the margins*

Using a sampling procedure, we get the Horvitz-Thompson estimators $\widehat{N}_{ij}$ from a contingency table (see Table 7.1). Now, the margins of this table are

Table 7.1. Table obtained through sampling: Exercise 7.1

80	170	150	400
90	80	210	380
10	80	130	220
180	330	490	1000

known for the entire population. The true totals of the rows are $(430, 360, 210)$, and the true totals of the columns $(150, 300, 550)$. Adjust the table obtained using sampling on the known margins of the population with the '*raking ratio*' method.

Solution

We start indiscriminately with an adjustment on the rows or on the columns. Here, we chose to start with an adjustment by row.

Calibration by row: iteration 1			
86.00	182.75	161.25	430.00
85.26	75.79	198.95	360.00
9.55	76.36	124.09	210.00
180.81	334.90	484.29	1000.00

Next, we adjust on the columns.

Calibration by column: iteration 2			
71.35	163.70	183.13	418.18
70.73	67.89	225.94	364.57
7.92	68.41	140.93	217.25
150.00	300.00	550.00	1000.00

We then repeat these two steps.

Calibration by row: iteration 3			
73.36 168.33 188.31	430.00
69.85 67.04 223.11	360.00
7.65 66.12 136.22	210.00
150.87 301.49 547.64	1000.00

Calibration by column: iteration 4			
72.94 167.50 189.12	429.56
69.45 66.71 224.07	360.23
7.61 65.79 136.81	210.22
150.00 300.00 550.00	1000.00

Calibration by row: iteration 5			
73.02 167.67 189.31	430.00
69.40 66.67 223.93	360.00
7.60 65.73 136.67	210.00
150.02 300.06 549.91	1000.00

Calibration by column: iteration 6			
73.01 167.64 189.34	429.98
69.39 66.65 223.97	360.01
7.60 65.71 136.69	210.01
150.00 300.00 550.00	1000.00

Calibration by row: iteration 7			
73.01 167.64 189.35	430.00
69.39 66.65 223.96	360.00
7.60 65.71 136.69	210.00
150.00 300.00 550.00	1000.00

Calibration by column: iteration 8			
73.01 167.64 189.35	430.00
69.39 66.65 223.96	360.00
7.60 65.71 136.69	210.00
150.00 300.00 550.00	1000.00

Calibration by row: iteration 9			
73.01 167.64 189.35	430.00
69.39 66.65 223.96	360.00
7.60 65.71 136.69	210.00
150.00 300.00 550.00	1000.00

Calibration by column: iteration 10			
73.01 167.64 189.35	430.00
69.39 66.65 223.96	360.00
7.60 65.71 136.69	210.00
150.00 300.00 550.00	1000.00

After 11 iterations, the adjustment is very accurate.

Calibration by row: iteration 11			
73.01 167.64 189.35	430.00
69.39 66.65 223.96	360.00
7.60 65.71 136.69	210.00
150.00 300.00 550.00	1000.00

Exercise 7.2 *Ratio estimation and adjustment*

We are interested in the population of 10000 students registered in their first year at a university. We know the total number of students whose parents have graduated from primary school, secondary school and higher education. We take a survey according to a simple random design without replacement of 150 students. We divide these 150 students according to the education level of their parents and their own marks (pass or fail) during the first year, and we get Table 7.2. The number of students whose parents have graduated from primary school, secondary school and higher education are respectively 5000, 3000 and 2000:

Table 7.2. Academic failure according to the education level of parents: Exercise 7.2

	Students results	
Education of parents	Fail	Pass
Primary	45	15
Secondary	25	25
Higher	10	30

1. Estimate the passing rate of students using the Horvitz-Thompson estimator and give a variance estimator and a 95% confidence interval for this rate.
2. Explain why it is *a priori* worthwhile to make an adjustment, and why this adjustment should decrease the value of the estimate from 1.
3. Estimate the passing rate of students using the post-stratified estimator and give a variance estimator and a 95% confidence interval for this rate.
4. Estimate the passing rate by the level of education of the parents using a raking ratio knowing that, in the total student population, the passing rate is in reality 40%.

Solution

1. The margins of Table 7.2 are given in Table 7.3. Since it is a simple random

Table 7.3. Table of academic failure with its margins: Exercise 7.2

	Fail	Pass	Total
Primary	45	15	60
Secondary	25	25	50
Higher	10	30	40
Total	80	70	150

sample, the Horvitz-Thompson estimator for the passing rate P (where P is the total number of passes divided by 10000) is given by

$$\widehat{P} = \frac{70}{150} = 0.467 = 46.7\%.$$

When a variable y_k takes as its value 0 (fail) or 1 (pass), we have

$$s_y^2 = \frac{1}{n-1} \sum_{k \in S} \left(y_k - \widehat{\overline{Y}}\right)^2 = \frac{n\widehat{P}(1-\widehat{P})}{n-1}.$$

Thus

$$
\begin{aligned}
\widehat{\mathrm{var}}(\widehat{P}) &= \frac{N-n}{N}\frac{s_y^2}{n} \\
&= \frac{N-n}{N}\frac{\widehat{P}(1-\widehat{P})}{n-1} \\
&= \frac{10000-150}{10000}\times\frac{70}{150}\times\frac{80}{150}\times\frac{1}{149} \\
&= 0.00164 \\
&= (0.0405)^2.
\end{aligned}
$$

The estimated 95% confidence interval is ($n = 150$ is sufficiently large):

$$
P \in \left[\widehat{P} \pm 1.96\sqrt{\widehat{\mathrm{var}}(\widehat{P})}\right] = [46.7\% \pm 8.0\%].
$$

2. The adjustment appears natural as the structure of the sample differs greatly from the expected structure 'on average'. Indeed, if we are interested for example in the first post-stratum (students whose parents have only a primary school education), we have:

$$
\mathrm{E}(n_1) = \frac{N_1}{N}n = \frac{5000}{10000}\times 150 = 75,
$$

while $n_1 = 60$. To judge the importance of the difference, let us calculate the interval for which n_1 has a 95% chance of being found (n_1 roughly follows a normal distribution):

$$
n_1 \in \left[\mathrm{E}(n_1) \pm 1.96\sqrt{\mathrm{var}(n_1)}\right],
$$

where

$$
\begin{aligned}
\mathrm{var}\left(\frac{n_1}{n}\right) &\approx \frac{N-n}{nN}\frac{N_1}{N}\left(1-\frac{N_1}{N}\right) \\
&= \frac{10000-150}{10000\times 150}\times\frac{1}{2}\times\left(1-\frac{1}{2}\right) \approx (0.0405)^2.
\end{aligned}
$$

Therefore,

$$
n_1 \in [75 \pm 11.91].
$$

Now, $n_1 = 60$ is outside of the interval! The adjustment must logically decrease the estimate from 1. In fact, we can already notice in Table 7.4 that the passing rates $\widehat{P}_h$ (proportions of passing by category h which are unbiased according to the theory of domain estimation) vary a lot from one category to another, thus showing a quite strong explanatory characteristic of the variable 'education level of parents'. We then notice that there are 'too many' in the category 'higher' in comparison to the

Table 7.4. Passing rates according to the education level of parents: Exercise 7.2

	n_h	$E(n_h)$	$\widehat{P}_h$
Primary	60	75	25%
Secondary	50	45	50%
Higher	40	30	75%
Total	150	150	46.7%

expected mean structure, and too few in the category 'primary'; that is, an over-representation of the category with the highest passing rate and correspondingly, a deficit in the category with the smallest passing rate. It is therefore logical that the simple estimator

$$\widehat{P} = \sum_{h=1}^{3} \frac{n_h}{n} \widehat{P}_h$$

is too high and that the post-stratification decreases the numerical estimate by correcting in a certain way the effect of the structure.

3. The post-stratified estimator is given by

$$\widehat{P}_{\text{post}} = \sum_{h=1}^{3} \frac{N_h}{N} \widehat{P}_h,$$

where the $\widehat{P}_h$ are the passing rates estimated for each post-stratum. Therefore,

$$\widehat{P}_{\text{post}} = \frac{1}{10000} \left\{ 5000 \times \frac{15}{60} + 3000 \times \frac{25}{50} + 2000 \times \frac{30}{40} \right\}$$

$$= \frac{1}{10000} \{1250 + 1500 + 1500\} = 0.425 = 42.5\%.$$

The post-stratification has indeed decreased the numerical estimate, as planned. The unbiased estimators of the population variance within the post-strata are the following:

$$s_{y1}^2 = \frac{\widehat{P}_1(1 - \widehat{P}_1)}{n_1 - 1} n_1 = \frac{15}{60} \times \frac{45}{60} \times \frac{60}{59} = 0.1906779,$$

$$s_{y2}^2 = \frac{\widehat{P}_2(1 - \widehat{P}_2)}{n_2 - 1} n_2 = \frac{25}{50} \times \frac{25}{50} \times \frac{50}{49} = 0.255102,$$

$$s_{y3}^2 = \frac{\widehat{P}_3(1 - \widehat{P}_3)}{n_3 - 1} n_3 = \frac{30}{40} \times \frac{10}{40} \times \frac{40}{39} = 0.1923076.$$

We can proceed with the (approximately) unbiased estimation of the variance:

$$\widehat{\text{var}}(\widehat{P}_{\text{post}})$$

$$= \frac{N-n}{nN^2} \sum_{h=1}^{3} N_h s_{yh}^2 + \frac{N-n}{n^2 N} \sum_{h=1}^{3} \frac{N-N_h}{N} s_{yh}^2$$

$$= \frac{10000-150}{150 \times 10000^2} \left\{ 5000 \times s_{y1}^2 + 3000 \times s_{y2}^2 + 2000 \times s_{y3}^2 \right\}$$

$$+ \frac{10000-150}{150^2(10000-1)} \left\{ \frac{5000}{10000} \times s_{y1}^2 + \frac{7000}{10000} \times s_{y2}^2 + \frac{8000}{10000} \times s_{y3}^2 \right\}$$

$$= \frac{10000-150}{150 \times 10000^2} \left\{ 953.3895 + 765.306 + 384.615 \right\}$$

$$+ \frac{10000-150}{150^2(10000-1)} \frac{1}{10000} \left\{ 953.3895 + 1785.714 + 1538.461 \right\}$$

$$= 0.0013812 + 0.0000187$$

$$= 0.0014$$

$$= (0.0374)^2.$$

Therefore, $\widehat{\text{var}}(\widehat{P}_{\text{post}}) < \widehat{\text{var}}(\widehat{P})$. We notice that the second term of $\widehat{\text{var}}(\widehat{P}_{\text{post}})$, being 0.0000187, is numerically negligible compared to the first term (0.0013812): this ratio of orders of magnitude is classical when the sample size is 'large'. The estimated 95% confidence interval is:

$$P \in \left[\widehat{P}_{\text{post}} \pm 1.96 \sqrt{\widehat{\text{var}}(\widehat{P}_{\text{post}})} \right] = [42.5\% \pm 7.3\%].$$

This confidence interval quite considerably covers that for the raw estimator (see 1.).

4. We have available two qualitative variables (level of education of the parents on the one hand and passing rate on the other hand), of which we know here the population sizes of each of the distinct values. If we consider the contingency table divided according to these two variables for the 150 students sampled, we know the 'theoretical' margins for the table but not the true values of the cases (see Table 7.5). The first three steps of the algorithm for calibrating on the margins are presented in Tables 7.6 to 7.8 and provide in case (i,j) the estimates $\widehat{N}_{ij}$ for the population size for the 10000 students, successively adjusted to the distinct values i and j of the variable by row and by column.

After 10 iterations, we obtain Table 7.9, nearly perfect by row and by column at the same time. Since we applied the algorithm directly on the population sizes, we get an estimated distribution of the total population for the 10000 students, from the 'asymptotically' unbiased estimators ($n = 150$ is sufficiently large in order for the bias to be negligible): it is then sufficient to read the passing rates by domain (each post-stratum technically constitutes a domain). The passing rates according to the level of education of the parents are therefore the following ratios (bias negligible):

Table 7.5. Table of academic failure with its margins in the population: Exercise 7.2

Start	Fail	Pass	Total	Margins
Primary	45	15	60	5000
Secondary	25	25	50	3000
Higher	10	30	40	2000
Total	80	70	150	
Margins	6000	4000		10000

Table 7.6. Adjustment on the margins, step 1: Exercise 7.2

Step 1	Fail	Pass	Total	Margins
Primary	3750	1250	5000	5000
Secondary	1500	1500	3000	3000
Higher	500	1500	2000	2000
Total	5750	4250	10000	
Margins	6000	4000		10000

Table 7.7. Adjustment on the margins, step 2: Exercise 7.2

Step 2	Fail	Pass	Total	Margins
Primary	3913.0	1176.5	5089.5	5000
Secondary	1565.2	1411.8	2977.0	3000
Higher	521.8	1411.7	1933.5	2000
Total	6000	4000	10000	
Margins	6000	4000		10000

Table 7.8. Adjustment on the margins, step 3: Exercise 7.2

Step 3	Fail	Pass	Total	Margins
Primary	3844.2	1155.8	5000	5000
Secondary	1577.3	1422.7	3000	3000
Higher	539.7	1460.3	2000	2000
Total	5961.2	4038.8	10000	
Margins	6000	4000		10000

- primary: $1138.9/5000.1 \approx 23\%$,
- secondary: $1408.4/3000 \approx 47\%$,
- higher: $1452.7/1999.9 \approx 73\%$.

These values must be compared to the three unbiased 'natural' estimators from the initial division of the sample, being respectively 25%, 50% and 75%. To choose the 'best' estimators, as all these estimators are unbiased or with negligible bias, it would remain to perform variance estimation.

Table 7.9. Table adjusted on the margins in 10 iterations: Exercise 7.2

Step 10	Fail	Pass	Total	Margins
Primary	3861.2	1138.9	5000.1	5000
Secondary	1591.6	1408.4	3000.0	3000
Higher	547.2	1452.7	1999.9	2000
Total	6000	4000	10000	
Margins	6000	4000		10000

Exercise 7.3 *Regression and unequal probabilities*

This exercise, theoretical enough, deals with regression estimation in the frame of sampling with unequal probabilities. It is composed of two independent parts.

First part:

The objective is to establish the expression of the regression estimator in the case where the regressors are the results x_k for a real variable, and the constant 1.

1. Recall the expression $\tilde{b}$ for the slope of the *true* regression line as a function of x_k and y_k, where k varies from 1 to N.

2. For a sampling design with unequal probabilities (π_k), what natural estimator of $\tilde{b}$ (denoted $\widehat{\tilde{b}}$) are we tempted to use in 'sticking' with the expression found in 1., by noticing that the numerator and the denominator of $\tilde{b}$ are sums?

3. With the mean 'regression' estimator set up by using the estimator of $\tilde{b}$ from 2., verify that the expected calibration (that is, the 'perfect' estimate of the total of x_k on the one hand, and of the population size N on the other hand) is no longer satisfied and that, as a result, the so-called 'regression' estimator is not the one that we think it is but is something else.

4. Set up the normal equations giving $\tilde{a}$ and $\tilde{b}$, the true regression coefficients from the relation:

$$y_k = a + bx_k + u_k, \quad \text{with} \quad \sum_{k \in U} u_k = 0.$$

We recall that these equations are obtained by writing the least square criteria and by differentiating it with respect to a and b.

5. Rewrite these equations by replacing all true sums (unknown) by their unbiased estimators and consider that the new system has for the solutions of a and b the estimated regression coefficients $\widehat{a}$ and $\widehat{b}$.

6. Finally, develop the regression estimator from $\widehat{a}$ and $\widehat{b}$ and verify that, this time, the estimator has the calibration properties required.

Second part:
In this part, it is a question of presenting a particular approach for the regression estimator. From the outcome of a sampling procedure (eventually very complex) we are led to use the estimator of the total:

$$\widehat{Y} = \sum_{k \in S} d_k y_k.$$

The weights $(d_k)_{k \in S}$ are real known values, determined by the sample.
The objective is to reweight the selected individuals by assigning them a new weight w_k in a context where we know two auxiliary variables x_k and z_k for each individual in the population (true totals X and Z known), in such a way to minimise

$$\sum_{k \in S} \frac{(w_k - d_k)^2}{d_k}, \quad \text{subject to} \quad \begin{cases} \sum_{k \in S} w_k x_k = X, \text{ and} \\ \sum_{k \in S} w_k z_k = Z. \end{cases}$$

1. Comment on this procedure.
2. Solve this and notice that the estimator obtained is the regression estimator on X and Z. We thus get a simple interpretation of this estimator.

Solution
First part:

1. The regression is written on the population:

$$y_k = a + b x_k + u_k, \quad (a,\ b) \in \mathbb{R}^2, \quad \text{with} \quad \sum_{k \in U} u_k = 0.$$

Attention: the u_k are not random (it is a matter here of rewriting y_k).
If we minimise $\sum_{k \in U} u_k^2$, we find:

$$\begin{cases} \tilde{b} = \dfrac{\sum_{k \in U} (y_k - \overline{Y})(x_k - \overline{X})}{\sum_{k \in U} (x_k - \overline{X})^2}, \\ \tilde{a} = \overline{Y} - \tilde{b}\overline{X}, \end{cases}$$

which are the true regression coefficients $\tilde{a}$ and $\tilde{b}$. These values are not calculable.

2. We are tempted to estimate all the sums with the Horvitz-Thompson estimators; that is to say, to use, quite naturally:

$$\widehat{\tilde{b}} = \frac{\sum_{k \in S} (y_k - \widehat{\overline{Y}})(x_k - \widehat{\overline{X}})/\pi_k}{\sum_{k \in S} (x_k - \widehat{\overline{X}})^2/\pi_k},$$

where

$$\widehat{\overline{Y}} = \frac{1}{N} \sum_{k \in S} \frac{y_k}{\pi_k},$$

is an unbiased estimator of $\overline{Y}$, and

$$\widehat{\overline{X}} = \frac{1}{N} \sum_{k \in S} \frac{x_k}{\pi_k},$$

is an unbiased estimator of $\overline{X}$. The divisions by π_k in the numerator and denominator proceed in the same way, as that for the classical estimator of a sum. Next, we determine $\widehat{a}$ by

$$\widehat{a} = \widehat{\overline{Y}} - \widehat{b}\widehat{\overline{X}}.$$

3. The mean 'regression' estimator (or supposedly like that) thus seems to be, at this stage:

$$\widetilde{\overline{Y}} = \widehat{\overline{Y}} + \widehat{b}(\overline{X} - \widehat{\overline{X}}) + \widehat{a}\left(1 - \frac{\widehat{N}}{N}\right),$$

where

$$\widehat{N} = \sum_{k \in S} \frac{1}{\pi_k}.$$

We must not omit the last term $\widehat{a}\left(1 - \widehat{N}/N\right)$ which must express the calibration on the constant.

Unfortunately, this formula *does not work*! Indeed, if we innocently select, for all $k \in U$: $y_k = 1$, then

$$\widehat{b} = \frac{\sum_{k \in S}\left(1 - \frac{\widehat{N}}{N}\right)(x_k - \widehat{\overline{X}})/\pi_k}{\sum_{k \in S}(x_k - \widehat{\overline{X}})^2/\pi_k} = \frac{\widehat{X}\left(1 - \frac{\widehat{N}}{N}\right)^2}{\sum_{k \in S}\left(x_k - \widehat{\overline{X}}\right)^2/\pi_k} \neq 0,$$

with

$$\widehat{X} = \sum_{k \in S} \frac{x_k}{\pi_k},$$

and, consequently, $\widetilde{\overline{Y}} \neq \overline{Y}$, where $\overline{Y} = 1$. There is therefore no calibration on the constant (which is to say, on the population size N if we think in terms of the total): the fundamental property of calibration is not satisfied! From this fact, $\widetilde{\overline{Y}}$ is not the regression estimator. The fundamental error originates from the use of N instead of $\widehat{N}$: the regression estimator must be constructed by estimating every total with its Horvitz-Thompson estimator, including the population size N, like how the following is shown.

4. The normal equations are given by the procedure:

$$\min_{a,b} \sum_{k \in U} (y_k - a - bx_k)^2,$$

which gives

$$\begin{cases} \displaystyle\sum_{k\in U} (y_k - \tilde{a} - \tilde{b}x_k)x_k = 0 \\ \displaystyle\sum_{k\in U} (y_k - \tilde{a} - \tilde{b}x_k) = 0. \end{cases}$$

5. The previous system, consisting of two equations for two unknowns $\tilde{a}$ and $\tilde{b}$ in $\mathbb{R}$, is translated in a new way if we are only interested in the sampled data (the other data being unknown, the solution for $\tilde{a}$ and $\tilde{b}$ to the previous system would not result in anything for the numerical design). We therefore solve

$$\begin{cases} \displaystyle\sum_{k\in S} \frac{(y_k - \widehat{a} - \widehat{b}x_k)\, x_k}{\pi_k} = 0 \\ \displaystyle\sum_{k\in S} \frac{(y_k - \widehat{a} - \widehat{b}x_k)}{\pi_k} = 0. \end{cases}$$

By denoting

$$\widehat{(XY)} = \sum_{k\in S} \frac{x_k y_k}{\pi_k}, \quad \widehat{Y} = \sum_{k\in S} \frac{y_k}{\pi_k}, \quad \text{and} \quad \widehat{(X^2)} = \sum_{k\in S} \frac{x_k^2}{\pi_k},$$

we arrive without difficulty at:

$$\widehat{b} = \frac{\widehat{(XY)} - \widehat{N}\dfrac{\widehat{X}}{N}\dfrac{\widehat{Y}}{N}}{\widehat{(X^2)} - \widehat{N}\left(\dfrac{\widehat{X}}{N}\right)^2}.$$

We deduce:

$$\widehat{b} = \frac{\sum_{k\in S}\left(x_k - \dfrac{\widehat{X}}{N}\right)\left(y_k - \dfrac{\widehat{Y}}{N}\right)/\pi_k}{\sum_{k\in S}\left(x_k - \dfrac{\widehat{X}}{N}\right)^2/\pi_k},$$

and

$$\frac{\widehat{Y}}{N} = \widehat{a} + \widehat{b}\,\frac{\widehat{X}}{N}.$$

6. • We are going to set:

$$\widehat{\overline{Y}}_{\text{reg}} = \widehat{\overline{Y}} + \widehat{b}(\overline{X} - \widehat{\overline{X}}) + \widehat{a}\left(1 - \frac{\widehat{N}}{N}\right).$$

• If $y_k = x_k$, we have

$$\begin{cases} \widehat{b} = 1 \\ \widehat{a} = 0, \end{cases}$$

and therefore $\widehat{\overline{Y}} = \widehat{\overline{X}}$. Hence

$$\widehat{\overline{X}}_{\text{reg}} = \widehat{\overline{X}} + 1(\overline{X} - \widehat{\overline{X}}) + 0 = \overline{X}.$$

- If $y_k = 1$, we obtain

$$\widehat{b} = \frac{\sum_{k \in S} \left(1 - \frac{\widehat{N}}{N}\right) \left(x_k - \frac{\widehat{X}}{N}\right)}{\sum_{k \in S} \left(x_k - \frac{\widehat{X}}{N}\right)^2}$$

being

$$\begin{cases} \widehat{b} = 0 \\ \widehat{a} = 1. \end{cases}$$

Hence

$$\widehat{\overline{Y}}_{\text{reg}} = \widehat{\overline{1}} + 0 + \left(1 - \frac{\widehat{N}}{N}\right),$$

with

$$\widehat{\overline{1}} = \frac{1}{N} \sum_{k \in S} \frac{1}{\pi_k} = \frac{\widehat{N}}{N},$$

which gives

$$\widehat{\overline{Y}}_{\text{reg}} = 1 = \overline{Y}.$$

There is therefore *double calibration* on each of the two auxiliary variables which are x_k and 1. Thus, $\widehat{\overline{Y}}_{\text{reg}}$ is indeed the regression estimator.

Note: we stress the role of the intercept in the 'model' at the start: to set a constant, it is said that we can perfectly estimate the total of the '1's, which is to say the population size N.

Second part:

1. We minimise a *distance* between 'raw weights' d_k and 'estimator weights from adjustment' w_k. The distance is of type chi-square (we can of course think of other distances but this one is often used in statistics).
   The constraints are the traditional properties of calibration. The minimisation of a distance is justified by the search for weights w_k as close as possible to the raw weights d_k: in fact, the weights d_k are generally established in order to define unbiased estimators. The calibration is going to destroy this property of the absence of bias, but the bias of the new estimator is *a priori* much smaller than the w_k are close to d_k.
2. The minimisation under the constraints lead to the Lagrangian calculation:

$$\mathcal{L} = \sum_{k \in S} \frac{(w_k - d_k)^2}{d_k} - 2\lambda \left(\sum_{k \in S} w_k x_k - X \right) - 2\mu \left(\sum_{k \in S} w_k z_k - Z \right).$$

By setting the partial derivatives of $\mathcal{L}$ equal to zero:

$$\frac{\partial \mathcal{L}}{\partial w_k} = 0, \text{ for all } k \text{ of } S,$$

we obtain

$$w_k = d_k + d_k \lambda x_k + d_k \mu z_k. \tag{7.2}$$

According to the constraints, we have

$$\begin{cases} X = \widehat{X} + \lambda \widehat{(X^2)} + \mu \widehat{(XZ)} \\ Z = \widehat{Z} + \mu \widehat{(Z^2)} + \lambda \widehat{(XZ)}, \end{cases}$$

where

$$\widehat{X} = \sum_{k \in S} d_k x_k, \quad \widehat{(X^2)} = \sum_{k \in S} d_k x_k^2, \quad \widehat{(XZ)} = \sum_{k \in S} d_k x_k z_k.$$

Therefore

$$\begin{pmatrix} \lambda \\ \mu \end{pmatrix} = T^{-1} \begin{pmatrix} X - \widehat{X} \\ Z - \widehat{Z} \end{pmatrix},$$

where

$$T = \begin{pmatrix} \widehat{(X^2)} & \widehat{(XZ)} \\ \widehat{(XZ)} & \widehat{Z^2} \end{pmatrix}.$$

Then, we establish the expression of the estimator adjusted according to Expression (7.2)

$$\sum_{k \in S} w_k y_k = \widehat{Y} + \lambda \widehat{(XY)} + \mu \widehat{(YZ)}$$

$$= \widehat{Y} + \left(\widehat{(XY)}, \ \widehat{(YZ)} \right) \begin{pmatrix} \lambda \\ \mu \end{pmatrix}$$

$$= \widehat{Y} + \left(\widehat{(XY)}, \ \widehat{(YZ)} \right) T^{-1} \begin{pmatrix} X - \widehat{X} \\ Z - \widehat{Z} \end{pmatrix}$$

$$= \widehat{Y} + (\widehat{A}, \ \widehat{B}) \begin{pmatrix} X - \widehat{X} \\ Z - \widehat{Z} \end{pmatrix}$$

$$= \widehat{Y} + \widehat{A}(X - \widehat{X}) + \widehat{B}(Z - \widehat{Z}),$$

where

$$\begin{pmatrix} \widehat{A} \\ \widehat{B} \end{pmatrix} = T^{-1} \begin{pmatrix} \widehat{(XY)} \\ \widehat{(ZY)} \end{pmatrix}.$$

We again find the *regression coefficient* estimated from the rewritten form:

$$y_k = Ax_k + Bz_k + u_k \ \left(\text{with} \sum_{k \in U} u_k = 0\right),$$

where A and B are estimated in the frame of sampling with unequal probabilities: it is in fact the (well-known) expression of the parameter of ordinary least squares, in which all the true sums (unknown) have been estimated by the unbiased Horvitz-Thompson estimator.

Conclusion: $\sum_{k \in S} w_k y_k$ is indeed the regression estimator of the total, which is therefore interpreted as the adjusted estimator 'closest' to the raw estimator, in the sense of the chi-square distance between the weights.

Exercise 7.4 *Possible and impossible adjustments*

Adjust Tables 7.10 and 7.11 to the margins labelled 'to adjust' using the 'raking ratio' method (we notice that the margins by row are satisfied immediately). Explain the problem posed by Table 7.11.

Table 7.10. Table to adjust on the margins: Exercise 7.4

	Data to adjust				Total	To adjust
	235	78	15	6	**334**	**334**
	427	43	17	12	**499**	**499**
	256	32	14	5	**307**	**307**
	432	27	32	2	**493**	**493**
Total	**1350**	**180**	**78**	**25**	**1633**	
To adjust	**25**	**78**	**180**	**1350**		**1633**

Table 7.11. Table to adjust on the margins: Exercise 7.4

	Data to adjust				Total	To adjust
	0	78	0	6	**84**	**84**
	427	0	17	0	**444**	**444**
	0	32	0	5	**37**	**37**
	432	0	32	0	**464**	**464**
Total	**859**	**110**	**49**	**11**	**1029**	
To adjust	**11**	**49**	**110**	**859**		**1029**

Solution

After the application on Table 7.10 of 10 iterations of the algorithm used in Exercises 7.1 and 7.2, we get Table 7.12. This example shows that even with an initial structure that is extremely far from the theoretical structure given by the margins (by column), we get to this adjusted table respecting the fixed margins.

By reorganising the rows and the columns of Table 7.11, we get Table 7.13. The initial structure respects the rows but not at all the columns. Indeed, the method of adjustment, by following the rules of three, obviously conserve the zeroes of the table. The method then comes back to separately adjusting the two 2×2 tables of the diagonal. Since $84 + 37 \neq 49 + 859$, and since $11 + 110 \neq 444 + 464$, it is impossible to adjust this table.

Table 7.12. Result of the adjustment of Table 7.10: Exercise 7.4

2.54	25.40	17.81	288.25	334
3.74	11.36	16.37	467.53	499
3.14	11.85	18.89	273.12	307
15.58	29.39	126.93	321.10	493
25	**78**	**180**	**1350**	**1633**

Table 7.13. Reorganisation of rows and columns of Table 7.11: Exercise 7.4

	Data to adjust				Total	To adjust
	78	6	0	0	**84**	**84**
	32	5	0	0	**37**	**37**
	0	0	427	17	**444**	**444**
	0	0	432	32	**464**	**464**
Total	**110**	**11**	**859**	**49**	**1029**	
To adjust	**49**	**859**	**11**	**110**		**1029**

Exercise 7.5 *Calibration and linear method*

Give the calibration equations for a problem of adjustment on two margins (of respective sizes H and I) in a simple random design with $q_k = 1$ (see course summaries) by selecting the adjustment function $F_k(u) = F(u) = 1+u$ (method is called linear).

Solution

The auxiliary variables are split into two groups (one for each qualitative variable)

$$x_{k1}, ..., x_{kh}, ..., x_{kH} \text{ (vertical margin of the table)},$$

and

$$z_{k1}, ..., z_{ki}, ..., z_{kI} \text{ (horizontal margin of the table)},$$

where, if U_{hi} indicated the population defined by the overlap of row h and column i of the contingency table:

$$x_{kh} = \begin{cases} 1 & \text{if } k \in U_{h.} = \bigcup_{i=1}^{I} U_{hi} \\ 0 & \text{otherwise}, \end{cases}$$

and

$$z_{ki} = \begin{cases} 1 & \text{if } k \in U_{.i} = \bigcup_{h=1}^{H} U_{hi} \\ 0 & \text{otherwise}. \end{cases}$$

We denote $\mathbf{x_k}$ as the vector of x_{kh} $(1 \le h \le H)$ and $\mathbf{z_k}$ as the vector of z_{ki} $(1 \le i \le I)$:

$$\mathbf{x_k} = (x_{k1}, x_{k2}, ..., x_{kH})', \quad \text{and} \quad \mathbf{z_k} = (z_{k1}, z_{k2}, ..., z_{kI})'.$$

The constraints linked to the H rows give way to the Lagrange multipliers $\boldsymbol{\lambda} = (\lambda_1, \lambda_2, \dots, \lambda_H)'$, likewise as the constraints linked to the I columns lead to multipliers $\boldsymbol{\mu} = (\mu_1, \mu_2, \dots, \mu_I)'$. The row constraints are of type:

$$\mathbf{X} = \sum_{k \in S} w_k \mathbf{x_k},$$

and the column constraints of type

$$\mathbf{Z} = \sum_{k \in S} w_k \mathbf{z_k},$$

where

$$\mathbf{X} = \sum_{k \in U} \mathbf{x}_k, \quad \text{and} \quad \mathbf{Z} = \sum_{k \in U} \mathbf{z}_k.$$

If the initial weight of k is denoted d_k, we get

$$w_k = d_k F\left(\mathbf{x_k'}\boldsymbol{\lambda} + \mathbf{z_k'}\boldsymbol{\mu}\right).$$

In the linear frame, we set $F(u) = 1 + u$. The calibration equations are therefore written, coordinate by coordinate:

$$X_h = \sum_{k \in S} x_{kh} d_k \left(1 + \sum_{h=1}^{H} \lambda_h x_{kh} + \sum_{i=1}^{I} \mu_i z_{ki} \right), \text{ with } 1 \leq h \leq H,$$

and

$$Z_i = \sum_{k \in S} z_{ki} d_k \left(1 + \sum_{h=1}^{H} \lambda_h x_{kh} + \sum_{i=1}^{I} \mu_i z_{ki} \right) \text{ with } 1 \leq i \leq I.$$

By distinguishing the rows and columns, as here $d_k = N/n$, we have, for all h,

$$N_{h.} = \#(U_{h.}) = \sum_{k \in S} x_{kh} \frac{N}{n} \left(1 + \lambda_{h(k)} + \mu_{i(k)} \right).$$

Furthermore, for all i,

$$N_{.i} = \#(U_{.i}) = \sum_{k \in S} z_{ki} \frac{N}{n} \left(1 + \lambda_{h(k)} + \mu_{i(k)} \right),$$

where $h(k)$ and $i(k)$ respectively indicate the row and the column in which k is situated, which gives

$$N_{h.} = \sum_{i=1}^{I} n_{hi} \frac{N}{n} \left(1 + \lambda_h + \mu_i \right), \text{ for all } h,$$

$$N_{.i} = \sum_{h=1}^{H} n_{hi} \frac{N}{n} \left(1 + \lambda_h + \mu_i \right), \text{ for all } i,$$

where n_{hi} is the sample size S intersecting row h and column i, and thus

$$N_{h.} = n_{h.} \frac{N}{n} \left(1 + \lambda_h \right) + \sum_{i=1}^{I} n_{hi} \frac{N}{n} \mu_i \quad (1 \leq h \leq H),$$

$$N_{.i} = n_{.i} \frac{N}{n} \left(1 + \mu_i \right) + \sum_{h=1}^{H} n_{hi} \frac{N}{n} \lambda_h \quad (1 \leq i \leq I).$$

Noticing that $n_{h.} N/n = \widehat{N}_{h.}$ estimates $N_{h.}$ without bias and $n_{hi} N/n = \widehat{N}_{hi}$ estimates N_{hi} without bias, the system is written:

$$\begin{cases} \widehat{N}_{h.} \lambda_h + \sum_{i=1}^{I} \widehat{N}_{hi} \mu_i = N_{h.} - \widehat{N}_{h.} \ (1 \leq h \leq H) \\ \sum_{h=1}^{H} \widehat{N}_{hi} \lambda_h + \widehat{N}_{.i} \mu_i = N_{.i} - \widehat{N}_{.i} \ (1 \leq i \leq I). \end{cases}$$

We must therefore solve a linear system of $H + I$ equations with $H + I$ unknowns. Still, as $\sum_{h=1}^{H} N_{h.} = \sum_{i=1}^{I} N_{.i} = N$, there are only $H + I - 1$ independent equations: we can therefore fix (to be chosen) one of the λ_h or one of the μ_i. The system being linear, there is no difficulty in the solution, and we finally obtain the λ_h and the μ_i, and then the adjusted weights w_k:

$$w_k = \frac{N}{n}(1 + \mathbf{x}_k'\boldsymbol{\lambda} + \mathbf{z}_k'\boldsymbol{\mu}).$$

Exercise 7.6 *Regression and strata*

In this exercise, the objective is to calculate the generalised regression estimator of the total in a stratified design. We assume that the totals of an auxiliary characteristic value x are known for each stratum and we use the weighting coefficient $q_k = 1/x_k$ to estimate the vector of regression coefficients, which are of general style:

$$\widehat{\mathbf{b}} = \left(\sum_{k \in S} \frac{\mathbf{x}_k \mathbf{x}_k' q_k}{\pi_k}\right)^{-1} \sum_{k \in S} \frac{\mathbf{x}_k y_k q_k}{\pi_k}.$$

The regression estimator of the total is conceived by alternatively using the following regressors:

1. We use a lone regressor given by the values x_k taken by the characteristic x (without intercept);
2. We use H regressors given by $x_k \delta_{kh}$, with h varying from 1 to H where H is the number of strata and δ_{kh} is 1 if k is in stratum h and 0 otherwise. We will verify that there is effectively calibration on the totals of x in each stratum.

In what way are these estimators distinguished? Which is most commendable?

Solution

In a preliminary way, we will denote that the presence of a weighting coefficient q_k appears naturally in regression theory. In this classical modelling approach where y_k is random, if we consider that its variance is proportional to x_k, the optimal estimator will reweight by the inverse of x_k.

1. If we use a lone regressor x, we have

$$\widehat{b} = \left(\sum_{k \in S} \frac{x_k x_k}{\pi_k x_k}\right)^{-1} \sum_{k \in S} \frac{x_k y_k}{\pi_k x_k} = \frac{\widehat{Y}_\pi}{\widehat{X}_\pi}.$$

We get

$$\widehat{Y}_{\text{reg}} = \widehat{Y}_\pi + (X - \widehat{X}_\pi)\frac{\widehat{Y}_\pi}{\widehat{X}_\pi} = X\frac{\widehat{Y}_\pi}{\widehat{X}_\pi}, \text{ where } X = \sum_{k \in U} x_k.$$

We encounter the ratio estimator.

2. If we use H regressors given by $x_k \delta_{kh}$, where H is the number of strata and δ_{kh} is 1 if k is in stratum h and 0 otherwise, we have $\mathbf{x}_k = (x_k \delta_{k1}, ..., x_k \delta_{kh}, ..., x_k \delta_{kH})'$. The matrix

$$\sum_{k \in S} \frac{\mathbf{x}_k \mathbf{x}_k'}{x_k \pi_k} = \operatorname{diag}\left(\sum_{k \in S_1} \frac{x_k}{\pi_k}, ..., \sum_{k \in S_h} \frac{x_k}{\pi_k}, ..., \sum_{k \in S_H} \frac{x_k}{\pi_k} \right)$$

is diagonal, as we have the equality $\delta_{ki} \delta_{kj} = 0$ as soon as $i \neq j$, where S_h indicates the sample of stratum h. Furthermore

$$\sum_{k \in S} \frac{\mathbf{x}_k y_k}{x_k \pi_k} = \left[\sum_{k \in S_1} \frac{y_k}{\pi_k}, ..., \sum_{k \in S_h} \frac{y_k}{\pi_k}, ..., \sum_{k \in S_H} \frac{y_k}{\pi_k} \right]'.$$

We therefore get $\widehat{\mathbf{b}} = \left[\widehat{b}_1, ..., \widehat{b}_h, ..., \widehat{b}_H \right]'$, with

$$\widehat{b}_h = \frac{\sum_{k \in S_h} y_k / \pi_k}{\sum_{k \in S_h} x_k / \pi_k} = \frac{\widehat{Y}_h}{\widehat{X}_h}.$$

Finally, if we denote X_h as the true total of x_k in stratum h,

$$\widehat{Y}_{\mathrm{reg}} = \widehat{Y}_\pi + \left[..., \sum_{k \in U_h} x_k - \sum_{k \in S_h} \frac{x_k}{\pi_k}, ... \right] \widehat{\mathbf{b}}$$

$$= \sum_{h=1}^{H} \widehat{Y}_h + \sum_{h=1}^{H} \widehat{b}_h (X_h - \widehat{X}_h)$$

$$= \sum_{h=1}^{H} \widehat{b}_h X_h = \sum_{h=1}^{H} \frac{\sum_{k \in S_h} y_k / \pi_k}{\sum_{k \in S_h} x_k / \pi_k} \sum_{k \in U_h} x_k = \sum_{h=1}^{H} X_h \frac{\widehat{Y}_h}{\widehat{X}_h}.$$

It is therefore the sum of ratio estimators in each stratum. Clearly, if we denote $y_k = \delta_{kh} x_k$, h fixed, we have $\widehat{Y}_{\mathrm{reg}} = X_h$. The estimator of the total of x_k in stratum h is calibrated. The first estimator ensures a calibration on X, the total on the set of the population. The second estimator ensures a calibration on the totals X_h, stratum by stratum (and therefore, as a result, on X). The second estimator, which uses more information, is *a priori* more efficient, especially if the relationship between x and y differs from one stratum to another: in fact, the variance from the first estimator involves the terms $y_k - (Y/X) x_k$ for all k of the population, whereas the variance of the second estimator involves the terms $y_k - (Y_h/X_h) x_k$ for all k of h, terms which are *a priori* smaller.

Exercise 7.7 *Calibration on sizes*

Consider H sub-populations U_h, where $h = 1, ..., H$, such that $U_h \cap U_\ell = \emptyset, h \neq \ell$. Show that, in a simple random design, if we apply a calibration method on the sizes $N_h = \#U_h$, without giving particular weighting q_k to the units and so that the pseudo-distance F_k used does not depend on units k, then the weights from calibration do not depend on the pseudo-distances used.

Solution

The H auxiliary characteristics naturally associated to the context take the values $x_{k1}, ..., x_{kh}, ..., x_{kH}$, for all $k \in U$ where $x_{kh} = 1$ if $k \in U_h$ and 0 otherwise. Furthermore, we have $x_{kh}x_{k\ell} = 0$, for all k whenever $h \neq \ell$. If $q_k = 1, k \in U$, and the pseudo-distance F does not depend on k, then the calibration equations are written

$$\sum_{k \in S} d_k x_{kh} F \left(\sum_{i=1}^{H} \lambda_i x_{ki} \right) = X_h = N_h,$$

for all h. Now

$$\sum_{i=1}^{H} \lambda_i x_{ki} = \lambda_{h(k)},$$

where $h(k)$ indicates the stratum for unit k, which gives, with the simple random design,

$$\sum_{k \in S} \frac{N}{n} x_{kh} F \left(\lambda_{h(k)} \right) = N_h,$$

for all h, and thus

$$\frac{N}{n} n_h F \left(\lambda_h \right) = N_h.$$

Finally

$$F \left(\lambda_h \right) = \frac{N_h n}{n_h N}.$$

The weights are therefore

$$w_k = \frac{N}{n} \frac{N_h n}{n_h N} = \frac{N_h}{n_h}, \quad \text{as soon as } k \in U_h,$$

and the estimator obtained is quite simply the 'classical' post-stratified estimator for whatever pseudo-distance F is used.

Exercise 7.8 *Optimal estimator*

We consider a stratified design with simple random sampling in each stratum. Furthermore, we know for the entire sampling frame an auxiliary characteristic value x. We are interested in the estimators of the total of type:

$$\widehat{Y}_b = \widehat{Y}_\pi + b(X - \widehat{X}_\pi),$$

for all $b \in \mathbb{R}, b$ fixed.

1. What is the best value of b? What are we going to finally keep as the 'optimal' *a priori* estimator?
2. Compare this optimal *a priori* estimator to the regression estimator obtained by using as regressors the characteristic x and the intercept.

Solution

1. Clearly, $\mathrm{E}(\widehat{Y}_b) = Y$ for all $b \in \mathbb{R}$. Furthermore:

$$\mathrm{var}(\widehat{Y}_b) = \mathrm{var}(\widehat{Y}_\pi - b\widehat{X}_\pi) = \mathrm{var}\widehat{Y}_\pi + b^2\mathrm{var}\widehat{X}_\pi - 2b\,\mathrm{cov}(\widehat{X}_\pi, \widehat{Y}_\pi).$$

The best value of b is that which minimises the mean square error, which is here equal to the variance. By differentiating the variance with respect to b and setting the derivative equal to zero, we find:

$$b_{\mathrm{opt}} = \frac{\mathrm{cov}(\widehat{X}_\pi, \widehat{Y}_\pi)}{\mathrm{var}(\widehat{X}_\pi)}.$$

Unfortunately, b_{opt} is incalculable: it is necessary to estimate it. We will choose, naturally,

$$\widehat{b}_{\mathrm{opt}} = \frac{\widehat{\mathrm{cov}}(\widehat{X}_\pi, \widehat{Y}_\pi)}{\widehat{\mathrm{var}}(\widehat{X}_\pi)},$$

where $\widehat{\mathrm{cov}}$ and $\widehat{\mathrm{var}}$ are the classical π-estimators of cov and var. The estimator loses, along the way, its optimality but we can think that it is 'almost as efficient' as the optimal estimator. The pseudo-optimal estimator is therefore:

$$\widehat{Y}_{\mathrm{opt}} = \widehat{Y}_\pi + (X - \widehat{X}_\pi)\frac{\widehat{\mathrm{cov}}(\widehat{X}_\pi, \widehat{Y}_\pi)}{\widehat{\mathrm{var}}(\widehat{X}_\pi)},$$

which gives, if we denote $\widehat{\overline{X}}_h$ and $\widehat{\overline{Y}}_h$ as the simple means respectively for x and y in the sample of stratum h,

$$\widehat{\mathrm{cov}}(\widehat{X}_\pi, \widehat{Y}_\pi) = \sum_{h=1}^{H} N_h^2 \widehat{\mathrm{cov}}(\widehat{\overline{X}}_h, \widehat{\overline{Y}}_h),$$

and

$$\widehat{\mathrm{var}}(\widehat{X}_\pi) = \sum_{h=1}^{H} N_h^2 \widehat{\mathrm{var}}(\overline{\overline{X}}_h),$$

which is

$$\widehat{Y}_{\mathrm{opt}} = \widehat{Y}_\pi + (X - \widehat{X}_\pi) \frac{\sum_{h=1}^{H} \frac{N_h^2}{n_h} \frac{N_h - n_h}{N_h} \frac{1}{n_h - 1} \sum_{k \in S_h} \left(x_k - \overline{\overline{X}}_h \right) \left(y_k - \overline{\overline{Y}}_h \right)}{\sum_{h=1}^{H} \frac{N_h - n_h}{N_h} \frac{N_h^2}{n_h} \frac{1}{n_h - 1} \sum_{k \in S_h} \left(x_k - \overline{\overline{X}}_h \right)^2}.$$

2. For the stratified design, the regression coefficient present in the expression of the regression estimator is $\mathbf{b} = \widehat{\mathbf{T}}^{-1}\widehat{\mathbf{t}}$, where

$$\widehat{\mathbf{T}} = \begin{bmatrix} \widehat{N}_\pi & \widehat{X}_\pi \\ \widehat{X}_\pi & \widehat{(X^2)}_\pi \end{bmatrix}, \quad \text{and} \quad \widehat{\mathbf{t}} = (\widehat{Y}_\pi, \widehat{(XY)}_\pi)'.$$

We set

$$\widehat{N}_\pi = \sum_{k \in S} \frac{1}{\pi_k}, \quad \widehat{(X^2)}_\pi = \sum_{k \in S} \frac{x_k^2}{\pi_k}, \quad \text{and} \quad \widehat{(XY)}_\pi = \sum_{k \in S} \frac{x_k y_k}{\pi_k}.$$

These expressions originate quite simply from the solution of the following equations:

$$\begin{cases} a\widehat{N}_\pi + b\widehat{X}_\pi = \widehat{Y}_\pi \\ a\widehat{X}_\pi + b\widehat{(X^2)}_\pi = \widehat{(XY)}_\pi. \end{cases}$$

Therefore,

$$\widehat{\mathbf{T}}^{-1} = \frac{1}{\widehat{N}_\pi \widehat{(X^2)}_\pi - \widehat{X}_\pi^2} \begin{bmatrix} \widehat{(X^2)}_\pi & -\widehat{X}_\pi \\ -\widehat{X}_\pi & \widehat{N}_\pi \end{bmatrix}.$$

Now, if we denote,

$$v_x^2 = \frac{\widehat{(X^2)}_\pi}{\widehat{N}_\pi} - \frac{\widehat{X}_\pi^2}{\widehat{N}_\pi^2} = \frac{1}{\widehat{N}_\pi} \sum_{k \in S} \frac{1}{\pi_k} \left(x_k - \frac{\widehat{X}_\pi}{\widehat{N}_\pi} \right)^2,$$

and

$$v_{xy} = \frac{\widehat{(XY)}_\pi}{\widehat{N}_\pi} - \frac{\widehat{X}_\pi \widehat{Y}_\pi}{\widehat{N}_\pi^2} = \frac{1}{\widehat{N}_\pi} \sum_{k \in S} \frac{1}{\pi_k} \left(x_k - \frac{\widehat{X}_\pi}{\widehat{N}_\pi} \right) \left(y_k - \frac{\widehat{Y}_\pi}{\widehat{N}_\pi} \right),$$

we get after a few calculations that:

$$\widehat{\mathbf{b}} = \widehat{\mathbf{T}}^{-1}\widehat{\mathbf{t}} = \frac{1}{v_x^2} \left[\widehat{\Delta}, v_{xy} \right]',$$

where $\widehat{\Delta}$ is a complex expression without interest. Finally, we have:

$$\widehat{Y}_{\mathrm{reg}} = \widehat{Y}_\pi + (X - \widehat{X}_\pi)\frac{v_{xy}}{v_x^2}.$$

As the sampling is stratified, with simple random sampling in each stratum,

$$\widehat{N}_\pi = N, \quad \widehat{X}_\pi = \sum_{h=1}^{H} N_h \widehat{\overline{X}}_h, \quad \widehat{Y}_\pi = \sum_{h=1}^{H} N_h \widehat{\overline{Y}}_h, \quad \widehat{\overline{X}}_\pi = \frac{\widehat{X}_\pi}{N}, \quad \widehat{\overline{Y}}_\pi = \frac{\widehat{Y}_\pi}{N}.$$

Hence

$$\widehat{Y}_{\mathrm{reg}} = \widehat{Y}_\pi + (X - \widehat{X}_\pi)\frac{\sum_{h=1}^{H}\frac{N_h}{n_h}\sum_{k\in S_h}\left(x_k - \widehat{\overline{X}}_\pi\right)\left(y_k - \widehat{\overline{Y}}_\pi\right)}{\sum_{h=1}^{H}\frac{N_h}{n_h}\sum_{k\in S_h}\left(x_k - \widehat{\overline{X}}_\pi\right)^2}.$$

Let us note that v_x^2 and v_{xy} do not estimate the population variances S_x^2 and S_{xy} without bias as they are ratio functions. The optimal estimator would be indisputably better if we knew the true regression coefficient, but it is penalised by the instability of the estimator for its regression coefficient. Indeed, this estimated regression coefficient is composed of a covariance ratio on a variance for a stratified design. In stratification, it is known that the higher the number of strata, the more the variance estimator is unstable. With many strata, the estimated regression coefficient of the optimal estimator $\widehat{Y}_{\mathrm{opt}}$ is therefore more unstable than that of the generalised regression estimator $\widehat{Y}_{\mathrm{reg}}$, which can have its theoretical advantage lost. In fact, in the coefficient of the optimal estimator, we must estimate the H means of the strata, which corresponds to the loss of H degrees of freedom. If the optimal estimator is asymptotically better, it can happen to be less effective when the sample sizes in the strata are small.

Exercise 7.9 *Calibration on population size*

We consider a Poisson sampling design with inclusion probabilities $\pi_k, k \in U$, and we are interested in the total Y of a characteristic of interest y. The objective consists of constructing a calibration estimator of Y using a sole auxiliary calibration variable $x_k = 1, k \in U$. We use the pseudo-distance G^α, parameterised by $\alpha \in \mathbb{R}$

$$G^\alpha(w_k, d_k) = \begin{cases} \dfrac{\dfrac{w_k^\alpha}{d_k^{\alpha-1}} + (\alpha-1)d_k - \alpha w_k}{\alpha(\alpha-1)} & \text{if } \alpha \in \mathbb{R}\setminus\{0,1\} \\[2ex] w_k \log\dfrac{w_k}{d_k} + d_k - w_k & \text{if } \alpha = 1 \\[2ex] d_k \log\dfrac{d_k}{w_k} + w_k - d_k & \text{if } \alpha = 0, \end{cases}$$

by taking $q_k = 1$, for all $k \in U$ (this parameterised expression integrates the principal distances used in practice).

Preliminary: Recall the unbiased estimator of the total population size N and express its variance. For a sample size that is on average $\bar{n}$, show that this variance is always higher than a threshold to be specified.

1. On which total are we going to calibrate this estimator?
2. Write the calibration equation.
3. Determine the value of the Lagrange multiplier λ (for all fixed $\alpha \in \mathbb{R}$).
4. Deduce the adjusted weights w_k of the calibration estimator.
5. Give the calibrated estimator. What type of estimator is it?

Solution

Preliminary: the unbiased estimator of a total Y in the Poisson case being

$$\widehat{Y} = \sum_{k \in S} \frac{y_k}{\pi_k},$$

we classically estimate the population size without bias by

$$\widehat{N} = \sum_{k \in S} \frac{1}{\pi_k} = \sum_{k \in U} \frac{I_k}{\pi_k}, \tag{7.3}$$

with a variance:

$$\text{var}(\widehat{N}) = \sum_{k \in U} \frac{1}{(\pi_k)^2} \text{var}(I_k) = \sum_{k \in U} \frac{1 - \pi_k}{\pi_k}. \tag{7.4}$$

The expected value of the (random) sample size being $\bar{n} = \sum_{k \in U} \pi_k$, the minimum variance threshold (convex function) is obtained by minimising (7.4) subject to $\sum_{k \in U} \pi_k = \bar{n}$ and $0 < \pi_k \leq 1$. We easily get π_k constant, equal as a result to $\bar{n}/N$. Hence

$$\min \left[\text{var}(\widehat{N}) \right] = \sum_{k \in U} \frac{1 - \frac{\bar{n}}{N}}{\frac{\bar{n}}{N}} = N \frac{N - \bar{n}}{\bar{n}}.$$

This threshold is high if $\bar{n}$ is 'sufficiently' small: this result is intuitive; it is obvious under these conditions that the variance can only be higher.

1. We calibrate on the population size. Indeed,

$$\sum_{k \in U} 1 = N.$$

This calibration is conceived, as a matter of fact, to 'thwart' the uncertainty dealt with in the previous question.

2. If we let $d_k = 1/\pi_k$, the calibration equation is then

$$N = \sum_{k \in S} d_k F_k^\alpha(\lambda \times 1),$$

where $d_k F_k^\alpha$ is the reciprocal of the function of w_k:

$$\frac{\partial G^\alpha(w_k, d_k)}{\partial w_k},$$

and λ is the Lagrange multiplier (real) associated with the constraint $\widehat{N} = N$.

3. As (derived without difficulty)

$$F_k^\alpha(u) = F^\alpha(u) = \begin{cases} \sqrt[\alpha-1]{1 + u(\alpha - 1)}, & \alpha \in \mathbb{R}\backslash\{1\} \\ \exp u, & \alpha = 1, \end{cases}$$

and since $\sum_{k \in S} d_k = \widehat{N}$, the calibration equation becomes $N = F^\alpha(\lambda)\widehat{N}$. Thus

$$\lambda = \begin{cases} \dfrac{\left(\frac{N}{\widehat{N}}\right)^{\alpha-1} - 1}{\alpha - 1} & \alpha \in \mathbb{R}\backslash\{1\} \\ \log \dfrac{N}{\widehat{N}} & \alpha = 1. \end{cases}$$

We notice that λ is a continuous function of α in $\mathbb{R}$, as

$$\lim_{\alpha \to 1} \frac{\left(\frac{N}{\widehat{N}}\right)^{\alpha-1} - 1}{\alpha - 1} = \log \frac{N}{\widehat{N}}.$$

4. Since

$$F^\alpha(\lambda) = \frac{N}{\widehat{N}},$$

for whatever value α is, the adjusted weights are written

$$w_k = d_k \frac{N}{\widehat{N}} = \frac{1}{\pi_k} \frac{N}{\widehat{N}}.$$

5. The calibrated estimator does not depend on the distance used and is

$$\sum_{k \in S} d_k \frac{N}{\widehat{N}} y_k = \frac{N}{\widehat{N}} \widehat{Y}.$$

It is a ratio estimator calibrated on the population size, called the 'Hájek ratio' (see also Exercise 3.24).

Exercise 7.10 *Double calibration*

Following a complex sampling design leading to individual weights d_k, we perform a first calibration on a vector of known totals $\mathbf{X}$ of $\mathbb{R}^p$, and then a second calibration on the pair $(\mathbf{X}, \mathbf{Z})$ of $\mathbb{R}^{p+q}$, where $\mathbf{Z}$ is a second vector of known totals of $\mathbb{R}^q$. Do we get *in fine* the same estimator as if we had disregarded the first step (which does not seem to contribute much *a priori*)?

Solution

- *Method a:*
  The first calibration leads to weights $\tilde{d}_k = d_k F(\mathbf{x}'_k \boldsymbol{\lambda})$ with

$$\sum_{k \in S} \tilde{d}_k \mathbf{x}_k = \mathbf{X},$$

  where $\mathbf{x}'_k$ indicates the transposed vector of $\mathbf{x}_k$. The second calibration starts from the weights $\tilde{d}_k$ and leads to

$$\tilde{\tilde{d}}_k = \tilde{d}_k F(\mathbf{x}'_k \boldsymbol{\mu} + \mathbf{z}'_k \boldsymbol{\delta}),$$

  where $\boldsymbol{\lambda}$, $\boldsymbol{\mu}$, $\boldsymbol{\delta}$ are the vectors of Lagrange multipliers associated with the constraints, respectively in $\mathbb{R}^p$, $\mathbb{R}^p$ and $\mathbb{R}^q$.

  Moreover,

$$\sum_{k \in S} \tilde{\tilde{d}}_k \mathbf{x}_k = \mathbf{X}, \quad \text{and} \quad \sum_{k \in S} \tilde{\tilde{d}}_k \mathbf{z}_k = \mathbf{Z}.$$

- *Method b:*
  With a direct calibration on $(\mathbf{X}, \mathbf{Z})$, we get *in fine* the weights

$$w_k = d_k F(\mathbf{x}'_k \boldsymbol{\alpha} + \mathbf{z}'_k \boldsymbol{\beta}),$$

  with

$$\sum_{k \in S} w_k \mathbf{x}_k = \mathbf{X}, \quad \text{and} \quad \sum_{k \in S} w_k \mathbf{z}_k = \mathbf{Z}.$$

Method *a* produces:

$$\tilde{\tilde{d}}_k = d_k F(\mathbf{x}'_k \boldsymbol{\lambda}) F(\mathbf{x}'_k \boldsymbol{\mu} + \mathbf{z}'_k \boldsymbol{\delta}).$$

A priori, the solution of the systems of equations related to the constraints leads to the weights $w_k \neq \tilde{\tilde{d}}_k$. On the other hand, there is a favourable case which leads to the same system of equations. Indeed, if $F(a)F(b) = F(a+b)$, we have:

$$\tilde{\tilde{d}}_k = d_k F[\mathbf{x}'_k (\boldsymbol{\lambda} + \boldsymbol{\mu}) + \mathbf{z}'_k \boldsymbol{\delta}].$$

The uniqueness of the solutions of the systems of equations (with F having 'good' properties of regularity) leads to

$$\lambda + \mu = \alpha \quad \text{and} \quad \delta = \beta.$$

which is to say that $\widetilde{\widetilde{d}}_k = w_k$, for all k (λ being determined by the first calibration, we select $\mu = \alpha - \lambda$). The estimator is then the same, that being with Method a or Method b. This case comes from the following property: $G(x) = \exp cx$, for any c in $\mathbb{R}$, is the only real continuous function satisfying $G(a + b) = G(a)G(b)$ for all (a, b) in $\mathbb{R}^2$. This is the method called 'raking ratio'. Under these conditions, in the case of marginal calibration ($\mathbf{x}_k$ and $\mathbf{z}_k$ contain the indicators), the eventual preliminary post-stratification steps on certain margins do not disrupt the ensuing calibration because they do not impact upon the weighting. If we do not use the *raking ratio*, the final estimates are, in theory, sensitive to the partial preliminary calibrations, even if it does not change a lot on the point of view of the bias and the variance, as soon as n is 'large'.

8

Variance Estimation

8.1 Principal techniques of variance estimation

There exist several approaches to estimate the variances of estimators. The two essential techniques are, on the one hand, the analytical approach, that is to say, the formatting of expressions for variance estimators, and on the other hand, replication methods that rely on re-samples conducted from the initially selected sample.

The analytical approach encounters two types of difficulties. On the one hand, it is necessary to manage the problem posed by the very complex calculation of double inclusion probabilities $\pi_{k\ell}$, which occurs in the majority of the sampling designs without replacement. On the other hand, it is necessary to bypass the difficulty posed by the manipulation of non-linear estimators. In fact, we know how to mathematically express the variance of a linear expression, but it is no longer possible to make exact calculations when products, ratios, powers and roots are involved. The treatment of the problem of second-order inclusion probabilities is quite complex, and requires us, on the one hand, to formulate simplifying assumptions on the design, and on the other hand, to completely explore the branching describing the sampling design. It is possible to use a recursive formula (see on this topic Raj, 1968) to construct variance estimators. This technique was used at the *Institut National de la Statistique et des Études Économiques*, (INSEE, France) in the POULPE software program used to estimate the variances in complex designs (see Caron, 1999). On the other hand, the treatment of the problem posed by the non-linearity of the estimators is more accessible due to the linearisation technique (see on this topic Deville, 1999), once the sample size is 'sufficiently large'.

Replication methods such as the *jackknife*, the *bootstrap* and balanced half-samples are used with 'sufficiently large' samples and permit the estimation of variances for non-linear estimators. Nevertheless, notable difficulties exist when the sampling is complex (multi-stage designs, unequal probability designs, multi-phase designs) and, above all, the properties of the variance estimator (bias, in particular) are not as well controlled as in the analytical approach when the sampling design is no longer simple random. The reader interested in these methods can refer to Wolter (1985); Efron and Tibshirani (1993) and Rao and Sitter (1995).

8.2 Method of estimator linearisation

The idea consists of linearising a complex estimator and assimilating its variance, under certain conditions, to that of its linear approximation. We then encounter, in a standard manner, the problem of variance estimation for a linear estimator. It is this approach that allows for the calculation of the precision of calibrated estimators, presented in Chapters 6 and 7 (ratio estimator, regression estimator, marginal calibration estimators), and of estimators with complex parameters, such as correlation coefficients, regression coefficients and inequality indicators. To estimate a parameter $\theta = f(Y^1, Y^2, ..., Y^p)$, where Y^i is the true total of a variable y^i ($i = 1$ to p), we use the substitution estimator modelled on the same functional form, that is:

$$\widehat{\theta} = f(\widehat{Y}^1, \widehat{Y}^2, ..., \widehat{Y}^p),$$

where $\widehat{Y}^i$ is a linear estimator of Y^i and therefore of type

$$\sum_{k \in S} w_k(S) y_k^i$$

(for example, the unbiased Horvitz-Thompson estimator), and f is a reasonably smooth function of $\mathbb{R}^p$ in $\mathbb{R}$, in practice of class C^2 (twice differentiable, the second-order derivative being continuous). If the mean estimators $\widehat{Y}^i/N$ have a mean square error that varies by $1/n$ (which is always the case in practice), and if n is sufficiently large so that $1/n^{3/2}$ is negligible compared to $1/n$ (it is therefore an 'asymptotic' vision where n and N become very large), then we show that $\mathrm{var}(\widehat{\theta}) \approx \mathrm{var}(\widehat{V})$, where $\widehat{V}$ is built on the model of $\widehat{Y}^i$ (thus with the same weights), being

$$\widehat{V} = \sum_{k \in S} w_k(S) v_k,$$

with, for all $k \in S$,

$$v_k = \sum_{j=1}^{p} y_k^j \frac{\partial f(a_1, a_2, ..., a_p)}{\partial a_j} \Big|_{(Y^1, Y^2, ..., Y^p)} .$$

The new variable v_k is called 'linearisation' of θ. The variance estimator of $\widehat{\theta}$ is naturally obtained from a variance estimator of $\widehat{V}$ by replacing v_k (incalculable) with:

$$\widehat{v}_k = \sum_{j=1}^{p} y_k^j \frac{\partial f(a_1, a_2, ..., a_p)}{\partial a_j} \bigg|_{(\widehat{Y}^1, \widehat{Y}^2, ..., \widehat{Y}^p)} .$$

We can show that this substitution is judicious (p remaining fixed when n increases). We can also proceed stepwise: if $\theta = f(Y^1, Y^2, ..., Y^p, \psi)$, where ψ is a function of totals $(Y^{p+1}, Y^{p+2}, ..., Y^q)$, for which we already calculated a linearised variable u_k, then the linearisation of θ is:

$$v_k = \sum_{j=1}^{p} y_k^j \frac{\partial f(a_1, a_2, ..., a_p, z)}{\partial a_j} \bigg|_{(Y^1, Y^2, ..., Y^p, \psi)}$$

$$+ u_k \frac{\partial f(a_1, a_2, ..., a_p, z)}{\partial z} \bigg|_{(Y^1, Y^2, ..., Y^p, \psi)} .$$

It is then sufficient to form $\widehat{v}_k$ by replacing all the unknown totals with their respective estimators.

EXERCISES

Exercise 8.1 *Variances in an employment survey*

The 1989 INSEE employment survey leads to Table 8.1, expressed in thousands of people. The sample size is larger than 10000, and the confidence intervals are given under the assumption of asymptotic normality of estimators.

Table 8.1. Labour force, employed and unemployed: Exercise 8.1

	Estimated size	95% confidence interval
Labour force	24062	± 129
Employed	21754	± 149
Unemployed	2308	± 76

1. Estimate the unemployment rate defined as the percentage of unemployed people among the labour force (the labour force is the sum of those employed and unemployed). What type of estimator is this?
2. Give the approximate mathematical expression for the estimated mean square error (MSE) of the estimated unemployment rate, as a function of:
   - the estimated variance of the estimator for the labour force,
   - the estimated variance of the estimator for the number of unemployed,
   - the estimated covariance between the estimators for the labour force and the number of unemployed,
   - the estimator of the labour force,
   - and the estimator of the unemployment rate.

3. Show that the MSE estimator of the unemployment rate can be calculated with the data from the table.

    Hint: to do this, we use the following general result. Let X and Y be any two random variables; then

$$\mathrm{cov}(X,Y) = \frac{\mathrm{var}(X+Y) - \mathrm{var}(X) - \mathrm{var}(Y)}{2}.$$

4. Use the previous expression to calculate the variance estimate for the unemployment rate and draw up an estimated 95% confidence interval.

Solution

1. Estimated unemployment rate:

$$\widehat{R} = \frac{\text{Unemployed}}{\text{Labour force}} = \frac{2308}{24062} \approx 9.6\%.$$

    This is a 'ratio' estimator (quotient of two estimators of the totals).

2. Since the sample size n is very large, the bias $1/n$ and the variance $1/n$ as well, the MSE is numerically similar to the variance (the squared bias becomes negligible). If X represents the labour force and Y the number of unemployed:

$$\widehat{\mathrm{MSE}} \approx \widehat{\mathrm{var}}\left(\frac{\widehat{Y}}{\widehat{X}}\right) = \frac{1}{\widehat{X}^2}\left\{\widehat{\mathrm{var}}\left(\widehat{Y}\right) + \widehat{R}^2\widehat{\mathrm{var}}\left(\widehat{X}\right) - 2\widehat{R}\widehat{\mathrm{cov}}\left(\widehat{X},\widehat{Y}\right)\right\},$$

    a well-known approximation for quotients.

3. The difficulty consists of the evaluation of the term $\widehat{\mathrm{cov}}\left(\widehat{X},\widehat{Y}\right)$, but since $\widehat{X} = \widehat{Y} + \widehat{Z}$, where Z is the number of employed people, we have

$$\widehat{\mathrm{cov}}\left(\widehat{Y},\widehat{X}\right) = \widehat{\mathrm{cov}}\left(\widehat{Y},\widehat{Y}+\widehat{Z}\right) = \widehat{\mathrm{var}}\left(\widehat{Y}\right) + \widehat{\mathrm{cov}}\left(\widehat{Y},\widehat{Z}\right).$$

    Furthermore

$$\widehat{\mathrm{cov}}\left(\widehat{Y},\widehat{Z}\right) = \frac{\widehat{\mathrm{var}}\left(\widehat{Y}+\widehat{Z}\right) - \widehat{\mathrm{var}}\left(\widehat{Y}\right) - \widehat{\mathrm{var}}\left(\widehat{Z}\right)}{2},$$

    which gives

$$\widehat{\mathrm{cov}}\left(\widehat{X},\widehat{Y}\right) = \frac{\widehat{\mathrm{var}}\left(\widehat{Y}+\widehat{Z}\right) + \widehat{\mathrm{var}}\left(\widehat{Y}\right) - \widehat{\mathrm{var}}\left(\widehat{Z}\right)}{2}.$$

Therefore:

$$
\widehat{\mathrm{var}} \left(\frac{\widehat{Y}}{\widehat{X}} \right)
$$

$$
= \frac{1}{\widehat{X}^2} \left\{ \widehat{\mathrm{var}} \left(\widehat{Y} \right) + \widehat{R}^2 \widehat{\mathrm{var}} \left(\widehat{X} \right) - 2\widehat{R} \left[\frac{\widehat{\mathrm{var}} \left(\widehat{X} \right) + \widehat{\mathrm{var}} \left(\widehat{Y} \right) - \widehat{\mathrm{var}} \left(\widehat{Z} \right)}{2} \right] \right\}
$$

$$
= \frac{1}{\widehat{X}^2} \left\{ \left(1 - \widehat{R} \right) \widehat{\mathrm{var}} \left(\widehat{Y} \right) - \widehat{R}(1 - \widehat{R}) \widehat{\mathrm{var}} \left(\widehat{X} \right) + \widehat{R} \widehat{\mathrm{var}} \left(\widehat{Z} \right) \right\}.
$$

4. Table 8.2 gives us the variance estimates for the three estimators of the totals $\widehat{X}, \widehat{Y}$ and $\widehat{Z}$ obtained from Table 8.1. It only remains to do the

Table 8.2. Estimated variances of the estimators: Exercise 8.1

$$
\begin{aligned}
\widehat{X} &= 24062 \quad \widehat{\mathrm{var}}(\widehat{X}) = 4332 \\
\widehat{Z} &= 21754 \quad \widehat{\mathrm{var}}(\widehat{Z}) = 5779 \\
\widehat{Y} &= 2308 \quad \widehat{\mathrm{var}}(\widehat{Y}) = 1504
\end{aligned}
$$

calculation, as $\widehat{R} \approx 0.09592$:

$$
\widehat{\mathrm{var}} \left(\frac{\widehat{Y}}{\widehat{X}} \right) = 2.66 \times 10^{-6},
$$

that is,

$$
\frac{Y}{X} \in [9.6\% \pm 0.3\%] \text{ (about) 95 times out of 100.}
$$

Exercise 8.2 *Tour de France*

Following a stage of the Tour de France, we complete a simple random survey without replacement of n cyclists (n fixed, supposedly 'large') among the N competitors. For each selected cyclist, we have available his average speed for the stage. Estimate the average speed of the entire group of cyclists and estimate the variance of this estimator.

Solution

To calculate an average of speeds on a route of given length L, it is necessary to calculate a harmonic mean. In fact, if we denote z_k as the speed of cyclist k and L as the length of the stage, we naturally define the average speed M of the group of cyclists by:

$$
M = \frac{\sum \text{kilometres travelled}}{\sum \text{time per cyclist}} = \frac{N \times L}{\sum \text{time per cyclist}},
$$

where the sums are for the group of N competitors. In fact, the time for cyclist k to complete the stage is equal to the length of the stage divided by the average speed z_k of this cyclist and thus

$$M = \frac{N \times L}{\sum_{k \in U} L/z_k} = \frac{N}{\sum_{k \in U} 1/z_k}.$$

We recognize the harmonic mean of z_k which can be written as a function of totals by setting $y_k = 1/z_k$:

$$M = \frac{N}{Y} = \frac{1}{\overline{Y}} = f(N, Y).$$

We thus estimate M by

$$\widehat{M} = \frac{N}{\widehat{Y}} = \frac{1}{\widehat{\overline{Y}}},$$

with

$$\widehat{Y} = N\widehat{\overline{Y}} = \frac{N}{n} \sum_{k \in S} \frac{1}{z_k}.$$

By noticing that N is a total, the linearisation of M is therefore

$$v_k = \left(\frac{1}{Y} - y_k \frac{N}{Y^2} \right) = \frac{1}{Y} \left(1 - y_k \frac{N}{Y} \right),$$

which can be estimated by

$$\hat{v}_k = \frac{1}{\widehat{Y}} \left(1 - y_k \frac{\widehat{N}}{\widehat{Y}} \right) = \frac{1}{\widehat{Y}} \left(1 - y_k \frac{N}{\widehat{Y}} \right) = \frac{1}{N\widehat{\overline{Y}}} \left(1 - \frac{y_k}{\widehat{\overline{Y}}} \right).$$

We get

$$\widehat{\mathrm{var}} \left(\widehat{M} \right) = N^2 \frac{N-n}{Nn} \frac{1}{n-1} \sum_{k \in S} (\hat{v}_k - \bar{\hat{v}})^2,$$

where

$$\bar{\hat{v}} = \frac{1}{n} \sum_{k \in S} \hat{v}_k = 0.$$

We easily deduce this:

$$\widehat{\mathrm{var}} \left(\widehat{M} \right) = \frac{1}{\widehat{\overline{Y}}^4} \frac{N-n}{Nn} s_y^2,$$

thus

$$\widehat{\mathrm{var}} \left(\frac{1}{\widehat{\overline{Y}}} \right) = \frac{\widehat{\mathrm{var}} \left(\widehat{\overline{Y}} \right)}{\widehat{\overline{Y}}^4}.$$

Exercise 8.3 *Geometric mean*

Show that the geometric mean G of values $y_k > 0, k \in U$ for a characteristic y can be written as a function of the total of a certain variable (to be determined). We recall:

$$G = \left(\prod_{k \in U} y_k \right)^{1/N}.$$

1. We assume that N is unknown. Give an estimator $\widehat{G}$ of G for any design. Then, estimate the variance of this estimator by way of the linearisation technique in the case of a design of fixed size n (n large).

2. We now assume that N is known. What estimator $\widehat{\widehat{G}}$ of G can we construct? What can we say about the sign of its bias (n large)? Give a variance estimator of $\widehat{\widehat{G}}$, and compare it to the variance estimator of $\widehat{G}$.

Solution

1. Since

$$G = \exp\left(\frac{1}{N} \sum_{k \in U} \log y_k \right),$$

letting $z_k = \log y_k, k \in U$, we have

$$G = \exp\left(\frac{1}{N} \sum_{k \in U} z_k \right) = \exp(\overline{Z}).$$

The geometric mean G is therefore written as a function of two totals: Z and N. We next estimate G with $\widehat{G} = \exp\left(\widehat{\overline{Z}}_R \right)$, where

$$\widehat{\overline{Z}}_R = \frac{1}{\widehat{N}_\pi} \sum_{k \in S} \frac{z_k}{\pi_k}.$$

We can also write:

$$\widehat{G} = \prod_{k \in S} \exp\left(\frac{1}{\pi_k \widehat{N}_\pi} \log y_k \right) = \left(\prod_{k \in S} y_k^{1/\pi_k} \right)^{1/\widehat{N}_\pi}.$$

This estimator is biased. The linearisation technique leads to the linearised variable v_k

$$v_k = \frac{z_k}{N} \exp \overline{Z} - \frac{\overline{Z}}{N} \exp \overline{Z} = \frac{1}{N}(z_k - \overline{Z})G,$$

which is estimated by

$$\hat{v}_k = \frac{1}{\widehat{N}_\pi}(z_k - \widehat{\overline{Z}}_R)\exp\widehat{\overline{Z}}_R = \frac{1}{\widehat{N}_\pi}(z_k - \widehat{\overline{Z}}_R)\widehat{G}.$$

Lastly, the variance estimator is written

$$\widehat{\mathrm{var}}\left[\widehat{G}\right] = \frac{1}{2}\frac{\widehat{G}^2}{\widehat{N}_\pi^2}\sum_{k\in S}\sum_{\substack{\ell\in S\\\ell\neq k}}\left(\frac{z_k - \widehat{\overline{Z}}_R}{\pi_k} - \frac{z_\ell - \widehat{\overline{Z}}_R}{\pi_\ell}\right)^2\frac{\pi_k\pi_\ell - \pi_{k\ell}}{\pi_{k\ell}}.$$

2. If N is known, we construct a new estimator $\widehat{\widehat{G}}$ on the model of $\widehat{G}$, without it being necessary to estimate N:

$$\widehat{\widehat{G}} = \exp(\widehat{\overline{Z}}_\pi)\text{ where }\widehat{\overline{Z}}_\pi = \frac{1}{N}\sum_{k\in S}\frac{z_k}{\pi_k}.$$

A limited development of $\widehat{\widehat{G}}$, an exponential function of $\widehat{\overline{Z}}_\pi$, around $\mathrm{E}(\widehat{\overline{Z}}_\pi) = \overline{Z}$ gives:

$$\widehat{\widehat{G}} = f(\widehat{\overline{Z}}_\pi) = f(\overline{Z}) + f'(\overline{Z})(\widehat{\overline{Z}}_\pi - \overline{Z}) + \frac{1}{2}f''(\overline{Z})(\widehat{\overline{Z}}_\pi - \overline{Z})^2 + R,$$

where R is of the same order of magnitude as $(\widehat{\overline{Z}}_\pi - \overline{Z})^3$. Therefore:

$$\widehat{\widehat{G}} = G + G(\widehat{\overline{Z}}_\pi - \overline{Z}) + \frac{G}{2}(\widehat{\overline{Z}}_\pi - \overline{Z})^2 + R.$$

In the end,

$$\mathrm{E}(\widehat{\widehat{G}} - G) = \frac{G}{2}\mathrm{E}(\widehat{\overline{Z}}_\pi - \overline{Z})^2 + \mathrm{E}(R),\text{ with }\mathrm{E}(R) = O\left(\frac{1}{n^{3/2}}\right)$$

$$\approx \frac{G}{2}\mathrm{E}(\widehat{\overline{Z}}_\pi - \overline{Z})^2\text{ for }n\text{ large}.$$

We have

$$\mathrm{E}(\widehat{\overline{Z}}_\pi - \overline{Z})^2 = \mathrm{var}(\widehat{\overline{Z}}_\pi) = O\left(\frac{1}{n}\right).$$

Therefore:

$$\mathrm{E}(\widehat{\overline{Z}}_\pi - \overline{Z})^3 = O\left(\frac{1}{n^{3/2}}\right).$$

Thus,

$$\mathrm{E}(\widehat{\widehat{G}} - G) > 0,$$

for n sufficiently large. The estimator $\widehat{\widehat{G}}$ overestimates (a little) G, but the bias is negligible if n is large. The sign of the bias is coherent with Jensen's inequality, which is well-known to probabilists and is applicable here because the exponential is convex. Indeed, Jensen's inequality states that if f is convex, throughout random variable X, we have: $\mathrm{E}f(X) \geq f(\mathrm{E}X)$. Here, we take $X = \widehat{\overline{Z}}_\pi$ and $f(X) = \exp(X)$, thus

$$\mathrm{E}\left[\exp(\widehat{\overline{Z}}_\pi)\right] \geq \exp\left[\mathrm{E}(\widehat{\overline{Z}}_\pi)\right],$$

and therefore

$$\mathrm{E}\widehat{G} \geq \exp(\overline{Z}) = G = \mathrm{E}(G).$$

The classical considerations for orders of magnitude (case where n is large) lead to the approximation:

$$\mathrm{var}(\widehat{\widehat{G}}) \approx G^2 \mathrm{var}(\widehat{\overline{Z}}_\pi) = \mathrm{var}\left(\sum_{k \in S} \frac{1}{\pi_k} \frac{Gz_k}{N}\right).$$

The linearisation technique would again exactly and logically give (as it follows from the limited development of $\widehat{G}$) the same result, which is to say $v_k = Gz_k/N$. Finally,

$$\widehat{\mathrm{var}}\left(\widehat{\widehat{G}}\right) = \frac{1}{2}\frac{\widehat{G}^2}{N^2}\sum_{k \in S}\sum_{\substack{\ell \in S \\ \ell \neq k}}\left(\frac{z_k}{\pi_k} - \frac{z_\ell}{\pi_\ell}\right)\frac{\pi_k\pi_\ell - \pi_{k\ell}}{\pi_{k\ell}}.$$

To compare the respective qualities of $\widehat{G}$ and $\widehat{\widehat{G}}$ (for all designs of fixed size and with equal probabilities, these two estimators correspond), we can compare the variances (or estimated variances). For a design with any probabilities π_k, we indeed see that everything depends on the value of $z_k = \log y_k$. If the z_k are 'rather constant' (therefore, if the y_k only vary a little), then $\widehat{G}$ is preferable to $\widehat{\widehat{G}}$. On the contrary, if the z_k are instead proportional to π_k (that is $y_k \approx \mu^{\pi_k}$), then we recommend $\widehat{\widehat{G}}$.

Exercise 8.4 *Poisson design and calibration on population size*

For a Poisson design with unequal probabilities, give the variance of the Horvitz-Thompson estimator of the total. Afterwards, give an unbiased estimator of this variance. For this same design (knowing the population size N), from now on we use the regression estimator, with only one auxiliary variable $x_k = 1, k \in U$.

1. Simplify the regression estimator. What well-known estimator is this? Is it unbiased?
2. Using the technique of linearisation, give the linearised variable associated with this estimator.
3. Give an approximation of the variance by means of the linearised variable. Formulate this variance with the variable y_k.

4. Give an estimator of this variance.
5. What probabilities π_k would we want to choose?
6. What happens to the previous results if the Poisson design is with equal probabilities, being $\pi_k = \bar{n}/N$, where $\bar{n}$ is the expected value of the total size of the sample?

Solution
We cannot use the Sen-Yates-Grundy variance, as the sample size is random. Therefore, we use the Horvitz-Thompson variance knowing that for a Poisson design, we have $\pi_{kl} = \pi_k \pi_\ell$ for all $k \neq \ell$, which gives:

$$\mathrm{var}(\widehat{Y}) = \sum_{k \in U} \frac{y_k^2}{\pi_k}(1 - \pi_k).$$

This expression is directly obtained by noting

$$\widehat{Y} = \sum_{k \in S} \frac{y_k}{\pi_k} = \sum_{k \in U} \frac{y_k}{\pi_k} I_k,$$

and by noticing that the random variables $I_k, k \in U$ are independent and follow Bernoulli distributions with parameters π_k. This variance can be estimated without bias by:

$$\widehat{\mathrm{var}}(\widehat{Y}) = \sum_{k \in S} \frac{y_k^2}{\pi_k^2}(1 - \pi_k).$$

1. The generalised regression estimator is

$$\widehat{Y}_{\mathrm{reg}} = \widehat{Y} + (N - \widehat{N})\hat{b},$$

where $\hat{b}$ estimates the 'true' regression coefficient of y_k on the constant 1, equal to:

$$\frac{\sum_{k \in U} y_k \times 1}{\sum_{k \in U} 1} = \frac{Y}{N} = \overline{Y},$$

and

$$\widehat{N} = \sum_{k \in S} \frac{1}{\pi_k}.$$

We select:

$$\hat{b} = \frac{\widehat{Y}}{\widehat{N}},$$

and therefore

$$\widehat{Y}_{\mathrm{reg}} = \widehat{Y} \frac{N}{\widehat{N}}.$$

The estimator calibrates well on N, since $\widehat{N}_{\mathrm{reg}} = N$ and an estimator based upon an estimated regression coefficient $\widehat{Y}/N$ would not lead to this fundamental property (see on this topic Exercise 7.3). It is that Hájek ratio that is a particular case of the ratio estimator where the auxiliary characteristic is $x_k = 1$. It is biased, with a $1/n$ bias, which is negligible when n is large.

2. The linearised variable of $\widehat{Y}_{\text{reg}} = N f(\widehat{Y}, \widehat{N})$ is

$$v_k = y_k - Y \frac{1}{N} = y_k - \overline{Y},$$

and its estimator

$$\hat{v}_k = y_k - \widehat{\overline{Y}}_H,$$

where $\widehat{\overline{Y}}_H$ is the Hájek ratio of the mean, which is

$$\frac{\widehat{Y}}{\widehat{N}} = \frac{1}{N} \times \widehat{Y}_{\text{reg}}.$$

3. The approximation of the variance is then

$$\text{var}(\widehat{Y}_{\text{reg}}) \approx \sum_{k \in U} \frac{v_k^2}{\pi_k}(1 - \pi_k) = \sum_{k \in U} \frac{1 - \pi_k}{\pi_k} \left(y_k - \overline{Y}\right)^2.$$

This variance, *a priori*, should be smaller than that of the Horvitz-Thompson estimator (see the preliminary question). Indeed, the y_k^2 of the Horvitz-Thompson variance are here replaced by $(y_k - \overline{Y})^2$, which are 'normally' smaller. Although the complex coefficients $(1 - \pi_k)/\pi_k$ disturb the comparisons, we remember that we still have

$$\sum_{k \in U} y_k^2 \geq \sum_{k \in U} (y_k - \overline{Y})^2.$$

4. Its estimator (slightly biased) is:

$$\widehat{\text{var}}(\widehat{Y}_{\text{reg}}) = \sum_{k \in S} \frac{1 - \pi_k}{\pi_k^2} \left(y_k - \widehat{\overline{Y}}_H\right)^2.$$

5. We try to minimise the convex function $\text{var}(\widehat{Y}_{\text{reg}})$ subject to:

$$\sum_{k \in U} \pi_k = \bar{n} \text{ and } \pi_k > 0,$$

where $\bar{n}$ is the expected value of the sample size (fixed by the survey taker). The Lagrangian technique finally leads to:

$$\text{for all } k \in U : \pi_k = \bar{n} \frac{|v_k|}{\sum_{k \in U} |v_k|} \text{ if } v_k \neq 0.$$

If $v_k = 0$, we consider π_k 'unimportant' in $]0, 1]$. In practice, v_k is unknown. It is necessary to estimate (even roughly) *a priori*.

6. Let us denote n_S as the sample size: this is a random variable, with expected value $\bar{n}$. Applying the previous expressions on $\pi_k = \bar{n}/N$, we easily get:

$$\widehat{Y} = N\frac{n_S}{\bar{n}}\widehat{\overline{Y}},$$

where

$$\widehat{\overline{Y}} = \frac{1}{n_S}\sum_{k \in S} y_k,$$

and

$$\widehat{Y}_{\text{reg}} = N\widehat{\overline{Y}},$$

$$\text{var}(\widehat{Y}_{\text{reg}}) = N^2\left(1 - \frac{\bar{n}}{N}\right)\frac{\sigma_y^2}{\bar{n}},$$

where

$$\sigma_y^2 = \frac{1}{N}\sum_{k \in U}(y_k - \overline{Y})^2.$$

At last, the estimator of the variance is

$$\widehat{\text{var}}(\widehat{Y}_{\text{reg}}) = N^2\left(1 - \frac{\bar{n}}{N}\right)\frac{\tilde{\sigma}_y^2}{\bar{n}},$$

where

$$\tilde{\sigma}_y^2 = \frac{1}{\bar{n}}\sum_{k \in S}\left(y_k - \widehat{\overline{Y}}\right)^2,$$

which can also be written

$$\widehat{\text{var}}(\widehat{Y}_{\text{reg}}) = \left(\frac{n_S}{\bar{n}}\right)^2 N^2\left(1 - \frac{\bar{n}}{N}\right)\frac{\widehat{\sigma}_y^2}{n_S},$$

where

$$\widehat{\sigma}_y^2 = \frac{1}{n_S}\sum_{k \in S}\left(y_k - \widehat{\overline{Y}}\right)^2.$$

Thus, the regression estimator on the constant is identical to the classical estimator of simple random sampling, and its variance resembles that of this same classical estimator (it is sufficient to replace the actual sample size n_S by its expected value $\bar{n}$). As for the estimator of the variance, we can say that it is 'nearly' the variance estimator of simple random sampling, multiplied by the corrective term $(n_S/\bar{n})^2$.

Exercise 8.5 *Variance of a regression estimator*

In a simple random design without replacement, estimate the variance of the regression estimator of the total when n is large

$$\widehat{Y}_{\text{reg}} = \widehat{Y} + \frac{s_{xy}}{s_x^2}(X - \widehat{X}).$$

Solution

In a simple random sampling of fixed size, the population size N needs to be known. The regression estimator is

$$\widehat{Y}_{\text{reg}} = \widehat{Y} + \frac{s_{xy}}{s_x^2}(X - \widehat{X}).$$

If we set

$$f(a_1, a_2, a_3, a_4) = a_1 + \frac{Na_3 - a_1 a_2}{Na_4 - a_2^2}(X - a_2),$$

we can write the regression estimator as a function of estimators of the totals

$$\widehat{Y}_{\text{reg}} = f(\widehat{Y}, \widehat{X}, \widehat{(XY)}, \widehat{(X^2)}),$$

where

$$\widehat{(XY)} = \frac{N}{n}\sum_{k \in S} x_k y_k, \quad \text{and} \quad \widehat{(X^2)} = \frac{N}{n}\sum_{k \in S} x_k^2.$$

In a simple design, the estimators of the totals $\widehat{Y}, \widehat{X}, \widehat{(XY)}$ and $\widehat{(X^2)}$ are all of $O_p(n^{-1/2})$, where $O_p(1/x)$ is a quantity which remains bounded in probability when multiplied by x. We start by calculating the partial derivatives:

$$\left.\frac{\partial f(a_1, a_2, a_3, a_4)}{\partial a_1}\right|_{a_1=Y, a_2=X, a_3=(XY), a_4=(X^2)} = 1$$

$$\left.\frac{\partial f(a_1, a_2, a_3, a_4)}{\partial a_2}\right|_{a_1=Y, a_2=X, a_3=(XY), a_4=(X^2)} = \frac{-S_{xy}}{S_x^2}$$

$$\left.\frac{\partial f(a_1, a_2, a_3, a_4)}{\partial a_3}\right|_{a_1=Y, a_2=X, a_3=(XY), a_4=(X^2)} = 0$$

$$\left.\frac{\partial f(a_1, a_2, a_3, a_4)}{\partial a_4}\right|_{a_1=Y, a_2=X, a_3=(XY), a_4=(X^2)} = 0,$$

where

$$(XY) = \sum_{k \in U} x_k y_k \quad \text{and} \quad (X^2) = \sum_{k \in U} x_k^2,$$

and S_{xy} and S_x^2 are respectively the corrected population covariance between x and y and the corrected population variance of x. We then get the linearised variable

$$v_k = y_k - \frac{S_{xy}}{S_x^2}x_k, \quad \text{estimated by} \quad \widehat{v}_k = y_k - \frac{s_{xy}}{s_x^2}x_k.$$

The (biased) variance estimator is therefore

$$\widehat{\mathrm{var}}\left[\widehat{Y}_{\mathrm{reg}}\right] = \frac{N(N-n)}{n}\frac{1}{n-1}\sum_{k\in S}(\hat{v}_k - \bar{\hat{v}})^2 \text{ where } \bar{\hat{v}} = \frac{1}{n}\sum_{k\in S}\hat{v}_k$$

$$= \frac{N(N-n)}{n}\left[s_y^2 + \frac{s_{xy}^2}{s_x^2} - 2\frac{s_{xy}^2}{s_x^2}\right]$$

$$= \frac{N(N-n)}{n}s_y^2\left(1 - \hat{\rho}^2\right),$$

where

$$\hat{\rho} = \frac{s_{xy}}{s_x s_y}.$$

Eventually,

$$\widehat{\mathrm{var}}\left[\widehat{Y}_{\mathrm{reg}}\right] = (1 - \hat{\rho}^2)\ \widehat{\mathrm{var}}\left(N\widehat{\overline{Y}}\right),$$

where $\widehat{\overline{Y}}$ is the simple mean in the sample.

Exercise 8.6 *Variance of the regression coefficient*

In some sampling design whose first two orders of inclusion probabilities are strictly positive, we consider the following parameter of interest:

$$\sigma_y^2 = \frac{1}{N}\sum_{k\in U}(y_k - \overline{Y})^2,$$

where $U = \{1, \ldots, k, \ldots, N\}$ indicates the population for which the size N is not supposedly known, and $\overline{Y}$ is the mean of the population.

1. Show that σ_y^2 is a function of totals (strictly).
2. Give the estimator of the parameter obtained by replacing the unknown totals of σ_y^2 with their Horvitz-Thompson estimators (called 'substitution estimator'). Simplify this expression to get a quadratic form in y_k.
3. Give the linearised variable associated to σ_y^2, then the approximate variance of the substitution estimator.
4. Give the variance estimator of the substitution estimator, in the case of simple random sampling. Give an expression as a function of moments about the mean.
5. By applying the same reasoning as before, give the Horvitz-Thompson estimator and the substitution estimator of the covariance σ_{xy}:

$$\sigma_{xy} = \frac{1}{N}\sum_{k\in U}(x_k - \overline{X})(y_k - \overline{Y}),$$

where

$$\overline{X} = \frac{1}{N}\sum_{k\in U}x_k, \quad \text{and} \quad \overline{Y} = \frac{1}{N}\sum_{k\in U}y_k.$$

6. Give the linearised variable associated to σ_{xy} and the estimator of the linearised variable.

7. Give the variance estimator of the substitution estimator of the covariance, in the case of simple random sampling. Give an expression as a function of moments about the mean.

8. By applying the technique of stepwise linearisation, give the linearised variable associated to the regression coefficient, in the regression of y on x. Come to a conclusion on the variance of the estimator of the regression coefficient.

Solution

1. The population variance of y_k in U can be written

$$\sigma_y^2 = \frac{1}{N}\sum_{k\in U} y_k^2 - \left(\frac{Y}{N}\right)^2 = \frac{(Y^2)}{N} - \left(\frac{Y}{N}\right)^2,$$

where

$$N = \sum_{k\in U} 1, \quad Y = \sum_{k\in U} y_k, \quad (Y^2) = \sum_{k\in U} y_k^2.$$

2. The substitution estimator is therefore

$$\widehat{\sigma}_{ysubs}^2 = \frac{\widehat{(Y^2)}}{\widehat{N}} - \left(\frac{\widehat{Y}}{\widehat{N}}\right)^2 = \frac{1}{\widehat{N}}\sum_{k\in S}\frac{\left(y_k - \widehat{\overline{Y}}_H\right)^2}{\pi_k},$$

where

$$\widehat{N} = \sum_{k\in S}\frac{1}{\pi_k}, \quad \widehat{Y} = \sum_{k\in S}\frac{y_k}{\pi_k}, \quad \widehat{(Y^2)} = \sum_{k\in S}\frac{y_k^2}{\pi_k}, \quad \text{and} \quad \widehat{\overline{Y}}_H = \frac{\widehat{Y}}{\widehat{N}}.$$

3. We can from that time calculate the linearised variable. Since

$$\widehat{\sigma}_{ysubs}^2 = f(\widehat{(Y^2)}, \widehat{Y}, \widehat{N}),$$

we obtain

$$v_k = y_k^2\frac{1}{N} - 2y_k\overline{Y}\frac{1}{N} + 2\overline{Y}\frac{Y}{N^2} - Y^2\frac{1}{N^2} = \frac{1}{N}\left\{\left(y_k - \overline{Y}\right)^2 - \sigma_y^2\right\}.$$

The approximate variance given by the linearisation is, for a simple design,

$$\text{var}\left(\widehat{\sigma}_{ysubs}^2\right) \approx \frac{N-n}{nN(N-1)}\sum_{k\in U}\left\{\left(y_k - \overline{Y}\right)^2 - \sigma_y^2\right\}^2$$

$$= \frac{N}{(N-1)}\frac{N-n}{Nn}\left(m_4 - \sigma_y^4\right),$$

where

$$m_4 = \frac{1}{N} \sum_{k \in U} \left(y_k - \overline{Y} \right)^4 .$$

This variance is very close to the exact variance obtained in Expression (2.13) of Exercise 2.21, page 56, particularly when N is large (see Expression (2.14)).

4. We estimate the linearised variable by

$$\hat{v}_k = \frac{1}{\widehat{N}} \left\{ \left(y_k - \widehat{\overline{Y}}_H \right)^2 - \hat{\sigma}_{ysubs}^2 \right\} .$$

For simple random sampling, we have, since N is then known:

$$\widehat{\mathrm{var}} \left(\hat{\sigma}_{ysubs}^2 \right) = \frac{N - n}{nN(n - 1)} \sum_{k \in S} \left\{ \left(y_k - \widehat{\overline{Y}}_H \right)^2 - \hat{\sigma}_{ysubs}^2 \right\}^2$$

$$= \frac{N - n}{N(n - 1)} \left(\hat{m}_4 - \hat{\sigma}_{ysubs}^4 \right) ,$$

where

$$\hat{m}_4 = \frac{1}{n} \sum_{k \in S} \left(y_k - \widehat{\overline{Y}}_H \right)^4 .$$

5. The true covariance is written:

$$\sigma_{xy} = \frac{1}{N} \sum_{k \in U} (x_k - \overline{X})(y_k - \overline{Y}) = \frac{1}{2N^2} \sum_{k \in U} \sum_{\substack{\ell \in U \\ \ell \neq k}} (x_k - x_\ell)(y_k - y_\ell).$$

The Horvitz-Thompson estimator of the covariance is:

$$\hat{\sigma}_{xy} = \frac{1}{2N^2} \sum_{k \in S} \sum_{\substack{\ell \in S \\ \ell \neq k}} \frac{(x_k - x_\ell)(y_k - y_\ell)}{\pi_{k\ell}} .$$

The covariance can also be written as a function of totals:

$$\sigma_{xy} = \frac{1}{N} \sum_{k \in U} x_k y_k - \frac{XY}{N^2} = \frac{(XY)}{N} - \frac{XY}{N^2} ,$$

where

$$N = \sum_{k \in U} 1, \quad X = \sum_{k \in U} x_k, \quad Y = \sum_{k \in U} y_k, \quad (XY) = \sum_{k \in U} x_k y_k ;$$

$$\hat{\sigma}_{xysubs} = \frac{\widehat{(XY)}}{\widehat{N}} - \frac{\widehat{X}\widehat{Y}}{\widehat{N}^2} = \frac{1}{\widehat{N}} \sum_{k \in S} \frac{\left(x_k - \widehat{\overline{X}}_H \right) \left(y_k - \widehat{\overline{Y}}_H \right)}{\pi_k} ,$$

where

$$\widehat{X} = \sum_{k \in S} \frac{x_k}{\pi_k}, \quad \widehat{(XY)} = \sum_{k \in S} \frac{y_k x_k}{\pi_k}, \quad \text{and} \quad \widehat{\overline{X}}_H = \frac{\widehat{X}}{\widehat{N}}.$$

6. By noticing that

$$\sigma_{xy} = f((XY), X, Y, N), \quad \text{and} \quad \widehat{\sigma}_{xysubs} = f(\widehat{(XY)}, \widehat{X}, \widehat{Y}, \widehat{N}),$$

we get the linearised variable for the covariance:

$$v_k = x_k y_k \frac{1}{N} - (XY)\frac{1}{N^2} - x_k \frac{Y}{N^2} - y_k \frac{X}{N^2} + 2\frac{XY}{N^3}$$

$$= \frac{1}{N}\left\{ (y_k - \overline{Y})(x_k - \overline{X}) - \sigma_{xy} \right\},$$

and therefore

$$\widehat{v}_k = \frac{1}{\widehat{N}}\left\{ \left(x_k - \widehat{\overline{X}}_H\right)\left(y_k - \widehat{\overline{Y}}_H\right) - \widehat{\sigma}_{xysubs} \right\}.$$

7. For a simple design, we have

$$\widehat{\text{var}}\,(\widehat{\sigma}_{xysubs}) = \frac{N-n}{nN(n-1)} \sum_{k \in S}\left\{ \left(x_k - \widehat{\overline{X}}_H\right)\left(y_k - \widehat{\overline{Y}}_H\right) - \widehat{\sigma}_{xysubs} \right\}^2$$

$$= \frac{N-n}{N(n-1)}\left(\widehat{m}_{22} - \widehat{\sigma}_{xysubs}^2 \right),$$

where

$$\widehat{m}_{22} = \frac{1}{n}\sum_{k \in S}\left(x_k - \widehat{\overline{X}}_H\right)^2\left(y_k - \widehat{\overline{Y}}_H\right)^2,$$

and

$$\widehat{\sigma}_{xysubs} = \frac{1}{n}\sum_{k \in S}(x_k - \widehat{\overline{X}}_H)(y_k - \widehat{\overline{Y}}_H).$$

8. The regression coefficient of y on x is:

$$b = \frac{\sigma_{xy}}{\sigma_x^2},$$

which we can estimate by

$$\widehat{b} = \frac{\widehat{\sigma}_{xysubs}}{\widehat{\sigma}_{xsubs}^2}.$$

The linearisation of $\widehat{b}$ is, by a stepwise reasoning:

$$w_k = \frac{1}{N}\left\{ (y_k - \overline{Y})(x_k - \overline{X}) - \sigma_{xy} \right\}\frac{1}{\sigma_x^2} - \frac{1}{N}\left\{ (x_k - \overline{X})^2 - \sigma_x^2 \right\}\frac{\sigma_{xy}}{\sigma_x^4}$$

$$= \frac{1}{N\sigma_x^2}(x_k - \overline{X})\left\{ (y_k - \overline{Y}) - (x_k - \overline{X})\frac{\sigma_{xy}}{\sigma_x^2} \right\}$$

$$= \frac{1}{N\sigma_x^2}(x_k - \overline{X})e_k,$$

where e_k is the true residual of the regression of y on x, which we naturally estimate by:

$$\widehat{w}_k = \frac{1}{\widehat{N}\widehat{\sigma}^2_{xsubs}}(x_k - \widehat{\overline{X}}_H)\left\{(y_k - \widehat{\overline{Y}}_H) - (x_k - \widehat{\overline{X}}_H)\widehat{b}\right\}.$$

The variance of $\widehat{b}$ is therefore approximated, if n is 'large', by the variance of $\sum_{k \in S} w_k / \pi_k$.

Exercise 8.7 *Variance of the coefficient of determination*

1. Using the technique of stepwise linearisation, give the linearised variable of the coefficient of determination (we assume N to be unknown) defined by

$$r^2 = \frac{\sigma^2_{xy}}{\sigma^2_x \sigma^2_y}.$$

   We will use the linearised variables of σ^2_x, σ^2_y and σ_{xy} obtained in Exercise 8.6. We note that the coefficient of determination is equal to the square of the linear correlation coefficient between x and y.
2. Show that the coefficient of determination can likewise be written as a function of regression coefficients.
3. Using the technique of stepwise linearisation, give the linearised variable of the coefficient of determination originating from the linearised variables of regression coefficients (see Question 8 of Exercise 8.6).
4. Do the two methods give the same result?

Solution

1. From Exercise 8.6, we have the linearisation of the population variance σ^2_y:

$$u_k(y) = \frac{1}{N}\left\{(y_k - \overline{Y})^2 - \sigma^2_y\right\},$$

and the covariance σ_{xy}:

$$v_k = \frac{1}{N}\left\{(y_k - \overline{Y})(x_k - \overline{X}) - \sigma_{xy}\right\}.$$

Since

$$r^2 = \frac{\sigma^2_{xy}}{\sigma^2_y \sigma^2_x},$$

the linearisation of r^2 is given by

$$w_k = 2v_k \frac{\sigma_{xy}}{\sigma_x^2 \sigma_y^2} - u_k(x) \frac{\sigma_{xy}^2}{\sigma_x^4 \sigma_y^2} - u_k(y) \frac{\sigma_{xy}^2}{\sigma_x^2 \sigma_y^4}$$

$$= r^2 \frac{1}{N} \left\{ \frac{2}{\sigma_{xy}} \left(y_k - \overline{Y} \right) \left(x_k - \overline{X} \right) - \frac{1}{\sigma_x^2} \left(x_k - \overline{X} \right)^2 - \frac{1}{\sigma_y^2} \left(y_k - \overline{Y} \right)^2 \right\}$$

$$= r^2 \frac{1}{N} \left\{ \frac{1}{\sigma_{xy}} \left(y_k - \overline{Y} \right) \left(x_k - \overline{X} \right) - \frac{1}{\sigma_x^2} \left(x_k - \overline{X} \right)^2 \right.$$
$$\left. + \frac{1}{\sigma_{xy}} \left(y_k - \overline{Y} \right) \left(x_k - \overline{X} \right) - \frac{1}{\sigma_y^2} \left(y_k - \overline{Y} \right)^2 \right\}$$

$$= \frac{r^2}{N \sigma_{xy}} \left\{ \left(x_k - \overline{X} \right) e_k + \left(y_k - \overline{Y} \right) f_k \right\},$$

where e_k and f_k are respectively the true residuals (unknown) of the regression of y on x and of x on y.

2. The coefficient of determination can be written as a function of regression coefficients. In fact, $r^2 = b_1 b_2$, where b_1 and b_2 are respectively the regression coefficients of y on x and of x on y.
3. If we denote $u_k(b_1)$ (or $u_k(b_2)$) as the linearisation of the regression coefficient of y on x (or of x on y), the linearisation of the coefficient of determination is

$$a_k = u_k(b_1) b_2 + u_k(b_2) b_1$$

$$= \frac{e_k(x_k - \overline{X})}{N \sigma_x^2} b_2 + \frac{f_k(y_k - \overline{Y})}{N \sigma_y^2} b_1$$

$$= \frac{r^2}{N \sigma_{xy}} \left\{ \left(x_k - \overline{X} \right) e_k + \left(y_k - \overline{Y} \right) f_k \right\}.$$

4. Yes, the two methods give the same result. The approximate variance of r^2, when n is rather large and if the design is of unequal probabilities π_k, is that of $\sum_{k \in S} a_k / \pi_k$.

Exercise 8.8 *Variance of the coefficient of skewness*

Consider any design of fixed size of which the first-order π_k and second-order $\pi_{k\ell}$ inclusion probabilities are strictly positive. The objective is to estimate the variance of the estimator of the coefficient of skewness

$$g = \frac{m_3}{\sigma_y^3},$$

where

$$\sigma_y^2 = \frac{1}{N} \sum_{k \in U} (y_k - \overline{Y})^2, \quad m_3 = \frac{1}{N} \sum_{k \in U} (y_k - \overline{Y})^3, \quad \overline{Y} = \frac{1}{N} \sum_{k \in U} y_k.$$

In everything that follows, we assume that the population size N is known and that the sample size n is large.

1. Give the substitution estimators of σ_y^2, m_3 and g (these estimators are obtained by writing σ_y^2, m^3 and g in the form of totals, with each of these totals being estimated without bias by the classical Horvitz-Thompson estimator).
2. Give the linearised variable of σ_y^2, and deduce the linearisation of σ_y^3.
3. Give the linearisation of m_3.
4. Deduce the linearisation of g from the two previous questions.
5. Give the estimator of the linearisation of g and lastly, estimate the variance of $\hat{g}$.

Solution

1. We denote

$$Y = \sum_{k \in U} y_k, \quad (Y^2) = \sum_{k \in U} y_k^2, \quad (Y^3) = \sum_{k \in U} y_k^3,$$

$$\widehat{Y} = \sum_{k \in S} \frac{y_k}{\pi_k}, \quad \widehat{(Y^2)} = \sum_{k \in S} \frac{y_k^2}{\pi_k}, \quad \widehat{(Y^3)} = \sum_{k \in S} \frac{y_k^3}{\pi_k}.$$

Since

$$\sigma_y^2 = \frac{(Y^2)}{N} - \frac{Y^2}{N^2}, \quad \text{and} \quad \widehat{\sigma}_y^2 = \frac{\widehat{(Y^2)}}{N} - \frac{\widehat{Y}^2}{N^2},$$

$$m_3 = \frac{(Y^3)}{N} - 3 \frac{Y(Y^2)}{N^2} + 2 \frac{Y^3}{N^3}, \quad \text{and} \quad \widehat{m}_3 = \frac{\widehat{(Y^3)}}{N} - 3 \frac{\widehat{Y}\widehat{(Y^2)}}{N^2} + 2 \frac{\widehat{Y}^3}{N^3},$$

we have

$$\widehat{g} = \frac{\frac{\widehat{(Y^3)}}{N} - 3 \frac{\widehat{Y}\widehat{(Y^2)}}{N^2} + 2 \frac{\widehat{Y}^3}{N^3}}{\left(\frac{\widehat{(Y^2)}}{N} - \frac{\widehat{Y}^2}{N^2} \right)^{3/2}} = \frac{N^2 \widehat{(Y^3)} - 3N\widehat{Y}\widehat{(Y^2)} + 2\widehat{Y}^3}{\left(N\widehat{(Y^2)} - \widehat{Y}^2 \right)^{3/2}}.$$

2. The linearisation of $\sigma_y^2 = f((Y^2), Y)$ is

$$u_k = y_k^2 \frac{1}{N} - 2y_k \frac{Y}{N^2} = \frac{y_k}{N} \left(y_k - 2\overline{Y} \right).$$

The linearisation of σ_y^3 is obtained by stepwise linearisation

$$v_k = u_k \frac{3}{2} \left(\sigma_y^2 \right)^{1/2} = \frac{3}{2} \sigma_y \frac{y_k}{N} \left(y_k - 2\overline{Y} \right).$$

3. The linearisation of $m_3 = g((Y^3), (Y^2), Y)$ is

$$w_k = \frac{1}{N}\left\{y_k^3 - 3y_k^2\frac{Y}{N} + y_k\left(\frac{-3(Y^2)}{N} + \frac{6Y^2}{N^2}\right)\right\}$$

$$= \frac{y_k}{N}\left(y_k^2 - 3y_k\frac{Y}{N} - 3\frac{(Y^2)}{N} + \frac{6Y^2}{N^2}\right).$$

4. The linearisation of g is (stepwise linearisation):

$$z_k = w_k\frac{1}{\sigma_y^3} - v_k\frac{m_3}{\sigma_y^6}$$

$$= g\left[w_k\frac{1}{m_3} - v_k\frac{1}{\sigma_y^3}\right]$$

$$= g\left[\frac{y_k}{N}\left(y_k^2 - 3y_k\frac{Y}{N} - 3\frac{(Y^2)}{N} + \frac{6Y^2}{N^2}\right)\frac{1}{m_3} - \frac{3}{2}\sigma_y\frac{y_k}{N}\left(y_k - 2\overline{Y}\right)\frac{1}{\sigma_y^3}\right]$$

$$= \frac{gy_k}{N}\left[\left(y_k^2 - 3y_k\frac{Y}{N} - 3\frac{(Y^2)}{N} + \frac{6Y^2}{N^2}\right)\frac{1}{m_3} - \frac{3}{2}\left(y_k - 2\overline{Y}\right)\frac{1}{\sigma_y^2}\right].$$

5. We estimate z_k by

$$\widehat{z}_k = \frac{\widehat{g}y_k}{N}\left[\left(y_k^2 - 3y_k\frac{\widehat{Y}}{N} - 3\frac{\widehat{(Y^2)}}{N} + \frac{6\widehat{Y}^2}{N^2}\right)\frac{1}{\widehat{m}_3} - \frac{3}{2}\left(y_k - 2\frac{\widehat{Y}}{N}\right)\frac{1}{\widehat{\sigma}_y^2}\right],$$

which allows for estimating the variance of $\hat{g}$ by:

$$\widehat{\mathrm{var}}(\hat{g}) = \frac{1}{2}\sum_{k\in S}\sum_{\substack{\ell\in S\\\ell\neq k}}\frac{\pi_k\pi_\ell - \pi_{k\ell}}{\pi_{k\ell}}\left(\frac{\widehat{z}_k}{\pi_k} - \frac{\widehat{z}_\ell}{\pi_\ell}\right)^2.$$

Exercise 8.9 *Half-samples*

The aim of this exercise is to present a method of variance estimation called the 'half-sample method' that is part of the class of methods of sample replication. It can be done when the initial drawing of a sample, denoted S, has been performed according to a stratification technique, with the drawing of $n_h = 2$ individuals by simple random sampling in each of the H strata initially constructed. We denote:

- $w_h = N_h/N$, where N_h is the size of stratum h, and N is the total population size.
- y_{h1} and y_{h2} are the values of y known for the two individuals selected in stratum h (denoted by the indicators $h1$ and $h2$).
- $d_h = y_{h1} - y_{h2}$.

Throughout this exercise, we neglect the sampling rates.

1. What is the unbiased estimator $\widehat{\overline{Y}}_{\text{strat}}$ used, from the sample S, to estimate the true mean $\overline{Y}$? What is its estimated variance $\widehat{\text{var}}(\widehat{\overline{Y}}_{\text{strat}})$ as a function of w_h and d_h?

2. In *each* of the H strata, we select, through simple random sampling in S, one of the two individuals $h1$ or $h2$. The sampling is independent from one stratum to another. We thus generate a random sampling design leading to a very simple probability distribution on the identifiers of S, denoted Pr^*. In relation to this distribution, we can calculate the expected values E^* and the variances var^* (the sample S is fixed when we manipulate this distribution).

   Vocabulary: This procedure, applied successively on the H strata, produces a sample S^* of size H called a 'half-sample'.

   a) With S being fixed, how many half-samples are possible?

   b) If we denote hi as the individual selected in stratum h according to Pr^* (two possible cases: $i = 1$ or $i = 2$), give the values of the probabilities $\text{Pr}^*(i = 1)$ and $\text{Pr}^*(i = 2)$.

   c) Deduce that the estimator:

   $$\widehat{\overline{Y}}_{1/2} = \sum_{h=1}^{H} w_h y_{hi} \quad \text{is such that} \quad \text{E}^*(\widehat{\overline{Y}}_{1/2}) = \widehat{\overline{Y}}_{\text{strat}}.$$

   d) We consider the dichotomous random variable ε_h, defined in stratum h according to:

   $$\varepsilon_h = \begin{cases} +1 \text{ if } & hi = h1, \\ -1 \text{ if } & hi = h2. \end{cases}$$

   Show that

   $$\widehat{\overline{Y}}_{1/2} - \widehat{\overline{Y}}_{\text{strat}} = \frac{1}{2} \sum_{h=1}^{H} w_h \varepsilon_h d_h.$$

   Then, find $(\widehat{\overline{Y}}_{1/2} - \widehat{\overline{Y}}_{\text{strat}})^2$.

   e) Calculate $\text{E}^*(\varepsilon_h)$, and show that:

   $$\text{var}^*(\widehat{\overline{Y}}_{1/2}) = \text{E}^*(\widehat{\overline{Y}}_{1/2} - \widehat{\overline{Y}}_{\text{strat}})^2 = \widehat{\text{var}}(\widehat{\overline{Y}}_{\text{strat}}).$$

3. Let us assume that we select X half-samples under the same conditions (which are those of 2.) and in an independent way, with X very large. The experiment x ($1 \leq x \leq X$) leads to the estimator denoted $\widehat{\overline{Y}}_{1/2}(x)$. Use the law of large numbers to estimate $\text{var}^*(\widehat{\overline{Y}}_{1/2})$, i.e., $\widehat{\text{var}}(\widehat{\overline{Y}}_{\text{strat}})$, and, *in fine*, $\text{var}(\widehat{\overline{Y}}_{\text{strat}})$.

   Note: This method is apparently without great interest if we want to estimate the variance of $\widehat{\overline{Y}}_{\text{strat}}$, as an exact analytical calculation is preferable.

Indeed, we verify that it works for means, and we apply it in more complex estimators such as, for example, ratios, for which we do not know how to get exact analytical expressions.

Solution

1. It is a classical stratified sampling, with simple random sampling in each stratum:

$$\widehat{\overline{Y}}_{\text{strat}} = \sum_{h=1}^{H} \frac{N_h}{N} \widehat{\overline{Y}}_h, \quad \text{where} \quad \widehat{\overline{Y}}_h = \frac{y_{h1} + y_{h2}}{2}.$$

Furthermore,

$$\widehat{\text{var}}(\widehat{\overline{Y}}_{\text{strat}}) = \sum_{h=1}^{H} \left(\frac{N_h}{N}\right)^2 (1 - f_h) \frac{s_{yh}^2}{n_h}.$$

Indeed, $n_h = 2$, and

$$s_{yh}^2 = \frac{1}{2-1} \sum_{\substack{k \in h \\ k \in S}} (y_k - \widehat{\overline{Y}}_h)^2 = (y_{h1} - \widehat{\overline{Y}}_h)^2 + (y_{h2} - \widehat{\overline{Y}}_h)^2 = \frac{d_h^2}{2}.$$

Therefore, since f_h is negligible,

$$\widehat{\text{var}}(\widehat{\overline{Y}}_{\text{strat}}) = \sum_{h=1}^{H} w_h^2 \frac{d_h^2}{4}.$$

2. The configuration of the sampling is as follows:
   a) In each of the H strata, there are two possibilities; hence, there are in total 2^H possible half-samples.
   b) $\text{Pr}^*(i = 1) = \text{Pr}^*(i = 2) = 1/2$, since we select at random one individual among the two.
   c) We recall that we reason here conditionally on S. Therefore

$$\text{E}^*(\widehat{\overline{Y}}_{1/2}) = \sum_{h=1}^{H} w_h \text{E}^*(y_{hi}).$$

Indeed,

$$\text{E}^*(y_{hi}) = y_{h1}\text{Pr}^*(i = 1) + y_{h2}\text{Pr}^*(i = 2) = \frac{y_{h1} + y_{h2}}{2} = \widehat{\overline{Y}}_h.$$

Conclusion: $\text{E}^*(\widehat{\overline{Y}}_{1/2}) = \widehat{\overline{Y}}_{\text{strat}}.$

d) We have

$$\widehat{\overline{Y}}_{1/2} - \widehat{\overline{Y}}_{\text{strat}} = \sum_{h=1}^{H} w_h(y_{hi} - \widehat{\overline{Y}}_h).$$

Indeed,

$$y_{hi} - \widehat{\overline{Y}}_h = \begin{cases} y_{h1} - \widehat{\overline{Y}}_h = \dfrac{y_{h1} - y_{h2}}{2} = \dfrac{d_h}{2}, & \text{with probability } \dfrac{1}{2} \\[2ex] y_{h2} - \widehat{\overline{Y}}_h = \dfrac{y_{h2} - y_{h1}}{2} = -\dfrac{d_h}{2}, & \text{with probability } \dfrac{1}{2}. \end{cases}$$

Thus,

$$y_{hi} - \widehat{\overline{Y}}_h = \varepsilon_h \frac{d_h}{2}.$$

Hence

$$\widehat{\overline{Y}}_{1/2} - \widehat{\overline{Y}}_{\text{strat}} = \sum_{h=1}^{H} w_h \varepsilon_h \frac{d_h}{2}.$$

We get

$$\left(\widehat{\overline{Y}}_{1/2} - \widehat{\overline{Y}}_{\text{strat}}\right)^2 = \frac{1}{4} \left[\sum_{h=1}^{H} w_h^2 \varepsilon_h^2 d_h^2 + \sum_{h=1}^{H} \sum_{\substack{\ell=1 \\ \ell \neq h}}^{H} w_h w_\ell \varepsilon_h \varepsilon_\ell d_h d_\ell \right].$$

e) The expected value of ε_h is:

$$\mathrm{E}^*(\varepsilon_h) = (+1) \times \frac{1}{2} + (-1) \times \frac{1}{2} = 0.$$

Now

$$\mathrm{var}^*(\widehat{\overline{Y}}_{1/2}) = \mathrm{E}^*(\widehat{\overline{Y}}_{1/2} - \widehat{\overline{Y}}_{\text{strat}})^2$$

$$= \frac{1}{4} \left[\sum_{h=1}^{H} w_h^2 d_h^2 + \sum_{h=1}^{H} \sum_{\substack{\ell=1 \\ \ell \neq h}}^{H} w_h w_\ell d_h d_\ell \mathrm{E}^*(\varepsilon_h \varepsilon_\ell) \right],$$

since $\mathrm{E}^*(\widehat{\overline{Y}}_{1/2}) = \widehat{\overline{Y}}_{\text{strat}}$ (according to c), for all h, $\varepsilon_h^2 = 1$, and d_h is not random for S fixed. In addition, for each pair (h, ℓ) where $h \neq \ell$,

$$\mathrm{E}^*(\varepsilon_h \varepsilon_\ell) = \mathrm{cov}(\varepsilon_h, \varepsilon_\ell) = 0,$$

for the drawings are independent from one stratum to another.

Conclusion:

$$\mathrm{var}^*(\widehat{\overline{Y}}_{1/2}) = \frac{1}{4} \sum_{h=1}^{H} w_h^2 d_h^2 = \widehat{\mathrm{var}}(\widehat{\overline{Y}}_{\text{strat}}).$$

3. By construction, the X estimators $\widehat{\overline{Y}}_{1/2}(x)$ are independent and identically distributed (*iid*). According to the law of large numbers, the empirical moments almost surely converge toward the 'true' moments according to the distribution (*). Therefore

$$\widetilde{V} = \frac{1}{X} \sum_{x=1}^{X} [\widehat{\overline{Y}}_{1/2}(x) - \widetilde{\overline{Y}}]^2 \text{ tends toward } \text{var}^*(\widehat{\overline{Y}}_{1/2}) = \widehat{\text{var}}(\overline{Y}_{\text{strat}}),$$

where

$$\widetilde{\overline{Y}} = \frac{1}{X} \sum_{x=1}^{X} \widehat{\overline{Y}}_{1/2}(x).$$

Conclusion:
Thus, using a system of X successive independent re-samples, we can very easily calculate an empirical statistic $\widetilde{V}$ which should not be 'too far' from the unbiased classical estimator $\widehat{\text{var}}(\overline{Y}_{\text{strat}})$ and thus from the true variance $\text{var}(\widehat{\overline{Y}}_{\text{strat}})$. This technique is in practice used to estimate the variance of *complex* estimators, i.e., non-linear estimators (ratios, regression coefficients, correlation coefficients).

9

Treatment of Non-response

Non-response is an inevitable phenomenon in surveys. We distinguish total non-response, which affects individuals for which we do not have available any workable collected information, and partial non-response, which corresponds to 'holes' in the information collected for a given individual (certain variables y_k are known, but others are not). In all cases, this phenomenon generates a bias and increases the variance that varies more or less explicitly as a function of the inverse of the sample size of the respondents. There exist two large classes of methods to correct the non-response: reweighting and imputation.

9.1 Reweighting methods

We denote ϕ_k as the probability of response of individual k: this entire approach rests on the idea that the decision of whether or not to respond is random and is formalised by a probability, which we consider here, to simplify, that it only depends on individual k (indeed, it could very likely depend on the set of identifiers sampled). If ϕ_k is known, before an eventual calibration, we estimate without bias the total Y by:

$$\widehat{Y}_\phi = \sum_{k \in r} \frac{y_k}{\pi_k \phi_k},$$

where π_k is the regular inclusion probability, and r indicates the sample of respondents ($r \subset S$). In practice, we try to model the probability ϕ_k (unknown) to be able to estimate it subsequently. The leads are then multiple, but often we try to partition the population U into sub-populations U_c inside of which the ϕ_k are supposedly constant:

$$\phi_k = \phi_c \text{ when } k \in U_c.$$

We are speaking of a homogeneous response model. We can also model ϕ_k by a logistic function (for example) if we have available quantitative or qualitative auxiliary information that is sufficiently reliable. Reweighting is essentially used to treat total non-response.

9.2 Imputation methods

Contrary to the case of the method of reweighting, we directly model the behaviour y_k by using a vector of auxiliary information $\mathbf{x}_k$. For example, we denote (model called 'superpopulation'):

$$y_k = \psi(\mathbf{x}'_k \mathbf{b}) + z_k,$$

where ψ is a known function and z_k is a random variable of null expected value and variance σ^2. We use the information on the respondents to estimate $\mathbf{b}$ and σ^2 and we predict y_k, for each non-respondent k, with y_k^*. Lastly, we calculate:

$$\widehat{Y}_I = \sum_{k \in r} \frac{y_k}{\pi_k} + \sum_{\substack{k \in S \\ k \notin r}} \frac{y_k^*}{\pi_k},$$

which allows for the conservation of the initial weights. If, within any sub-population, we believe in the model $y_k = b + z_k$, we can impute $y_k^* = y_\ell$, where ℓ is an identifier selected at random in the respondent sub-population: this is a technique called 'hot deck'. The study of the quality of $\widehat{Y}_I$ is performed by bringing into play the random variable z_k. Imputation is essentially used to treat partial non-response.

EXERCISES

Exercise 9.1 *Weight of an aeroplane*

We wish to estimate the total weight of 250 passengers on a charter flight. For that, we select a simple random sample of 25 people for whom we intend to ask their height (in centimetres) and their weight (in kilograms). Five people refuse to respond, but we can all the same note their gender (1: male and 2: female). Among the others, five have given their height but did not want to say their weight. The collected data is finally presented in Table 9.1.

1. What methods can we use to correct the effects of non-response? Justify your decisions in a precise way, by explaining the models that you use. Perform the numerical applications.
2. You learn that 130 passengers are men and 120 are women. Would you modify your estimation method? Why?
3. Among the 10 non-responses for weight, we select a simple random sample comprised of individuals b, g, w, x. Using a particularly persuasive interviewer, we get them to admit their height and their weight. This complementary information is given in Table 9.2. How can we take this into consideration?

Table 9.1. Sample of 25 selected individuals: Exercise 9.1

Individual	Gender	Height	Weight
a	1	170	60
b	1	170	
c	1	180	70
d	1	190	80
e	1	190	80
f	1	170	70
g	1	170	
h	1	180	80
i	1	180	80
j	1	180	80
k	1	180	
l	1	190	
m	1	190	90
n	2	150	40
o	2	160	50
p	2	170	60
q	2	150	50
r	2	160	60
s	2	180	70
t	2	180	
u	1		
v	1		
w	2		
x	2		
y	2		

Table 9.2. Complementary information for four individuals: Exercise 9.1

Individual	Gender	Weight	Height
b	1	*80*	*170*
g	1	*100*	*170*
w	2	*90*	*180*
x	2	*60*	*150*

Solution

1. Two types of non-response appear: total non-response for individuals u to y and partial non-response for b, g, k, l and t. The total non-response is treated in general by modifying the weights of the respondents (technique of 'reweighting'). Since only the gender variable is known, we can construct, at best, cells based on the gender variable. To justify this practice, we can have two points of view:

- A 'probabilistic' point of view, which postulates that the non-respondents of one given gender in fact account for a simple random subsample of the initially selected sample (gender by gender), whose size is equal to the number of respondents for the gender considered. A second approach, equivalent in terms of the estimator, depends on a Bernoulli type of response model: all individuals of a given gender have the same probability of response, estimated by the response rate characterising the gender (maximum likelihood estimator). A third way, equivalent in terms of the estimator, of adhering to this point of view, consists of saying that, conditionally on the gender, the weight variable and the 'response' variable are independent (the fact of deciding not to respond does not depend on the weight). With these three approaches, the reweighting estimator is:

$$\widehat{\overline{Y}}_\phi = \sum_{h=1,2} \frac{n_h}{n} \widehat{\overline{Y}}_{hr},$$

where n_h is the number of selected people of gender h ($h = 1, 2$) and $\widehat{\overline{Y}}_{hr}$ is the average weight of the respondents of gender h. If we treat the partial non-responses as total non-responses, it is theoretically unbiased if the probabilistic model is exact.

- A more 'modellistic' point of view, which is less interested in the process of selecting the non-respondents but which postulates a statistical model of type:

$$y_{hi} = \mu_h + \varepsilon_{hi},$$

where y_{hi} is the weight of individual i of gender h, μ_h is 'mean' of the weight characteristic of gender h and ε_{hi} is a random variable whose expected value is 0 (it is a classical approach in statistics: everything happens as if a random process had generated y_{hi} according to this model). The estimator is still $\widehat{\overline{Y}}_\phi$, but this time we are interested in its expected value under the model:

$$\mathcal{E}(\widehat{\overline{Y}}_\phi) = \sum_{h=1,2} \frac{n_h}{n} \mathcal{E}(\widehat{\overline{Y}}_{hr}) = \sum_{h=1,2} \frac{n_h}{n} \mu_h.$$

Therefore,

$$\mathrm{E}\, \mathcal{E}(\widehat{\overline{Y}}_\phi) = \sum_{h=1,2} \frac{N_h}{N} \mu_h = \mathcal{E}(\overline{Y}) = \mathrm{E}\, \mathcal{E}(\overline{Y}).$$

We have $\mathrm{E}\, \mathcal{E}(\widehat{\overline{Y}}_\phi - \overline{Y}) = 0$, and therefore $\widehat{\overline{Y}}_\phi$ remains 'unbiased' if we bring into play the expected value under the model.

The partial non-response is treated in general by imputation, using a behaviour model. In every case, we use the auxiliary information given by the variable 'size', which is strongly linked to weight. To treat the partial non-response, we can for example use imputation by depending on a linear regression:

$$\text{weight} = a + b \times \text{height} + \text{residual},$$

and by estimating the parameters a and b from the respondents. This model can be repeated gender by gender. We find:

- for men (9 observations):

$$\text{weight} = -83.80 + 0.89 \times \text{height} + \text{residual} \quad (R^2 = 0.64),$$

- for women (6 observations):

$$\text{weight} = -75.10 + 0.80 \times \text{height} + \text{residual} \quad (R^2 = 0.80).$$

We could equally construct cells of 'homogeneous behaviour' by using the height variable and imputing the non-responses of one cell with the mean of respondents of the indicated cell. The problem is that of the composition of the cells. As a matter of course, many compositions are possible. A natural option consists of regrouping the individuals of the same height and same gender, which would lead to:

$$\text{cell } 1 = \{a, b, f, g\} \quad \text{cell } 2 = \{c, h, i, j, k\} \quad \text{cell } 3 = \{d, e, \ell, m\}$$
$$\text{cell } 4 = \{n, q\} \qquad \text{cell } 5 = \{o, r\} \qquad \text{cell } 6 = \{p\}$$
$$\text{cell } 7 = \{s, t\}.$$

Numerical applications:

The reweighting estimation depends on the calculation of $\widehat{\overline{Y}}_{hr}$, which can be conceived in two ways:

- either from the lone respondents of the variable 'weight', in which case there is no recognition of the partial non-response (the variable 'height' is at no time of assistance, so the information is lost),
- or from the respondents of the variable 'weight' for which we add the partial non-respondents after imputing a value for 'weight'.

In the first case, we have:

$$\widehat{\overline{Y}}_{1r} = \frac{690}{9} \approx 76.7 \text{ and } \widehat{\overline{Y}}_{2r} = \frac{330}{6} = 55,$$

thus

$$\widehat{\overline{Y}}_1 = \frac{15}{25} \times 76.7 + \frac{10}{25} \times 55 = 68 \text{ kg}.$$

In the second case, we impute five values. If we choose the imputation originally from a regression by gender:

$$y_b^* = y_g^* = -83.80 + 0.89 \times 170 = 67.5$$
$$y_k^* = -83.80 + 0.89 \times 180 = 76.4$$
$$y_\ell^* = -83.80 + 0.89 \times 190 = 85.3$$
$$y_t^* = -75.10 + 0.80 \times 180 = 68.9.$$

Then, we calculate:

$$\widehat{\overline{Y}}_{1r} = \frac{690 + 67.5 \times 2 + 76.4 + 85.3}{9 + 4} \approx 75.9,$$

and

$$\widehat{\overline{Y}}_{2r} = \frac{330 + 68.9}{6 + 1} \approx 57,$$

thus

$$\widehat{\overline{Y}}_2 = \frac{15}{25} \times 75.9 + \frac{10}{25} \times 57 = 68.3 \text{ kg}.$$

If we choose imputation by homogeneous cell:

$$y_b^* = y_g^* = \frac{60 + 70}{2} = 65,$$

$$y_k^* = \frac{70 + 80 \times 3}{4} = 77.5,$$

$$y_\ell^* = \frac{80 \times 2 + 90}{3} = 83.3.$$

$$y_t^* = 70.$$

Then, we calculate:

$$\widehat{\overline{Y}}_{1r} = \frac{690 + 65 \times 2 + 77.5 + 83.3}{9 + 4} \approx 75.4,$$

and

$$\widehat{\overline{Y}}_{2r} = \frac{330 + 70}{6 + 1} \approx 57.1,$$

thus

$$\widehat{\overline{Y}}_3 = \frac{15}{25} \times 75.4 + \frac{10}{25} \times 57.1 = 68.1 \text{ kg}.$$

By way of comparison, we notice that the simple mean of the weights of 15 respondents is equal to $\widehat{\overline{Y}}_4 = 66.7$ kg. This estimate is the most natural if we believe in a model in which men and women have the same probability of response.

Conclusion of this question:
Everything is dependent on the behaviour model in which we believe. There is not therefore, among the four previous estimators, an approach that is indisputably better than the others.

2. This supplementary information allows for post-stratification on the entire population and thus leads to a modification in weights. In fact, the proportions of men in the sample and in the population differ. It is the same for women. We recall:
   - reweighting estimator by gender:

$$\widehat{\overline{Y}}_\phi = \sum_{h=1,2} \frac{n_h}{n} \widehat{\overline{Y}}_{hr},$$

- post-stratified estimator by gender:

$$\widehat{\overline{Y}}_{\text{post}} = \sum_{h=1,2} \frac{N_h}{N} \widehat{\overline{Y}}_{hr}.$$

We have

$$\frac{n_1}{n} = \frac{15}{25} = 60\% \quad \text{and} \quad \frac{N_1}{N} = \frac{130}{250} = 52\%,$$

and

$$\frac{n_2}{n} = \frac{10}{25} = 40\% \quad \text{and} \quad \frac{N_2}{N} = \frac{120}{250} = 48\%.$$

The post-stratified estimator uses 'exact' weights: its variance is smaller than that of the reweighting estimators. If we redo the first three approaches of Question 1, the post-stratified estimates become:

$$\widehat{\overline{Y}}_{1,\text{post}} = 0.52 \times 76.7 + 0.48 \times 55 = 66.3 \text{ kg},$$

$$\widehat{\overline{Y}}_{2,\text{post}} = 0.52 \times 75.9 + 0.48 \times 57 = 66.8 \text{ kg},$$

$$\widehat{\overline{Y}}_{3,\text{post}} = 0.52 \times 75.4 + 0.48 \times 57.1 = 66.6 \text{ kg}.$$

Each estimate appears to be smaller than its counterpart from Question 1: this is naturally due to an over-representation of men in the sample, which is rectified by the post-stratification (men have a higher average weight than women).

3. The new values obtained for partial non-respondents b and g are indeed larger than those obtained by imputation (regression or homogeneous cells), which shows that the non-response is very much related to the weight variable (people for whom weight is high in comparison to their height refuse to respond). We could use these new values to perform more pertinent imputations. If we had available sufficient values in the supplementary table obtained due to the persuasive interviewer, we could for example conceive a regression model uniquely from the initial non-respondent sub-population, to impute individual values to the partial non-respondents (by postulating a link of the same nature between height and weight among initial non-respondents). This would allow for the limiting of the bias generated by the non-response. Alas, this is not the case here: with so few values, we can 'only' add the values of the new respondents in the initial calculations to produce the model parameters. This is somewhat a last resort, which is going to certainly reduce the bias but does not resolve the underlying problem linked to the dependence between weight and non-response. The regression equations are modified by the recognition of two supplementary points for each gender:

- for men (11 observations):

$$\text{weight} = 29.60 + 0.28 \times \text{height} + \text{residual} \ (R^2 = 0.05),$$

- for women (8 observations):

$$\text{weight} = -95.40 + 0.96 \times \text{height} + \text{residual} \ (R^2 = 0.66).$$

We notice the poor quality of the adjustment for men. The means $\widehat{Y}_{hr}$ are found to be modified as a consequence, which is the same for imputations by homogeneous cell. We get

$$\widehat{\overline{Y}}_1 = \frac{15}{25} \times \frac{690 + 80 + 100}{11} + \frac{10}{25} \times \frac{330 + 90 + 60}{8} = 71.5 \text{ kg.}$$

With regression imputation:

$$y_k^* = 80, \qquad y_\ell^* = 82.8, \qquad y_t^* = 77.4,$$

thus

$$\widehat{\overline{Y}}_2 = \frac{15}{25} \times 79.4 + \frac{10}{25} \times 61.9 = 72.4 \text{ kg.}$$

With mean imputation by class (homogeneous cell):

$$y_k^* = 77.5, \qquad y_\ell^* = 83.3, \qquad y_t^* = \frac{70 + 90}{2} = 80,$$

thus

$$\widehat{\overline{Y}}_3 = \frac{15}{25} \times \frac{690 + 80 + 100 + 77.5 + 83.3}{13}$$
$$+ \frac{10}{25} \times \frac{330 + 90 + 60 + 80}{9} = 72.5 \text{ kg.}$$

Finally, the mean of the 19 respondents leads to $\widehat{\overline{Y}}_4 = 71.1$ kg.

Exercise 9.2 *Weighting and non-response*

It is a matter here of presenting two concurrent estimators in the presence of non-response. We consider a simple random sampling of size n in a population of size N and we are interested in the true mean $\overline{Y}$ of a variable y. We know in addition, for each individual in the sampling frame, a qualitative auxiliary variable x which takes C modalities (which leads to defining C 'cells' in the population). We denote n_r as the respondent sample size and $\widehat{\overline{Y}}_r$ as the mean of y in the sample of respondents. We assume from now on that in each cell c $(1 \leq c \leq C)$, there is at least one responding individual.

1. If we make the assumption that all the individuals of the population have the same probability of response, what reweighting estimator for the respondents are we going to use? When do we use this approach?

2. If we apply the preceding model, but only for each of the modalities of x, what estimator $\widehat{\overline{Y}}_\phi$ are we going to choose? We denote $\widehat{\overline{Y}}_{cr}$ as the mean of y_i of the respondents, for which x takes the value c and n_{cr} is the number of corresponding respondents. Compare the structures of $\widehat{\overline{Y}}_r$ and $\widehat{\overline{Y}}_\phi$.

3. *Numerical application:* In a survey on income for 300 people (simple random sampling), we have available the variable x 'place of residence', which allows distinguishing of three types of habitats: rural ($c = 1$), urban fringe and suburban ($c = 2$) and downtown ($c = 3$). The data is presented by category in Table 9.3. Calculate $\widehat{\overline{Y}}_r$ and $\widehat{\overline{Y}}_\phi$.

Table 9.3. Non-response according to category: Exercise 9.2

	Subpopulation		
	$c = 1$	$c = 2$	$c = 3$
Number of respondents	80	70	50
Mean annual income	9 800	11 600	13 600
Sample size	100	100	100

4. If we choose to proceed with case-by-case mean imputation of the respondents, what estimator are we going to get?

5. We are now interested in the bias and the variance of the two preceding estimators. Count all the random variables involved in the survey process, by taking into account the *ex-post* splitting into C cells (we consider that the behaviour of an individual i can be modelled by a random variable R_i that is 1 if i responds and 0 otherwise). Next, we consider that all these variables are fixed, except the sample S: under these simplified conditions, how must we comprehend the random nature of the estimator?

6. Under the previous conditions, give the bias and the conditional variance of $\widehat{\overline{Y}}_r$ and $\widehat{\overline{Y}}_\phi$. We will differentiate, in the expressions of bias, between one part that tends towards zero when n increases (denoted B^0) and one part that is insensitive to n (denoted B^∞).

7. If n is 'large', what are the favourable conditions to limit the conditional biases of $\widehat{\overline{Y}}_r$ and $\widehat{\overline{Y}}_\phi$? (We can go back to the numerical example of 3.)

8. We are trying to compare the conditional variances of $\widehat{\overline{Y}}_r$ and $\widehat{\overline{Y}}_\phi$, in the case where the sampling rates involved are negligible and where the variances of y_i among the respondents do not depend on the cell. Noting that these two estimators are written in the form $\sum_{c=1}^{C} w_c \widehat{\overline{Y}}_{cr}$, where the (w_c) are the non-random weights of the sum equal to 1, find the weighting scheme (w_c) which minimises the variance. Come to a conclusion on the 'better' of the two estimators $\widehat{\overline{Y}}_r$ and $\widehat{\overline{Y}}_\phi$ from the point of view of the variance.

9. If the 'reality' effectively corresponds to an independence between the value of y, being y_i, and the fact of whether or not to respond (modelled by the variable R_i), but only cell by cell and not in the total population,

which estimator are we going to finally retain in the case where the sample size is large?

10. We assume here that the 'reality' this time corresponds to an independence on the whole population between the variable y and the fact of whether or not to respond. We are placed in the case where N is very large compared to n.

a) To judge the respective biases of the two estimators, what terms must be compared?

b) Calculate the expected values for the squares of the two terms in question, by making use of the fact that (n_c) and (n_{cr}) approximately follow multinomial distributions $(n \ll N)$ (in this question, we retain only the conditioning with respect to R_i and n_r).

c) If we now assume that the probability of response does not depend upon the individual, considering the expected values for the squares of the two biases, which of the estimators $\widehat{\overline{Y}}_r$ and $\widehat{\overline{Y}}_\phi$ appears on average to be less biased?

d) Come to a conclusion on the respective quality of $\widehat{\overline{Y}}_r$ and $\widehat{\overline{Y}}_\phi$, under the assumptions from 8. and 10.(c).

Solution

1. The probability of response is manipulated like an inclusion probability (the sample of respondents is considered in theory to be selected with equal probabilities in the 'primitive' sample). The probability of response is going to be estimated by the global response rate, being n_r/n, which has the property of being the maximum likelihood estimator in a Bernoulli model. Hence:

$$\widehat{\overline{Y}}_\phi = \frac{1}{N} \sum_{i \in r} \frac{y_i}{\frac{n}{N} \frac{n_r}{n}} = \widehat{\overline{Y}}_r,$$

where r is the set of respondents. We use this assumption when we consider that the 'decision' to not respond does not even depend on the subject of the survey, i.e., neither on the value of y nor on any known auxiliary variable (in particular, not on x).

2. The variable x possesses C modalities: that leads to differentiating C sub-populations (cells 1, 2,..., C). In cell c, we denote n_{cr} as the number of respondents among the n_c sampled. The response probability is estimated, in cell c, by the ratio n_{cr}/n_c, the maximum likelihood estimator. Therefore, if we denote $r_c = r \cap c$:

$$\widehat{\overline{Y}}_\phi = \frac{1}{N} \sum_{c=1}^{C} \sum_{i \in r_c} \frac{y_i}{\frac{n}{N} \frac{n_{cr}}{n_c}} = \sum_{c=1}^{C} \frac{n_c}{n} \widehat{\overline{Y}}_{cr}.$$

We compare the structure of $\widehat{\overline{Y}}_\phi$ with that of $\widehat{\overline{Y}}_r$ written like this:

$$\widehat{\overline{Y}}_r = \sum_{c=1}^{C} \frac{n_{cr}}{n_r} \widehat{\overline{Y}}_{cr},$$

which differs in the weighting system of means $\widehat{\overline{Y}}_{cr}$: we use the weights of the categories in the sample for $\widehat{\overline{Y}}_\phi$ and the weights of the categories in the respondent sample for $\widehat{\overline{Y}}_r$.

3. We notice that the response rates obviously differ from one category to another (80% rural, 70% suburban and 50% downtown), which explains the numerical difference between $\widehat{\overline{Y}}_r$ and $\widehat{\overline{Y}}_\phi$. We have

$$\widehat{\overline{Y}}_\phi = \frac{100}{300} \times 9\,800 + \frac{100}{300} \times 11\,600 + \frac{100}{300} \times 13\,600 \approx 11\,670 \text{ Francs,}$$

and

$$\widehat{\overline{Y}}_r = \frac{80}{200} \times 9\,800 + \frac{70}{200} \times 11\,600 + \frac{50}{200} \times 13\,600 = 11\,380 \text{ Francs.}$$

The categories with the smallest mean income (rural zone) respond best: this is why the correction (going from $\widehat{\overline{Y}}_r$ to $\widehat{\overline{Y}}_\phi$) makes the estimate increase.

4. In case c, we keep y_i if individual i is a respondent ($i \in r_c$) and $\widehat{\overline{Y}}_{cr}$ otherwise (which affects $n_c - n_{cr}$ individuals). The estimator originally from the imputation is thus:

$$\widehat{\overline{Y}}_I = \frac{1}{N} \sum_{c=1}^{C} \frac{\sum_{i \in r_c} y_i + (n_c - n_{cr}) \widehat{\overline{Y}}_{cr}}{n/N}$$

$$= \frac{1}{n} \sum_{c=1}^{C} \left[(n_{cr} \widehat{\overline{Y}}_{cr} + (n_c - n_{cr}) \widehat{\overline{Y}}_{cr} \right] = \widehat{\overline{Y}}_\phi.$$

We hold that, case by case, mean imputation or reweighting by the inverse of the response probability leads to the same estimator.

5. We can distinguish four random variables (more or less interdependent):
   - The sample S, as a list of sampled identifiers;
   - The responding or non-responding characteristic of every individual in the population, which can be formed by a random variable R_i that is 1 is individual i responds and 0 otherwise;
   - The vector of sample sizes intersecting the C cells, being

$$\mathbf{n} = (n_1, n_2, \ldots, n_C);$$

   - The vector of respondent sample sizes by cell, being

$$\mathbf{n}_r = (n_{1r}; n_{2r}; \ldots; n_{Cr}).$$

If, for example, we fix S and R_i, then $\mathbf{n}$ and $\mathbf{n}_r$ are determined. If we are placed in the situation where we fix R_i, $\mathbf{n}$ and $\mathbf{n}_r$, there remains the random variable on S: this random variable leaves *a priori* a large number of combinations possible for the sample of respondents. The fixing of R_i comes back to designating non-respondents in the population. The fixing of n_c comes back, in the frame of a simple random survey of size n, to considering that we eventually complete a simple random survey of size n_c in cell c (a well-known result that is often used in the frame of domain estimation: consider here that a cell is a domain). The supplementary fixing of n_{cr} (still by virtue of the fundamental result of the domain estimation in the case of simple sampling) eventually consists in summarising the situation as such: in a given cell c distinguishing *a priori* a population of respondents and a population of non-respondents, we perform a simple random survey of size n_{cr} among the respondents and a simple random survey of size $(n_c - n_{cr})$ among the non-respondents (the latter obviously not yielding any information y).

6. *Attention!* The expected values and variances are simply denoted E(.) and var(.), but it is indeed a question, in the entire series, of conditional moments about $\mathbf{R}$ (vector of R_i), $\mathbf{n}$ and $\mathbf{n}_r$

$$\mathrm{E}(\widehat{\overline{Y}}_r) - \overline{Y} = \sum_{c=1}^{C} \frac{n_{cr}}{n_r}\, \overline{Y}_{cr} - \overline{Y},$$

where $\overline{Y}_{cr}$ is the true mean among the respondents of cell c (this parameter has here a significance, since the respondents in the population of the cell are fixed, from the fact of conditioning upon R_i). We know in fact that $\mathrm{E}(\widehat{\overline{Y}}_{cr} \mid \mathbf{n}, \mathbf{n_r}, \mathbf{R}) = \overline{Y}_{cr}$. We notice that in the absence of conditioning by R_i we cannot define $\overline{Y}_{cr}$, as there is no *a priori* respondent population. Since

$$\overline{Y}_r = \sum_{c=1}^{C} \frac{N_{cr}}{N_r}\, \overline{Y}_{cr},$$

where N_{cr} counts the respondents in the total population of cell c, we have:

$$\mathrm{E}(\widehat{\overline{Y}}_r) - \overline{Y} = \sum_{c=1}^{C} \left(\frac{n_{cr}}{n_r} - \frac{N_{cr}}{N_r} \right) \overline{Y}_{cr} + (\overline{Y}_r - \overline{Y}) = B^0 + B^\infty.$$

It is clear that if n is large, n_{cr}/n_r is close to N_{cr}/N_r (classical estimation theory of proportions in the case of simple random sampling, since n_{cr}/n_r is an unbiased estimator of N_{cr}/N_r and its variance varies by $1/n_r$), and B^0 indeed approaches zero. On the other hand, it is very likely that $\overline{Y}_r \neq \overline{Y}$, the difference between the two magnitudes (that is B^∞) being nothing depending on n. Furthermore,

$$\text{var}(\widehat{\overline{Y}}_r) = \sum_{c=1}^{C} \left(\frac{n_{cr}}{n_r}\right)^2 \text{var}(\widehat{\overline{Y}}_{cr}) = \sum_{c=1}^{C} \left(\frac{n_{cr}}{n_r}\right)^2 \left(1 - \frac{n_{cr}}{N_{cr}}\right) \frac{S_{cr}^2}{n_{cr}}$$

$$= \frac{1}{n_r} \sum_{c=1}^{C} \left(\frac{n_{cr}}{n_r}\right) \left(1 - \frac{n_{cr}}{N_{cr}}\right) S_{cr}^2,$$

where S_{cr}^2 is the population variance of y_i for the N_{cr} respondents of cell c. The variance of $\widehat{\overline{Y}}_r$ thus varies *in fine* by $1/n_r$. We have:

$$\text{E}(\widehat{\overline{Y}}_\phi) - \overline{Y} = \sum_{c=1}^{C} \frac{n_c}{n} \text{E}(\widehat{\overline{Y}}_{cr}) - \overline{Y}$$

$$= \sum_{c=1}^{C} \left(\frac{n_c}{n} - \frac{N_c}{N}\right) \overline{Y}_{cr} + \sum_{c=1}^{C} \frac{N_c}{N} (\overline{Y}_{cr} - \overline{Y}_c)$$

$$= B^0 + B^\infty.$$

For reasons similar to those mentioned for $\widehat{\overline{Y}}_r$, B^0 approaches zero if n increases, but B^∞ does not depend on n and is not null unless $\overline{Y}_{cr} = \overline{Y}_c$, for all c.

$$\text{var}(\widehat{\overline{Y}}_\phi) = \sum_{c=1}^{C} \left(\frac{n_c}{n}\right)^2 \left(1 - \frac{n_{cr}}{N_{cr}}\right) \frac{S_{cr}^2}{n_{cr}}.$$

We can write the variance as such:

$$\text{var}(\widehat{\overline{Y}}_\phi) = \frac{1}{n_r} \sum_{c=1}^{C} \frac{n_r/n}{n_{cr}/n_c} \left(1 - \frac{n_{cr}}{N_{cr}}\right) \frac{n_c}{n} S_{cr}^2.$$

The variance of $\widehat{\overline{Y}}_\phi$ therefore varies *in fine* by $1/n_r$.

7. If n is large, the problem is concentrated on B^∞. To reasonably use $\widehat{\overline{Y}}_r$, it is necessary to assume that $\overline{Y} = \overline{Y}_r$. That concretely returns to making the assumption that in the whole population the phenomenon of non-response does not at all depend on the value of y: technically, if we assume that y is a random variable for which the y_i constitute N realisations, we get the equality if there is independence between y and R, i.e., if $f(y \mid R) = f(y)$, where $f(y)$ is the distribution of y, or, which is equivalent, if $\Pr[R_i = 1 \mid y] = \Pr(R_i = 1)$. On the other hand, we use $\widehat{\overline{Y}}_\phi$ when we believe that $\overline{Y}_{cr} = \overline{Y}_c$ in each cell, i.e., that the non-response is not related to the value of y within each cell. The application of the assumption at a finer level sometimes lets us approach the reality in a more acceptable way. If we go back to the numerical example of 3., we indeed see that it is not reasonable to use $\widehat{\overline{Y}}_r$ because the mean income increases while the response rate decreases. An individual with high income (living downtown) responds less readily than an individual with lower income (characterising

the rural zone). However, for a given type of habitat and in the absence of other information, we can believe that y does not influence (or only slightly) on the non-response.

8. We try to minimise

$$\text{var}\left(\sum_{c=1}^{C} w_c \widehat{\overline{Y}}_{cr}\right) \quad \text{subject to} \quad \sum_{c=1}^{C} w_c = 1,$$

in the assumption where n_{cr} is small compared to N_{cr} and S_{cr}^2 is a constant denoted S^2 (this last assumption is a little simplifying, but the validity of the result naturally covers the case where these population variances are 'a little bit different' from one another). We have:

$$\text{var}\left(\sum_{c=1}^{C} w_c \widehat{\overline{Y}}_{cr}\right) \approx \sum_{c=1}^{C} w_c^2 \frac{S^2}{n_{cr}}.$$

Using the Lagrangian method, we find $w_c = n_{cr}/n_r$. The estimator of minimal variance is therefore $\widehat{\overline{Y}}_r$. Under the assumptions from the start, we have $\text{var}(\widehat{\overline{Y}}_\phi) > \text{var}(\widehat{\overline{Y}}_r)$.

9. The overall quality of the estimator $\widehat{\overline{Y}}$ is measurable by the criterion of the mean square error (MSE):

$$\text{MSE}(\widehat{\overline{Y}}) = \text{E}[\widehat{\overline{Y}} - \overline{Y}]^2 = \text{var}(\widehat{\overline{Y}}) + \text{Bias}^2.$$

Comparing from this point of view the two estimators $\widehat{\overline{Y}}_r$ and $\widehat{\overline{Y}}_\phi$, we see that the variances vary by $1/n_r$ (see 6.) and therefore approach zero when n becomes large, that the B^0 parts of the bias also approach 0 and that the B^∞ parts consequently remain the prominent terms of the numerical point of view. In the conditions stated, B^∞ is 0 for $\widehat{\overline{Y}}_\phi$ but not for $\widehat{\overline{Y}}_r$. We thus without hesitation keep $\widehat{\overline{Y}}_\phi$.

10. a) If there is independence between y and R in the whole population (therefore, in particular cell by cell), the B^∞ components are null, for $\widehat{\overline{Y}}_r$ as well as for $\widehat{\overline{Y}}_\phi$. To judge the impact of the biases, it therefore remains to compare the squares of the following terms:

$$B_1 = \sum_{c=1}^{C} \left(\frac{n_{cr}}{n_r} - \frac{N_{cr}}{N_r}\right) \overline{Y}_{cr}, \text{ for } \widehat{\overline{Y}}_r,$$

and

$$B_2 = \sum_{c=1}^{C} \left(\frac{n_c}{n} - \frac{N_c}{N}\right) \overline{Y}_{cr}, \text{ for } \widehat{\overline{Y}}_\phi.$$

The direct comparison of squares of these values is not possible; hence, the approach of part b).

b) The distribution of n_c is hypergeometric, for whatever c

$$n_c \sim \mathcal{H}\left(n, \frac{N_c}{N}\right),$$

as the sampling is simple random without replacement and of fixed size n. However, if n is very small compared to N, the hypergeometric distribution can be approximated by a multinomial distribution, which we will do hereafter. We therefore have, conditionally on R_i:

$$\mathrm{E}\left(\frac{n_c}{n}\right) = \frac{N_c}{N},$$

$$\mathrm{var}\left(\frac{n_c}{n}\right) = \frac{1}{n}\left(\frac{N_c}{N}\right)\left(1 - \frac{N_c}{N}\right),$$

and

$$\mathrm{cov}\left(\frac{n_c}{n}, \frac{n_d}{n}\right) = -\frac{1}{n}\frac{N_c}{N}\frac{N_d}{N}, \qquad \text{for all } c \neq d,$$

Now

$$
\begin{aligned}
\mathrm{E}(B_2^2) &= \mathrm{var}\,(B_2) \\
&= \sum_{c=1}^{C} \overline{Y}_{cr}^2 \,\mathrm{var}\left(\frac{n_c}{n}\right) + \sum_{c=1}^{C}\sum_{\substack{d=1\\d\neq c}}^{C} \overline{Y}_{cr}\overline{Y}_{dr}\mathrm{cov}\left(\frac{n_c}{n}, \frac{n_d}{n}\right) \\
&= \frac{1}{n}\sum_{c=1}^{C} \frac{N_c}{N}\,(\overline{Y}_{cr} - \tilde{Y}_r)^2,
\end{aligned}
$$

with

$$\tilde{Y}_r = \sum_{c=1}^{C} \frac{N_c}{N}\,\overline{Y}_{cr}.$$

Likewise, conditionally on n_r and R_i:

$$n_{cr} \sim \mathcal{H}\left(n_r, \frac{N_{cr}}{N_r}\right).$$

Therefore, by a calculation similar to the preceding, in approaching the hypergeometric by a multinomial,

$$\mathrm{E}(B_1^2) = \mathrm{var}(B_1) = \frac{1}{n_r}\sum_{c=1}^{C} \frac{N_{cr}}{N_r}\,(\overline{Y}_{cr} - \tilde{Y}_r)^2.$$

c) This assumption of constant probability places us in the scope where y and R are independent on the entire population. From this fact, the response probability satisfies:

$$\Pr[R_i = 1 \mid y] = \Pr[R_i = 1] = \text{constant}.$$

The response rate is therefore pretty much identical in every cell, thus:

$$\frac{N_{cr}}{N_c} \approx \frac{N_r}{N} \quad \Leftrightarrow \quad \frac{N_{cr}}{N_r} \approx \frac{N_c}{N}.$$

The equality is only approximate: we would rigorously get this by considering the expected values of the two members with respect to the distribution of R_i. However, if N_c is 'large' (we assume this), the approximation must be good or very good. In this case, $\widetilde{Y}_r \approx \overline{Y}_r$ and we finally get:

$$\mathrm{E}(B_2^2) \approx \frac{n_r}{n} \mathrm{E}(B_1^2),$$

where n_r/n is the response rate, less than 1. On average, B_2^2 is less than B_1^2.

d) From the point of view of the conditional variance, $\widehat{\overline{Y}}_r$ is preferable to $\widehat{\overline{Y}}_\phi$ (see 8.). From the point of view of the squared conditional bias, $\widehat{\overline{Y}}_\phi$ is instead preferable to $\widehat{\overline{Y}}_r$. Thus, there is not any evidence to say that overall (conditional criterion of the MSE) $\widehat{\overline{Y}}_\phi$ is preferable to $\widehat{\overline{Y}}_r$, all the more so as the advantage of $\widehat{\overline{Y}}_\phi$ in terms of bias is indeed slim if the response rate is good. Even if we cannot come up, at this stage, with a general rule, we hold out as, if the reality is indeed that of a constant response probability, the splitting of the population into cells c and the concerted use of $\widehat{\overline{Y}}_\phi$ instead of $\widehat{\overline{Y}}_r$ can very well carry a false sense of security: it is not because we are based on a finer splitting with an estimator adapted on this splitting that the estimate is on average better!

Exercise 9.3 *Precision and non-response*

For this exercise, it is a question of calculating the accuracy of an estimator in the presence of non-response, when we consider the sample of respondents as resulting from a two-phase survey. The variable of interest is denoted y.

1. *Preliminary:* We consider, in a sample S of given size n, that the individuals are all likely to respond with a probability ϕ, and they act independently from one another. We denote R_i as the random variable linked to individual i, which is 1 if he responds and 0 otherwise. What is the distribution of R_i? What is the distribution of R, the total number of respondents? Using two different methods, estimate ϕ and notice that we reach the response rate m/n (m indicates the value taken by R).

2. From now on, we are going to model the process leading to the sample of respondents. We begin from a sample of size n selected by simple random sampling. In addition, in the sample, we distinguish two sub-populations (respectively indicated by $h = 1$ and $h = 2$): the first is that of individuals likely to respond, independent from one another, with a probability ϕ_1, and the second is that of individuals likely to respond, independent from one another, with a probability ϕ_2. From the point of view of individual information, we consider that we are capable of replacing every non-responding sampled individual in his category h, but not every individual in the complete population. If the sample of respondents consists in fine of m_h individuals in sub-population h, what estimator $\widehat{Y}_\phi$ of the total Y are we going to use? (We assume that we always have at least one respondent in each category h.) We will verify that it is unbiased, after having seen the four types of random variables related to the modelling.

3. Using the appropriate conditioning, express the true variance $\mathrm{var}(\widehat{Y}_\phi)$. In some terms, there remain expected values, which we will not try to calculate.

4. With a quick calculation that assimilates the expected value of $1/m_h$ to $1/\mathrm{E}(m_h)$, which is justified only if n is 'quite large', give a more legible version of $\mathrm{var}(\widehat{Y}_\phi)$, as a function of ϕ_h.

5. Propose an unbiased variance estimator for $\widehat{Y}_\phi$.

   Hint: If we denote $\overline{\widehat{Y}}_{hr}$ as the mean of y_i calculated on the m_h respondents of h, n_h as the sample size intersecting the sub-population h, and $\overline{\widehat{Y}}_\phi$ as the unbiased mean estimator $\overline{Y}$, calculate the expected value of

$$\sum_{h=1,2} \frac{n_h}{n} \, (\overline{\widehat{Y}}_{hr} - \overline{\widehat{Y}}_\phi)^2,$$

   using the conditioning according to the different random variables.

Solution

1. The random variable R_i follows a Bernoulli distribution: $R_i \sim \mathcal{B}(1, \phi)$, therefore R has a binomial distribution

$$R = \sum_{i \in S} R_i \sim \mathcal{B}(n, \phi).$$

In fact, the R_i are independent (fundamental assumption).
Estimation of ϕ:

- Search for an unbiased estimator (method 1):
  As $\mathrm{E}(R) = n\phi$, we have

$$\mathrm{E}\left(\frac{R}{n}\right) = \phi.$$

We therefore choose:

$$\widehat{\phi} = \frac{m}{n},$$

where m is the realisation of the random variable R.

- Search for the maximum likelihood estimator (method 2):
  The distribution of R has as a density function:

$$\Pr[R = m] = \binom{n}{m} \phi^m (1 - \phi)^{n-m}.$$

It remains to maximise $\phi^m (1 - \phi)^{n-m}$, for given m. By differentiating with respect to ϕ, we easily find $\widehat{\phi} = m/n$.

2. There are two sub-populations of interest ($h = 1$ and $h = 2$). The sample S consists of a part of size n_1 (denoted S_1) crossing population 1 and another part of size n_2 (denoted S_2) crossing population 2. The response mechanism gives a sample r_1 of size m_1 in S_1 and a sample r_2 of size m_2 in S_2. Thus, there are four types of random variables:

- the sizes n_1 and n_2,
- the sample S,
- the sizes m_1 and m_2,
- the samples r_1 and r_2.

Fig. 9.1. Respondent and non-respondent samples: Exercise 9.3

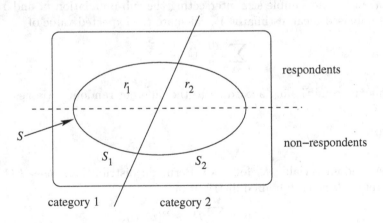

In the presence of non-response, we reweight by the inverse of the estimated response probability, in order to limit the biases. It is therefore going to be necessary to distinguish between the two sub-populations. In addition, the true response probabilities ϕ_1 and ϕ_2 being known, it is necessary to replace them with their respective estimators. Conditionally on n_1 and n_2, we know following from 1) that the 'good' estimators of ϕ_1 and ϕ_2 are m_1/n_1 and m_2/n_2. We therefore use:

$$\widehat{Y}_\phi = \sum_{h=1,2} \frac{\sum_{i \in r_h} y_i}{\frac{n}{N} \frac{m_h}{n_h}} = N \sum_{h=1,2} \frac{n_h}{n} \left(\frac{1}{m_h} \sum_{i \in r_h} y_i \right).$$

We can write

$$\widehat{Y}_\phi = N \sum_{h=1,2} \frac{n_h}{n} \widehat{\overline{Y}}_{rh},$$

where

$$\frac{1}{m_h} \sum_{i \in r_h} y_i = \widehat{\overline{Y}}_{rh}$$

is the mean of y_i in sample r_h. If the modelling is exact, the estimator $\widehat{Y}_\phi$ is unbiased on Y.

We have:

$$\mathrm{E}(\widehat{Y}_\phi) = \underset{S}{\mathrm{E}}[\underset{m_h|S}{\mathrm{E}}(\underset{r_h|S,m_h}{\mathrm{E}}\widehat{Y}_\phi)].$$

The three conditional expected values correspond to the 'fitting' of successive random variables: S foremost, then m_h (S fixed) and then finally r_h (m_h and S fixed). Indeed, a fundamental theorem (shown below) says that if we fix m_h, since the non-response comes from a Bernoulli scheme, everything occurs as if r_h came from a simple random sample of size m_h in S_h (fixed), which is written

$$\underset{r_h|S,m_h}{\mathrm{E}}(\widehat{Y}_\phi) = N \sum_{h=1,2} \frac{n_h}{n} \underset{r_h|S,m_h}{\mathrm{E}}(\widehat{\overline{Y}}_{rh}) = N \sum_{h=1,2} \frac{n_h}{n} \widehat{\overline{Y}}_h,$$

where

$$\widehat{\overline{Y}}_h = \frac{1}{n_h} \sum_{i \in S_h} y_i.$$

Thus

$$\underset{r_h|S,m_h}{\mathrm{E}}(\widehat{Y}_\phi) = N \widehat{\overline{Y}},$$

where $\widehat{\overline{Y}}$ indicates the simple mean of n values y_i for i in S. Since $\widehat{\overline{Y}}$ does not depend on m_h, we have $\underset{m_h|S}{\mathrm{E}}(\underset{r_h|S,m_h}{\mathrm{E}}\widehat{Y}_\phi) = N \widehat{\overline{Y}}$. Eventually,

$$\mathrm{E}(\widehat{Y}_\phi) = N \underset{S}{\mathrm{E}}(\widehat{\overline{Y}}) = Y.$$

Complements

Let us show that in some population of size n, if we select a sample according to a Bernoulli process of probability ϕ and whose size is m, then conditionally on m the selection is carried out by simple random sampling of size m. We have: $m \sim \mathcal{B}(n, \phi)$, the binomial distribution. Then, for every s sample,

$$\Pr[s \mid m] = \begin{cases} \dfrac{\Pr(s \text{ and } m)}{\Pr(m)} = \dfrac{\Pr(s)}{\Pr(m)} & \text{if } \#s = m \\ 0 & \text{if } \#s \neq m. \end{cases}$$

For all s of size m,

$$\Pr[s \mid m] = \frac{\phi^m (1-\phi)^{n-m}}{\binom{n}{m} \phi^m (1-\phi)^{n-m}} = \binom{n}{m}^{-1}.$$

This probability characterises the simple random sample of size m in a population of size n.

3. We have, by the decomposition formula of the variance,

$$\operatorname{var}(\widehat{Y}_\phi) = \operatorname*{var}_{S} \operatorname*{E}_{m_h|S} \operatorname*{E}_{r_h|S,m_h} (\widehat{Y}_\phi) + \operatorname*{E}_{S} \operatorname*{var}_{m_h|S} \operatorname*{E}_{r_h|S,m_h} (\widehat{Y}_\phi) + \operatorname*{E}_{S} \operatorname*{E}_{m_h|S} \operatorname*{var}_{r_h|S,m_h} (\widehat{Y}_\phi).$$

Let us examine each of the three terms on the right-hand side:

- *Term 3:*

$$\operatorname*{var}_{r_h|S,m_h} (\widehat{Y}_\phi) = N^2 \sum_{h=1,2} \left(\frac{n_h}{n}\right)^2 \operatorname*{var}_{r_h|S,m_h} (\widehat{\overline{Y}}_{rh}),$$

since S is fixed, the n_h are fixed as well. The conditioning allows assimilating the sampling to a stratified sampling by category h: therefore, there is no covariance. Thus, according to the theorem from Question 2., we have

$$\operatorname*{var}_{r_h|S,m_h} (\widehat{Y}_\phi) = N^2 \sum_{h=1,2} \left(\frac{n_h}{n}\right)^2 \left(1 - \frac{m_h}{n_h}\right) \frac{s_{yh}^2}{m_h},$$

where s_{yh}^2 is the variance of y_i in the subsample S_h, which gives

$$\operatorname*{E}_{S} \operatorname*{E}_{m_h|S} \operatorname*{var}_{r_h|S,m_h} (\widehat{Y}_\phi) = \operatorname*{E}_{S} \left\{ N^2 \sum_{h=1,2} \left(\frac{n_h}{n}\right)^2 \left(\operatorname*{E}_{m_h|S} \left(\frac{1}{m_h}\right) - \frac{1}{n_h}\right) s_{yh}^2 \right\}.$$

To go a bit further, we can consider a conditioning of S by n_h, being:

$$\operatorname*{E}_{S} = \operatorname*{E}_{n_h} \operatorname*{E}_{S|n_h}.$$

We notice that this distinction was not imposed in the calculation of the bias. Now, still by virtue of the same fundamental theorem, we know that if we condition by n_h everything happens as if we did simple random sampling of size n_h in the sub-population h. Thus $\operatorname*{E}_{S|n_h} (s_{yh}^2) = S_{yh}^2$, the population variance of y_i throughout sub-population h. Since $1/m_h$ evidently only depends on n_h (and not on S), we at last obtain a third term equal to:

$$N^2 \mathop{\mathrm{E}}_{n_h} \left\{ \sum_{h=1,2} \left(\frac{n_h}{n}\right)^2 \left[\mathop{\mathrm{E}}_{m_h|n_h} \left(\frac{1}{m_h}\right) - \frac{1}{n_h} \right] \right\} S_{yh}^2,$$

thus:

$$\frac{N^2}{n^2} \sum_{h=1,2} \left\{ \mathop{\mathrm{E}}_{n_h} \left[n_h^2 \mathop{\mathrm{E}}_{m_h|n_h} \left(\frac{1}{m_h}\right) \right] - \mathop{\mathrm{E}}_{n_h}(n_h) \right\} S_{yh}^2.$$

We can certainly write

$$\mathop{\mathrm{E}}_{n_h}(n_h) = n \, \frac{N_h}{N},$$

(N_h is the size of sub-population h), but the other expected values cannot be calculated in an exact manner, even if it is not possible to simplify this term any more.

- *Term 2:* Since $\mathop{\mathrm{E}}_{r_h|S,m_h} (\widehat{Y}_\phi) = N\overline{Y}$ and since this term does not depend on m_h, we have: $\mathop{\mathrm{var}}_{m_h|S} (N\overline{Y}) = 0$. The second term is therefore null.

- *Term 1:*

$$\mathop{\mathrm{E}}_{m_h|S} (\mathop{\mathrm{E}}_{r_h|S,m_h} (\widehat{Y}_\phi)) = \mathop{\mathrm{E}}_{m_h|S} (N\overline{Y}) = N\overline{Y}.$$

The first term is

$$N^2 \mathop{\mathrm{var}}_{S}(\widehat{\overline{Y}}) = N^2 \frac{1-f}{n} S_y^2.$$

Finally, by bringing together the three terms:

$$\mathrm{var}(\widehat{Y}_\phi)$$

$$= N^2 \left[\frac{1-f}{n} \, S_y^2 + \frac{1}{n^2} \sum_{h=1,2} \left\{ \mathop{\mathrm{E}}_{n_h} \left[n_h^2 \mathop{\mathrm{E}}_{m_h|n_h} \left(\frac{1}{m_h}\right) \right] - \mathop{\mathrm{E}}_{n_h}(n_h) \right\} S_{yh}^2 \right].$$

4. $m_h \sim \mathcal{B}(n_h, \phi_h)$. If n is large, the coefficient of variation of m_h is small, and therefore the approximation $\mathrm{E}(1/m_h) \approx 1/\mathrm{E}(m_h)$, although incorrect on the theoretical point of view, can be proven to be numerically acceptable. Under this approximation:

$$\mathrm{E}\left(\frac{1}{m_h}\right) \approx \frac{1}{n_h \phi_h},$$

and therefore, because $\mathrm{E}(n_h) = n\frac{N_h}{N}$:

$$\mathrm{var}(\widehat{Y}_\phi) \approx N^2 \frac{1}{n} \left[(1-f) \, S_y^2 + \sum_{h=1,2} \left(\frac{1}{\phi_h} - 1\right) \frac{N_h}{N} S_{yh}^2 \right].$$

We see in particular that if $\phi_h = 1$, for all h, we indeed find the classical scope of the theory without non-response. The second term of $\mathrm{var}(\widehat{Y}_\phi)$, which is positive, therefore constitutes the loss in accuracy (in order of magnitude) due to the lone phenomenon of non-response. Clearly, if we can act on ϕ_h (policy of training the interviewers, to go back to the non-respondents, etc.), we have complete interest in doing this (ϕ_h large), especially where $N_h S_{yh}^2$ is large.

5. We have:

$$\text{var}(\widehat{Y}_\phi) = N^2 \left[\frac{1-f}{n} S_y^2 \right] + \underset{S}{E} \underset{m_h|S}{E} \left[N^2 \sum_{h=1,2} \left(\frac{n_h}{n} \right)^2 \left(1 - \frac{m_h}{n_h} \right) \frac{s_{yh}^2}{m_h} \right],$$

and likewise:

$$s_{yh}^2 = \underset{r_h|S,m_h}{E} (s_{hr}^2),$$

where s_{hr}^2 is the sample variance of y_i in the sample r_h (therefore calculable).

Estimation of S_y^2: The sampling design being complex, it is necessary to try to estimate, on the one hand the 'inter' sub-population variance, and on the other hand the 'intra' population variance. We are therefore brought to calculate the expected value of:

$$\sum_{h=1,2} \frac{n_h}{n} (\widehat{\overline{Y}}_{hr} - \widehat{\overline{Y}}_\phi)^2 = \sum_{h=1,2} \frac{n_h}{n} \widehat{\overline{Y}}_{hr}^2 - \widehat{\overline{Y}}_\phi^2.$$

where $\widehat{\overline{Y}}_\phi$ indicates the unbiased estimator of $\overline{Y}$, being

$$\widehat{\overline{Y}}_\phi = \sum_{h=1,2} \frac{n_h}{n} \widehat{\overline{Y}}_{hr}.$$

We get

$$\text{E}\left(\sum_{h=1,2} \frac{n_h}{n} \widehat{\overline{Y}}_{hr}^2 \right)$$

$$= \underset{S}{E} \underset{m_h|S}{E} \left(\sum_{h=1,2} \frac{n_h}{n} \underset{r_h|m_h,S}{E} (\widehat{\overline{Y}}_{hr}^2) \right)$$

$$= \underset{S}{E} \underset{m_h|S}{E} \left(\sum_{h=1,2} \frac{n_h}{n} \underset{r_h|m_h,S}{\text{var}} (\widehat{\overline{Y}}_{hr}) + \sum_{h=1,2} \frac{n_h}{n} \widehat{\overline{Y}}_h^2 \right)$$

$$= \underset{S}{E} \underset{m_h|S}{E} \underset{r_h|m_h,S}{E} \left(\sum_{h=1,2} \frac{n_h}{n} \left(1 - \frac{m_h}{n_h} \right) \frac{s_{hr}^2}{m_h} \right) + \underset{n_h}{E} \underset{S|n_h}{E} \left(\sum_{h=1,2} \frac{n_h}{n} \widehat{\overline{Y}}_h^2 \right).$$

Now,

$$\mathop{\mathrm{E}}_{S|n_h}\left(\sum_{h=1,2}\frac{n_h}{n}\,\widehat{\overline{Y}}_h^2\right)=\sum_{h=1,2}\frac{n_h}{n}[\mathop{\mathrm{var}}_{S|n_h}(\widehat{\overline{Y}}_h)+\overline{Y}_h^2].$$

Thus,

$$\mathop{\mathrm{E}}_{n_h}\mathop{\mathrm{E}}_{S|n_h}\left(\sum_{h=1,2}\frac{n_h}{n}\,\widehat{\overline{Y}}_h^2\right)$$

$$=\mathop{\mathrm{E}}_{n_h}\left(\sum_{h=1,2}\frac{n_h}{n}\left(1-\frac{n_h}{N_h}\right)\frac{S_{yh}^2}{n_h}\right)+\sum_{h=1,2}\frac{N_h}{N}\,\overline{Y}_h^2$$

$$=\mathop{\mathrm{E}}_{n_h}\mathop{\mathrm{E}}_{S|n_h}\mathop{\mathrm{E}}_{m_h|S}\mathop{\mathrm{E}}_{r_h|m_h,S}\left(\sum_{h=1,2}\frac{n_h}{n}\left(1-\frac{n_h}{N_h}\right)\frac{s_{hr}^2}{n_h}\right)+\sum_{h=1,2}\frac{N_h}{N}\,\overline{Y}_h^2.$$

On the other hand,

$$\mathrm{E}(\widehat{\overline{Y}}_\phi^2)=\mathrm{var}(\widehat{\overline{Y}}_\phi)+\overline{Y}^2.$$

Finally,

$$\mathrm{E}\left(\sum_{h=1,2}\frac{n_h}{n}\,(\widehat{\overline{Y}}_{hr}-\widehat{\overline{Y}}_\phi)^2\right)$$

$$=\mathrm{E}\left(\sum_{h=1,2}\frac{n_h}{n}\left(1-\frac{m_h}{n_h}\right)\frac{s_{hr}^2}{m_h}\right)+\mathrm{E}\left(\sum_{h=1,2}\frac{n_h}{n}\left(1-\frac{n_h}{N_h}\right)\frac{s_{hr}^2}{n_h}\right)$$

$$-\mathrm{var}(\widehat{\overline{Y}}_\phi)+\left(\sum_{h=1,2}\frac{N_h}{N}\,\overline{Y}_h^2-\overline{Y}^2\right).$$

Therefore

$$\sum_{h=1,2}\frac{N_h}{N}\,(\overline{Y}_h-\overline{Y})^2$$

$$=\mathrm{E}\left[\sum_{h=1,2}\frac{n_h}{n}\,(\widehat{\overline{Y}}_{hr}-\widehat{\overline{Y}}_\phi)^2-\sum_{h=1,2}\frac{n_h}{n}\left(1-\frac{m_h}{N_h}\right)\frac{s_{hr}^2}{m_h}\right]+\mathrm{var}(\widehat{\overline{Y}}_\phi).$$

In addition, it is easy to verify that:

$$\mathop{\mathrm{E}}_{n_h}\mathop{\mathrm{E}}_{S|n_h}\mathop{\mathrm{E}}_{m_h|S}\mathop{\mathrm{E}}_{r_h|m_h,S}\left(\sum_{h=1,2}\frac{n_h}{n}\,s_{hr}^2\right)=\sum_{h=1,2}\frac{N_h}{N}\,S_{yh}^2,$$

which is the intra-population variance. Therefore:

$$S_y^2 = \sum_{h=1,2} \frac{N_h}{N}(\overline{Y}_h - \overline{Y})^2 + \sum_{h=1,2} \frac{N_h}{N} S_{yh}^2 \quad (N \text{ large})$$

$$= \text{var}(\widehat{\overline{Y}}_\phi)$$

$$+ \text{E}\left[\sum_{h=1,2} \frac{n_h}{n}(s_{hr}^2 + (\widehat{\overline{Y}}_{hr} - \widehat{\overline{Y}}_\phi)^2) - \sum_{h=1,2} \frac{n_h}{n}\left(1 - \frac{m_h}{N_h}\right)\frac{s_{hr}^2}{m_h}\right]$$

$$= \text{var}(\widehat{\overline{Y}}_\phi) + \text{E}(\delta),$$

where δ represents the complex term between square brackets.
Return to the estimation of $\text{var}(\widehat{Y}_\phi)$:

$$\text{var}(\widehat{Y}_\phi)$$

$$= N^2 \frac{1-f}{n}\left(\frac{\text{var}(\widehat{Y}_\phi)}{N^2} + \text{E}(\delta)\right) + \text{E}\left(N^2 \sum_{h=1,2}\left(\frac{n_h}{n}\right)^2\left(1 - \frac{m_h}{n_h}\right)\frac{s_{hr}^2}{m_h}\right).$$

Therefore,

$$\left(1 - \frac{1-f}{n}\right)\text{var}(\widehat{Y}_\phi) = N^2\text{E}\left(\frac{1-f}{n}\delta + \sum_{h=1,2}\left(\frac{n_h}{n}\right)^2\left(1 - \frac{m_h}{n_h}\right)\frac{s_{hr}^2}{m_h}\right).$$

An unbiased estimator of $\text{var}(\widehat{Y}_\phi)$ is thus:

$$\widehat{\text{var}}(\widehat{Y}_\phi)$$

$$= \left(1 - \frac{1-f}{n}\right)^{-1} N^2 \left(\frac{1-f}{n}\delta + \sum_{h=1,2}\left(\frac{n_h}{n}\right)^2\left(1 - \frac{m_h}{n_h}\right)\frac{s_{hr}^2}{m_h}\right).$$

In the most frequent conditions, we have n_h large, with f negligible compared to 1 (and therefore m_h negligible compared to N_h). After the developments that are derived, we come to an approximately unbiased estimator

$$\widehat{V}' = N^2\left(\frac{\widetilde{S}_y^2}{n} + \frac{1}{n}\sum_{h=1,2}\frac{n_h}{n}(n_h(1 - \widetilde{\phi}_h) - 1)\frac{s_{hr}^2}{m_h}\right),$$

with

$$\widetilde{S}_y^2 = \sum_{h=1,2}\frac{n_h}{n}(s_{hr}^2 + (\widehat{\overline{Y}}_{hr} - \widehat{\overline{Y}}_\phi)^2),$$

and

$$\widetilde{\phi}_h = \frac{m_h}{n_h}.$$

The term $\widetilde{S}_y^2$ is of an overall population variance nature, traced to the decomposition of S_y^2 into 'inter-variance' and 'intra-variance', and $\widetilde{\phi}_h$ is the response rate observed in category h.

It is difficult to simplify anymore, the order of magnitude (and even the sign) of $n_h(1 - \widetilde{\phi}_h) - 1$ possibly being of any nature. That being said, we can be satisfied with the term in $\widetilde{S}_y^2/n$ provided that, for $h = 1$ and $h = 2$, $|\, n_h(1 - \widetilde{\phi}_h) - 1 \,|$ is negligible compared to $m_h = \widetilde{\phi}_h n_h$, thus in practice $n_h(1 - \widetilde{\phi}_h)$ is negligible compared to $\widetilde{\phi}_h n_h$, or moreover $\widetilde{\phi}_h \geq 90\ \%$.

Exercise 9.4 *Non-response and variance*

We consider a sampling plan with unequal probabilities π_k (the π_k do not depend upon S) producing a sample S. The phenomenon of non-response leads to a sample r (consequently included in S); we model the behaviour of non-response using the Bernoulli approach, by distinguishing the categories c within which the response probability of each of the individuals is ϕ_c. We consider that the behaviours of response are independent from one individual to another and that there is independence between 'deciding whether or not to respond' and 'being in S' or not.

First part:
In this part, we consider that the response probabilities ϕ_c are known.

1. What natural unbiased estimator $\widehat{Y}_\phi$ are we going to use to estimate the total Y? Verify that its bias is effectively null.
2. Write the decomposition formula of the variance by distinguishing the randomness of producing S and the randomness of producing r given S.
3. Deduce the variance of $\widehat{Y}_\phi$ (we denote V as the variance that the Horvitz-Thompson estimator would have if there had not been any non-response). Ascertain that the supplementary imprecision brought by the non-response can be easily formalised.
4. How should we estimate (without bias) this variance? We will write the variance V under the following form:

$$V = \sum_{k \in U} \sum_{\ell \in U} w_{k\ell} y_k y_\ell,$$

where $w_{k\ell}$ depends on k and ℓ.

Second part:
In this part, we consider that the response probabilities ϕ_c are unknown. We call n_c the number of selected individuals that belong to category c (n_c random) and m_c the number of respondents among these n_c individuals.

1. In the Bernoulli model, how do we estimate ϕ_c?
2. Give the estimator $\widehat{Y}_\phi$ that is used.
3. Using the fact that conditioning by m_c amounts to a simple random sampling of size m_c in a population of size n_c, show that $\widehat{Y}_\phi$ is unbiased.
4. How would we naturally estimate the variance of $\widehat{Y}_\phi$? Is the estimator biased?
5. The variance expression obtained above is comprised, alas, of very complex terms that are the double inclusion probabilities $\pi_{k\ell}$. Assuming that we have available 'ready-made' software that can estimate the accuracy using a design with unequal probabilities π_k but cannot treat the non-response, how can we estimate the accuracy of the estimator obtained in 2.?

Solution
First part:

1. Naturally, the non-response procedure is treated like a supplementary sampling stage, where each individual k of S is kept with a probability ϕ_k which is appropriate for this. The estimator is only concerned with the individuals k of r, the only identifiers that we know y_k:

$$\widehat{Y}_\phi = \sum_{k \in r} \frac{y_k}{\pi_k \phi_k}, \tag{9.1}$$

here with

$$\phi_k = \phi_c \quad \text{if} \quad k \in c.$$

We can write:

$$\widehat{Y}_\phi = \sum_{k \in U} \frac{y_k}{\pi_k \phi_k} \times R_k \times I_k,$$

where

$$I_k = \begin{cases} 1 \text{ if } k \text{ is selected} \\ 0 \text{ otherwise,} \end{cases}$$

and

$$R_k = \begin{cases} 1 \text{ if } k \text{ responds} \\ 0 \text{ otherwise.} \end{cases}$$

R_k follows a Bernoulli distribution $B(1, \phi_k)$ and I_k follows a Bernoulli distribution $B(1, \pi_k)$.

$$\mathrm{E}(R_k) = \Pr[k \text{ responds}] = \phi_k,$$

$$\mathrm{E}(I_k) = \Pr[k \text{ is selected}] = \pi_k.$$

Thus,

$$\mathrm{E}(\widehat{Y}_\phi) = \sum_{k \in U} \frac{y_k}{\pi_k \phi_k} \mathrm{E}(R_k \times I_k).$$

Indeed
$$E(R_k \times I_k) = ER_k \times EI_k = \phi_k \times \pi_k.$$

The equality above is due to the assumption of independence between 'sampling' and 'response': the fact of knowing that an individual is selected gives absolutely no information about its behaviour in terms of response. In other words, being sampled is neither a particular incentive nor a restraint in accepting to respond in a survey. The independence of these two events is far from being evident in practice, as the fact of being selected in the sample can prompt, in itself, a response. With this assumption:
$$E(\widehat{Y}_\phi) = Y.$$

2. The variance decomposition gives:
$$\text{var}(\widehat{Y}_\phi) = \underset{S}{\text{var}}[\underset{r|S}{E}(\widehat{Y}_\phi \mid S)] + \underset{S}{E}[\underset{r|S}{\text{var}}(\widehat{Y}_\phi \mid S)],$$

where $r \mid S$ is the randomness from the Bernoulli model, and S is the 'classical' randomness producing the sample.
3. In the first place, it is necessary to express each of the conditional terms.

$$\underset{r|S}{E}[\widehat{Y}_\phi \mid S] = \underset{r|S}{E}\left[\sum_{k \in S} \frac{y_k}{\pi_k \phi_k} R_k \mid S\right] = \sum_{k \in S} \frac{y_k}{\pi_k \phi_k}\phi_k = \sum_{k \in S} \frac{y_k}{\pi_k},$$

where $\sum_{k \in S} y_k/\pi_k$ is the Horvitz-Thompson estimator used in the absence of non-response. Thus,

$$\underset{S}{\text{var}}[\underset{r|S}{E}(\widehat{Y}_\phi \mid S)] = \underset{S}{\text{var}}\left[\sum_{k \in S} \frac{y_k}{\pi_k}\right] = V,$$

$$\underset{r|S}{\text{var}}[\widehat{Y}_\phi \mid S] = \underset{r|S}{\text{var}}\left[\sum_{k \in S} \frac{y_k}{\pi_k \phi_k} R_k\right] = \sum_{k \in S}\left(\frac{y_k}{\pi_k \phi_k}\right)^2 \underset{r|S}{\text{var}}(R_k).$$

The covariances between the R_k are null, as the 'response' behaviours are independent from one individual to another. It is again based on an assumption for which the relevance in practice remains questionable. We can indeed imagine that the decision for individual i to respond (or not) has an 'influence' on the decision of another individual ℓ to respond (or not). This phenomenon routinely occurs for surveys in clusters. Now,

$$\text{var}(R_k) = \phi_k(1 - \phi_k) \text{ (Bernoulli distribution)},$$

and

$$\underset{r|S}{\text{var}}[\widehat{Y}_\phi \mid S] = \sum_{k \in S} \frac{1 - \phi_k}{\phi_k \pi_k^2} y_k^2 = \sum_{k \in U} \frac{1 - \phi_k}{\phi_k \pi_k} \frac{y_k^2}{\pi_k} I_k.$$

Therefore,

$$\underset{S\ r|S}{\mathrm{E}}[\mathrm{var}(\widehat{Y}_\phi \mid S)] = \sum_{k \in U} \frac{1 - \phi_k}{\phi_k} \frac{y_k^2}{\pi_k}.$$

Conclusion:
We have

$$\mathrm{var}(\widehat{Y}_\phi) = V + \sum_{k \in U} \frac{1 - \phi_k}{\phi_k} \frac{y_k^2}{\pi_k},$$

where V is thus the accuracy that we would get in the absence of non-response, and

$$\sum_{k \in U} \frac{1 - \phi_k}{\phi_k} \frac{y_k^2}{\pi_k}$$

is the loss in supplementary accuracy specifically due to the non-response. In particular, we verify that for all k, if $\phi_k = 1$ then $\mathrm{var}(\widehat{Y}_\phi) = V$. Since $\phi_k = \phi_c$ if $k \in c$, we can write:

$$\mathrm{var}(\widehat{Y}_\phi) = V + \sum_{c=1}^{C} \frac{1 - \phi_c}{\phi_c} \left(\sum_{k \in c} \frac{y_k^2}{\pi_k} \right),$$

where $(1 - \phi_c)/\phi_c$ is a decreasing function of ϕ_c.

4. We notice that V is a quadratic form, which justifies the adopted composition. The $w_{k\ell}$ are complex functions of k and ℓ, involving double inclusion probabilities $\pi_{k\ell}$. Then, reusing the approach from 1., we have:

$$\widehat{V} = \sum_{k \in r} \sum_{\substack{\ell \in r \\ \ell \neq k}} \frac{w_{k\ell}}{\pi_{k\ell}} \frac{y_k y_\ell}{\phi_k \phi_\ell} + \sum_{k \in r} \frac{w_{kk}}{\pi_k \phi_k} y_k^2. \qquad (9.2)$$

where $\widehat{V}$ estimates V without bias. The calculation of the expected value is done as in 1. and again uses the independences expressed in the assumption through:

$$\mathrm{E}(I_k I_\ell R_k R_\ell) = \mathrm{E}(I_k I_\ell)(\mathrm{E}R_k)(\mathrm{E}R_\ell) = \pi_{k\ell} \phi_k \phi_\ell.$$

In addition, the second part of $\mathrm{var}(\widehat{Y}_\phi)$ is estimated (without bias) by:

$$\sum_{k \in r} \frac{1 - \phi_k}{\phi_k^2} \frac{y_k^2}{\pi_k^2}.$$

Finally,

$$\widehat{\mathrm{var}}(\widehat{Y}_\phi) = \widehat{V} + \sum_{k \in r} \frac{1 - \phi_k}{\phi_k^2} \frac{y_k^2}{\pi_k^2},$$

where $\widehat{V}$ is the 'classical' variance estimator, weighted to take into consideration the non-response, and

$$\sum_{k \in r} \frac{1 - \phi_k}{\phi_k^2} \frac{y_k^2}{\pi_k^2}$$

is a 'supplementary' term specifically due to the existence of non-response.

Second part:

1. The natural estimator is:

$$\widehat{\phi}_c = \frac{m_c}{n_c}$$

(empirical response rate in category c). Here, n_c is a random variable, but if we think conditionally on n_c, $\widehat{\phi}_c$ is the maximum likelihood estimator of ϕ_c in the Bernoulli model. Actually, according to the non-response mechanism, $m_c \sim \mathcal{B}(n_c, \phi_c)$. Furthermore, conditionally on n_c, the estimator $\widehat{\phi}_c$ is unbiased for ϕ_c.

2. Since ϕ_c is unknown, we take a page from Expression (9.1) and we replace ϕ_c by $\widehat{\phi}_c$:

$$\widehat{Y}_\phi = \sum_{c=1}^{C} \sum_{k \in r_c} \frac{y_k}{\pi_k \widehat{\phi}_c} = \sum_{c=1}^{C} \sum_{k \in r_c} \frac{y_k}{\pi_k \, m_c/n_c} \quad \text{(with } r_c = r \cap c\text{)}.$$

3. In the first place, we are placed in S (we therefore condition with respect to S). We look for:

$$\mathop{\mathrm{E}}_{r|S}(\widehat{Y}_\phi \mid S) = \mathop{\mathrm{E}}_{m_c|S}[\mathrm{E}(\widehat{Y}_\phi \mid m_c, S)],$$

where $\mathop{\mathrm{E}}_{m_c|S}$ indicates the expected value with respect to the distribution of m_c conditionally on S. Indeed,

$$\mathrm{E}(\widehat{Y}_\phi \mid m_c, S) = \sum_{c=1}^{C} \mathrm{E}\left(\sum_{k \in r_c} \frac{y_k}{\pi_k \, m_c/n_c} \, \middle| \, m_c, S \right)$$

$$= \sum_{c=1}^{C} n_c \mathrm{E}\left[\sum_{k \in r_c} \frac{y_k/\pi_k}{m_c} \, \middle| \, m_c, S \right],$$

and

$$\mathrm{E}\left[\sum_{k \in r_c} \frac{y_k/\pi_k}{m_c} \, \middle| \, m_c, S \right] = \sum_{k \in S_c} \frac{y_k/\pi_k}{n_c},$$

for all c, with $S_c = S \cap c$. In effect, this expected value being conditional on the sample size m_c resulting from a Bernoulli sampling, we calculate this as if we had dealt with a simple random sample of size m_c in a population of size n_c:

$$\mathrm{E}(\widehat{Y}_\phi \mid m_c, S) = \sum_{c=1}^{C} \sum_{k \in S_c} \frac{y_k}{\pi_k} = \sum_{k \in S} \frac{y_k}{\pi_k}.$$

Therefore,

$$\mathop{\mathrm{E}}_{r|S}(\widehat{Y}_\phi \mid S) = \mathop{\mathrm{E}}_{m_c|S}\left(\sum_{k\in S}\frac{y_k}{\pi_k}\right) = \sum_{k\in S}\frac{y_k}{\pi_k},$$

because $\sum_{k\in S} y_k/\pi_k$ does not depend on m_c. Eventually, by deconditioning by S at the ultimate step, we obtain

$$\mathrm{E}(\widehat{Y}_\phi) = \mathop{\mathrm{E}}_{S}\mathrm{E}(\widehat{Y}_\phi \mid S) = \sum_{k\in U} y_k.$$

4. We go from Expression (9.2), and we replace the ϕ_k by their estimator (maximum likelihood) $\widehat{\phi}_k$:

$$\widehat{\widehat{V}} = \sum_{k\in r}\sum_{\substack{\ell\in r\\ \ell\neq k}}\frac{w_{k\ell}}{\pi_{k\ell}}\frac{y_k y_\ell}{\widehat{\phi}_k\widehat{\phi}_\ell} + \sum_{k\in r}\frac{w_{kk}}{\pi_k\widehat{\phi}_k}y_k^2 + \sum_{k\in r}\frac{(1-\widehat{\phi}_k)y_k^2}{\widehat{\phi}_k^2\,\pi_k^2}.$$

Even if $\widehat{\phi}_k$ estimates ϕ_k without bias (and if in knowing that n_c is large, we can remember that $\widehat{\phi}_k$ has the 'good properties' that all maximum likelihood estimators have), $\widehat{V}$ being a complex expression in ϕ_k (presence of squares, roots, products), this substitution operation renders $\widehat{\widehat{V}}$ slightly biased.

5. In fact, for a sample S and in the absence of non-response, the 'ready-made' software knows how to calculate the following variance estimator:

$$\widehat{\mathrm{var}}(\widehat{Y}_\phi) = \sum_{k\in S}\sum_{\substack{\ell\in S\\ \ell\neq k}}\frac{w_{k\ell}}{\pi_{k\ell}}y_k y_\ell + \sum_{k\in S}w_{kk}\frac{y_k^2}{\pi_k}.$$

We notice that if at the start of running the software we give the variable $z_k = y_k/\widehat{\phi}_k$, instead of y_k and if we do the calculation on the respondents only (for which z_k is perfectly known), the software is going to calculate:

$$\widehat{V}_2 = \sum_{k\in r}\sum_{\substack{\ell\in r\\ \ell\neq k}}\frac{w_{k\ell}}{\pi_{k\ell}}\frac{y_k}{\widehat{\phi}_k}\frac{y_\ell}{\widehat{\phi}_\ell} + \sum_{k\in r}w_{kk}\frac{y_k^2}{\pi_k\widehat{\phi}_k^2}$$

$$= \widehat{\widehat{V}} - \sum_{k\in r}\frac{w_{kk}}{\pi_k\widehat{\phi}_k}y_k^2 - \sum_{k\in r}\frac{1-\widehat{\phi}_k}{\widehat{\phi}_k^2}\frac{y_k^2}{\pi_k^2} + \sum_{k\in r}w_{kk}\frac{y_k^2}{\pi_k\widehat{\phi}_k^2}.$$

Therefore,

$$\widehat{V} = \widehat{V}_2 + \sum_{k \in r} \frac{w_{kk}}{\pi_k \widehat{\phi}_k} y_k^2 \left\{ 1 + \frac{1 - \widehat{\phi}_k}{\widehat{\phi}_k \pi_k w_{kk}} - \frac{1}{\widehat{\phi}_k} \right\}$$

$$= \widehat{V}_2 + \sum_{k \in r} \frac{(1 - \widehat{\phi}_k)(1 - w_{kk}\pi_k)}{\widehat{\phi}_k^2 \pi_k^2} y_k^2$$

$$= \widehat{V}_2 + \sum_{c=1}^{C} \sum_{k \in r_c} \frac{1 - \widehat{\phi}_c}{\widehat{\phi}_c^2} \frac{1 - w_{kk}\pi_k}{\pi_k^2} y_k^2.$$

Since the inclusion probabilities π_k do not depend on S, we have

$$w_{kk} = \frac{1 - \pi_k}{\pi_k}.$$

Therefore,

$$\widehat{V} = \widehat{V}_2 + \sum_{c=1}^{C} \sum_{k \in r_c} \frac{1 - \widehat{\phi}_c}{\widehat{\phi}_c^2} \frac{y_k^2}{\pi_k}.$$

Summary of the approach:
- Calculate, for each category c and for each respondent k of c, the variable $y_k / \widehat{\phi}_c$.
- Give the values thus obtained at the start of the 'ready-made' software and note the output value $\widehat{V}_2$.
- Add the (positive) value:

$$\sum_{c=1}^{C} \frac{1 - \widehat{\phi}_c}{\widehat{\phi}_c^2} \sum_{k \in r_c} \frac{y_k^2}{\pi_k}.$$

Exercise 9.5 *Non-response and superpopulation*

In this exercise, we introduce a model called the 'superpopulation', by examining the randomness of a completely different nature than for survey randomness. Thus, we consider adding to the survey randomness a randomness term governed by a superpopulation model. Without this approach (that of sampling 'models'), it is difficult to treat the non-responses through imputation. We consider that each value y_k of the finite population of size N is, indeed, the result of a random variable generated by the following simple model:

$$y_k = a + z_k,$$

where a is a real number (unknown) and z_k is a random variable of the expected value $\mathcal{E}(z_k) = 0$ and of the variance $\mathcal{V}(z_k) = \sigma^2$. The z_k are independent among one another. The notations $\mathcal{E}$ and $\mathcal{V}$ are voluntarily differentiated from the notations E and var for the traditional expected value and variance because it is a matter of randomness of a different nature. As a result of a simple random sample, we obtain m responses for a selected sample of size n. We denote S as the selected sample, r as the sample of respondents, $\widehat{\overline{Y}}_r$ as the mean (known) for the r respondents and $\overline{Y}$ as the true mean (unknown) for the population of size N. We assume that the response behaviour is independent from one individual to another.

I) *Reweighting with the 'classical' view*

It is common to come across, for the estimator of $\overline{Y}$, the value $\widehat{\overline{Y}}_r$, i.e., the simple mean of the respondents.

1. Justify this estimator with a simple probabilistic model, in a reweighting point of view.
2. With the previous model, show that if the size m is fixed, everything comes along as if we had produced a simple random sampling of size m in a population of size N (*hint:* calculate the conditional probability of selecting r knowing m and S, then 'decondition' by S).
3. Deduce that, if the model from 1. is true, $\widehat{\overline{Y}}_r$ is conditionally unbiased on m; then, calculate its true variance (conditional on m) and give an unbiased estimator for it (still for m fixed).
4. What problems would we have if we wanted to calculate a bias or a variance unconditionally on m?

II) *Mean imputation, 'superpopulation' view*

We are going to verify that, in a completely different point of view, we find the estimator $\widehat{\overline{Y}}_r$ and that we are able to calculate a bias and a variance, in the 'superpopulation' sense.

1. Having available information on the lone respondents, i.e., $\{y_k \mid k \in r\}$, how do we estimate ('at best', in a way of specifying) the known parameters a and σ^2 of the superpopulation model?
2. Under these conditions, what 'optimal' value are we going to naturally impute for the selected but non-respondent individual ℓ?
3. Verify that then the mean imputation estimator $\overline{Y}$ can only be $\widehat{\overline{Y}}_r$.
4. The bias in a classical sense is: $\mathrm{E}(\widehat{\overline{Y}}_r) - \overline{Y}$. In the sense of the model, the bias is obtained by taking the expected value $\mathcal{E}$ (compared to the model) of the classical bias, being:

$$\mathcal{E}[\mathrm{E}(\widehat{\overline{Y}}_r) - \overline{Y}].$$

We remember that the expected value E is conceived with respect to the randomness generating r. We consider that the sampling leading to r produces a randomness completely independent from that of the model, which goes back to saying that the two expected values $\mathcal{E}$ and E interchange. Then, calculate the bias of $\widehat{\overline{Y}}_r$.

5. The variance in a classical sense is:

$$\text{var}(\widehat{\overline{Y}}_r) = \text{E}[\widehat{\overline{Y}}_r - \text{E}(\widehat{\overline{Y}}_r)]^2.$$

In the sense of the model, we now define the variance like this:

$$\mathcal{V}(\widehat{\overline{Y}}_r) = \mathcal{E}\text{E}(\widehat{\overline{Y}}_r - \overline{Y})^2.$$

Calculate this variance as a function of m, N and σ^2.

6. What expression can we use to estimate $\mathcal{V}(\widehat{\overline{Y}}_r)$ in a way to obtain an unbiased estimator $\widehat{V}$ under the model, i.e., such that $\mathcal{E}\widehat{V} = \mathcal{V}(\widehat{\overline{Y}}_r)$? We distinguish two cases:
   a) We know how to locate the individual respondents.
   b) After imputation, we no longer know who has responded and who has not responded and consequently, we no longer know which are the imputed values.

7. Finally, what are the 'benefits' and the 'drawbacks' of the two points of view, addressing respectively I) and II), which both lead to the same estimator $\widehat{\overline{Y}}_r$?

III) *Imputation by drawing of individuals (quick overview of the method)*
We are placed in the event where that are more respondents than non-respondents (response rate higher than 50%). To impute the values of the $(n - m)$ non-respondents, we randomly assign them selected values *without replacement* among the m responses. The 'donors' therefore make up a simple random sample S^* taken from r (this is a sort of hot deck, but without replacement). We denote as y_ℓ^* the value imputed in this way for unit ℓ of $S - r$.

1. In an approach by modelling behaviour, what is the justification of this method? Write the mean estimator that is imposed and specify under what condition and in what sense it is unbiased.
2. Verify that, when the composition of r is known and fixed, the weights are random and can take two values that we will specify.
3. Deduce that this estimator is unbiased for $\overline{Y}$, in the 'traditional' sense of bias, under the conditions of Part I).

We could carry on with the exercise by calculating the accuracy (traditional or in the 'model' sense). We could show that the accuracy in the traditional sense is worse with this method III) than with that developed in II).

Solution
I) *Reweighting with the 'classical' view*

1. This approach comes back to adopting a Bernoulli model: individual k responds with probability ϕ or does not respond with probability $(1 - \phi)$. Therefore, the distribution of the indicator variable 'k responds' (denoted

R_k) is a Bernoulli distribution with parameter ϕ: $R_k \sim \mathcal{B}(1, \phi)$. We know that with the inclusion probability π_k and the response probability ϕ_k, associated with individual k, the classical reweighting estimator of the total is:

$$\widehat{Y}_\phi = \sum_{k \in r} \frac{y_k}{\pi_k \times \phi_k}.$$

It is calculable and unbiased if (and only if) $\phi_k > 0$ and ϕ_k are known for all $k = 1, \ldots, N$, but this approach is unrealistic, since in practice the ϕ_k stay unknown. Here, the model needs $\phi_k = \phi$, for all $k = 1, \ldots, N$, which in practice produces a bias as this model does not reflect the reality. With ϕ being unknown, it is necessary to estimate it. We easily verify that, S being known, the maximum likelihood estimator of parameter ϕ is:

$$\sum_{k \in S} \frac{R_k}{n} = \frac{m}{n} = \text{Empirical response rate.}$$

Actually, the likelihood function is (conditionally on S):

$$\prod_{k \in S} \phi^{R_k} (1 - \phi)^{1 - R_k}$$

and it suffices to maximise it on ϕ, with R_k being 'known'. We recall that here, $\pi_k = n/N$. At last,

$$\widehat{Y}_\phi = \sum_{k \in r} \frac{y_k}{\frac{n}{N}\frac{m}{n}} = N \frac{\sum_{k \in r} y_k}{m} = N \widehat{\overline{Y}}_r.$$

Thus, $\widehat{\overline{Y}}_\phi = \widehat{\overline{Y}}_r$, and the use of $\widehat{\overline{Y}}_r$ is justified with such a model.

2. Let us set S and m. The randomness rests on the m identifiers of the respondents among the individuals of S, i.e., on the composition of r:

$$\Pr(r \mid S, m) = \frac{\Pr(r \text{ and } m \mid S)}{\Pr(m \mid S)}.$$

As it is well understood that we consider only samples r consisting of exactly m individuals, we have:

$$\Pr(r \mid S, m) = \frac{\Pr(r \mid S)}{\Pr(m \mid S)}.$$

In the Bernoulli model, m follows a binomial distribution $\mathcal{B}(n, \phi)$, as $m = \sum_{k \in S} R_k$ and the variables R_k are independent among one another by assumption of behaviour

$$\Pr(m \mid S) = \binom{n}{m} \phi^m (1 - \phi)^{n-m}.$$

In addition,

$$\Pr(r \mid S) = \phi^m (1 - \phi)^{n-m}.$$

Actually, r is a well-determined sample of size m. To get it, it is necessary to 'select' with probability ϕ exactly the m individuals which comprise r (that justifies the term ϕ^m), while the other individuals of S (i.e., the non-respondents) are selected with probability $1 - \phi$ (hence the term $(1 - \phi)^{n-m}$). Therefore,

$$\Pr(r \mid S, \; m) = \frac{1}{\binom{n}{m}},$$

an expression which characterises a simple random sample of size m in S. At last, we decondition by S:

$$\Pr(r \mid m) = \sum_{S \supset r} \Pr(r \mid S, \; m) \Pr(S \mid m) = \sum_{S \supset r} \Pr(r \mid S, \; m) \Pr(S). \quad (9.3)$$

Effectively,

$$\Pr(S \mid m) = \frac{\Pr(m \mid S)}{\Pr(m)} \Pr(S),$$

and $\Pr(m \mid S)$ does not depend on S (S of size fixed on n), therefore

$$\Pr(m) = \Pr(m \mid S), \quad \text{and} \quad \Pr(S = s \mid m) = \Pr(S = s) = p(s).$$

Indeed

$$p(s) = \frac{1}{\binom{N}{n}},$$

and there are $\binom{N-m}{n-m}$ terms in the sum. We get

$$\Pr(r \mid m) = \frac{\binom{N-m}{n-m}}{\binom{n}{m} \binom{N}{n}} = \frac{1}{\binom{N}{m}}.$$

That characterises a simple random sample of size m in a population of size N.

3. If the Bernoulli model is true (and only in this case), we have, using the fundamental result of the previous question and for all $m > 0$:

- $\mathrm{E}[\widehat{\overline{Y}}_r \mid m] = \overline{Y}$ (fundamental property of simple random sampling of size m). We notice that the knowledge of ϕ is (fortunately) useless.

- $\mathrm{var}[\widehat{\overline{Y}}_r \mid m] = \left(1 - \dfrac{m}{N}\right) \dfrac{S_y^2}{m}$ (property of simple random sampling of size m), with

$$S_y^2 = \frac{1}{N-1} \sum_{k \in U} (y_k - \overline{Y})^2,$$

- $\widehat{\text{var}}[\widehat{\overline{Y}}_r \mid m] = \left(1 - \dfrac{m}{N}\right) \dfrac{s_r^2}{m}$, with

$$s_r^2 = \frac{1}{m-1} \sum_{k \in r} (y_k - \widehat{\overline{Y}}_r)^2$$

(calculated on the sample of respondents). The estimator $\widehat{\text{var}}[\widehat{\overline{Y}}_r \mid m]$ is unbiased for $\text{var}[\widehat{\overline{Y}}_r \mid m]$ conditionally on m, for $\text{E}(s_r^2 \mid m) = S_y^2$ (property of simple random sampling).

4. There are in fact two problems: a problem for the expected value and the variance and a calculating problem for the variance.
   - The problem of burden is derived from the situation where $m = 0$. That remains possible with probability $\Pr(m = 0) = (1 - \phi)^n$. This probability can in addition be non-negligible if ϕ is small. In this unfavourable case, $\widehat{\overline{Y}}_r$ is very obviously incalculable since there are no respondents. From this fact, the 'deconditioning' by m can only be developed by preserving the condition $m > 0$, i.e., by considering:

$$\text{E}(\widehat{\overline{Y}}_r \mid m > 0) \quad \text{and} \quad \text{var}(\widehat{\overline{Y}}_r \mid m > 0).$$

This condition implies a modified distribution of m, with $m > 0$,

$$\Pr'(m) = \frac{\Pr(m)}{\Pr(m > 0)}.$$

Since $\overline{Y} = \text{E}(\widehat{\overline{Y}}_r \mid m)$ does not depend on m for $m > 0$, we have $\text{E}(\widehat{\overline{Y}}_r \mid m > 0) = \text{E}\left[\text{E}(\widehat{\overline{Y}}_r \mid m) \mid m > 0\right] = \overline{Y}$. It is not the same for the variance.
   - For the variance, independently from the difficulty that comes from being raised, the deconditioning by m leads to serious calculation difficulties: it is necessary to calculate $\underset{m}{\text{E}} \left(\text{var}(\widehat{\overline{Y}}_r \mid m)\right)$ while $\text{var}(\widehat{\overline{Y}}_r \mid m)$ has a $1/m$ expression. Indeed, we do not know how to exactly calculate $\underset{m}{\text{E}}(1/m)$, and it would then be necessary to develop an approximate formula (which would only make sense for n large).

II) *Mean imputation, 'superpopulation' view*

1. It is a quite classical problem in mathematical statistics: we have available r values y_k that are independent and identically distributed (iid). The optimum linear estimator in the least squares sense is:

$$\widehat{a} = \widehat{\overline{Y}}_r \quad \text{(called the Gauss-Markov estimator.)}$$

The criterion is one of minimal variance among the linear estimators and without bias: $\widehat{a}$ is the estimator of type $\sum_{k \in r} \lambda_k y_k$ such that $\mathcal{E}(\widehat{a}) = a$

that minimises $\mathcal{E}(\hat{a} - a)^2$. According to this same criterion of optimality, we find:

$$\hat{\sigma}^2 = \frac{\text{SSR}}{m-1},$$

where

$$\text{SSR} = \sum_{k \in r} \hat{z}_k^2 \text{ (sum of squared residuals)},$$

with $\hat{z}_k = y_k - \widehat{\overline{Y}}_r$, thus $\hat{\sigma}^2 = s_r^2$.

2. We are going to estimate (or, more precisely, predict) the unknown value y_ℓ for the selected but non-respondent individual ℓ by the optimal value:

$$\mathcal{E}[y_\ell \mid \{y_k \mid k \in r\}] = a + \mathcal{E}(z_\ell \mid \{z_k \mid k \in r\}) = a,$$

where a must be at best estimated, thus by $\hat{a}$, i.e., that $y_\ell^* = \widehat{\overline{Y}}_r$. We therefore impute, for each non-respondent, the mean value of the respondents. Even without the theoretical arsenal shown above, this practice is intuitive and natural.

3. We denote $\widehat{\overline{Y}}_I$ as the final mean estimator $\overline{Y}$ after imputation.

$$\widehat{\overline{Y}}_I = \frac{1}{N} \left[\sum_{k \in r} \frac{y_k}{n/N} + \sum_{k \in S \backslash r} \frac{y_k^*}{n/N} \right],$$

where $S \backslash r$ is the set of selected but non-respondent individuals, n/N is the inverse of the sampling weight, and y_k^* is the imputed value for individual k. We get

$$\widehat{\overline{Y}}_I = \frac{1}{N} \left[N \frac{m}{n} \widehat{\overline{Y}}_r + (n-m) \frac{\widehat{\overline{Y}}_r}{n} N \right] = \frac{m \widehat{\overline{Y}}_r + (n-m) \widehat{\overline{Y}}_r}{n} = \widehat{\overline{Y}}_r.$$

4. The bias is

$$\text{Bias} = \mathcal{E}[\text{E}(\widehat{\overline{Y}}_r - \overline{Y})],$$

and the expected value is $\text{E}(\overline{Y}) = \overline{Y}$, however $\mathcal{E}(\overline{Y}) \neq \overline{Y}$, because $\overline{Y}$ consists of values y_k, and each y_k is a random variable. Therefore,

$$\text{Bias} = \text{E}[\mathcal{E}(\widehat{\overline{Y}}_r - \overline{Y})].$$

Since the two types of randomness are by hypothesis of independent nature, we can interchange the operators E and $\mathcal{E}$.

Note on this point: Concretely, this signifies that the sampling of individuals to construct S and the response process, i.e., the transition from S to

r, is carried out for both independently from the values y that the individuals take. In other words, we require that the 'behaviour of response/non-response' not depend on the realisations of the random variables y_k, or, otherwise stated, the fact of whether or not to respond does not depend on the value y (which all the same consists of a rather strong assumption, contrary to its appearance).

$$\text{Bias} = \text{E}[\mathcal{E}(\widehat{\overline{Y}}_r - \overline{Y})] = \text{E}[\mathcal{E}(\widehat{\overline{Y}}_r) - \mathcal{E}(\overline{Y})].$$

Indeed,

$$\mathcal{E}(\widehat{\overline{Y}}_r) = \mathcal{E}\left(\frac{\sum_{k\in r} y_k}{m}\right) = \sum_{k\in r}\frac{\mathcal{E}(y_k)}{m} = m\frac{a}{m} = a,$$

and

$$\mathcal{E}(\overline{Y}) = \mathcal{E}\left(\frac{\sum_{k\in U} y_k}{N}\right) = \sum_{k\in U}\frac{\mathcal{E}(y_k)}{N} = N\frac{a}{N} = a.$$

Therefore, the bias is: $B = \text{E}(0) = 0$.

5. For the variance, we are going to use the same property and interchange the operators E and $\mathcal{E}$

$$\mathcal{V}(\widehat{\overline{Y}}_r) = \text{E}[\mathcal{E}(\widehat{\overline{Y}}_r - \overline{Y})^2].$$

We have:

$$\widehat{\overline{Y}}_r = a + \widehat{\overline{Z}}_r \text{ and } \overline{Y} = a + \overline{Z},$$

where

$$\widehat{\overline{Z}}_r = \frac{1}{n}\sum_{k\in r} z_k \quad \text{and} \quad \overline{Z} = \frac{1}{N}\sum_{k\in U} z_k.$$

Therefore,

$$\mathcal{V}(\widehat{\overline{Y}}_r) = \text{E}[\mathcal{E}(\widehat{\overline{Z}}_r - \overline{Z})^2].$$

We have, as well,

$$\mathcal{E}(\widehat{\overline{Z}}_r - \overline{Z})^2$$

$$= \mathcal{E}\left(\frac{\sum_{k\in r} z_k}{m} - \frac{\sum_{k\in U} z_k}{N}\right)^2 = \mathcal{E}\left[\sum_{k\in r} z_k\left(\frac{1}{m} - \frac{1}{N}\right) - \sum_{k\notin r}\frac{z_k}{N}\right]^2$$

$$= \mathcal{V}\left(\sum_{k\in r} z_k\left(\frac{1}{m} - \frac{1}{N}\right) - \sum_{k\notin r}\frac{z_k}{N}\right) \quad \text{because} \quad \mathcal{E}z_k = 0$$

$$= \sum_{k\in r}\left(\frac{1}{m} - \frac{1}{N}\right)^2 \mathcal{V}(z_k) + \sum_{k\notin r}\frac{1}{N^2}\mathcal{V}(z_k) \quad \text{for the } z_k \text{ are } independent$$

$$= \left[m\left(\frac{1}{m} - \frac{1}{N}\right)^2 + (N - m)\frac{1}{N^2}\right]\sigma^2 = \frac{N - m}{Nm}\sigma^2.$$

Indeed, this expression only depends on *fixed* quantities, thus it is not sensitive to S, which leads to

$$E(\mathcal{E}(\widehat{\overline{Z}}_r - \overline{Z})^2) = \mathcal{E}(\widehat{\overline{Z}}_r - \overline{Z})^2 = \left(1 - \frac{m}{N}\right)\frac{\sigma^2}{m}.$$

Note: The expression resembles that of the variance for a simple random sample of size m in a population of size N. The lone difference is that the traditional population variance defined on the population S_y^2 has been 'replaced' here by the variance from the model σ^2.

6. a) We recalled in II.1 that the traditional expression of the variance in the sample of respondents denoted s_r^2 was unbiased under the model (classical theory of the linear model), that is:

$$\mathcal{E}(s_r^2) = \sigma^2.$$

Hence the natural estimator,

$$\widehat{V} = \left(1 - \frac{m}{N}\right)\frac{s_r^2}{m} \quad \text{such that} \quad \mathcal{E}\widehat{V} = V(\widehat{\overline{Y}}_r).$$

b) After imputation and in the absence of remembering the imputed data, the simple expression that we try to naturally calculate is the overall variance in S, being:

$$s^2 = \frac{1}{n-1}\sum_{k \in S}(y_k^* - \widehat{\overline{Y}}_r)^2,$$

with

$$y_k^* = \begin{cases} y_k & \text{if } k \text{ responded} \\ \widehat{\overline{Y}}_r & \text{otherwise.} \end{cases}$$

We recall that

$$\widehat{\overline{Y}}_r = \frac{\sum_{k \in S} y_k^*}{n}.$$

The estimator s^2 can seem at first glance impossible to calculate since it brings into play all the individuals of S and not only the respondents. Actually, this is not the case, as it is sufficient to observe that for all $k \notin r$, $y_k^* = \widehat{\overline{Y}}_r$ and therefore the terms corresponding to the non-respondents disappear. Hence,

$$s^2 = \frac{1}{n-1}\sum_{k \in r}(y_k^* - \widehat{\overline{Y}}_r)^2 = \frac{1}{n-1}\sum_{k \in r}(y_k - \widehat{\overline{Y}}_r)^2,$$

and therefore

$$s^2 = \frac{m-1}{n-1}s_r^2.$$

It is thus necessary to use, to estimate $\mathcal{V}(\widehat{\overline{Y}}_r)$ without bias:

$$\widehat{V} = \underbrace{\left(1 - \frac{m}{N}\right) \frac{1}{m} \left(\frac{n-1}{m-1} s^2\right)}_{\beta \text{ term}} \underbrace{\frac{n-1}{m-1} \left(1 - \frac{m}{N}\right) \frac{s^2}{m}}_{\alpha \text{ term}}.$$

The α term is the expression that we are 'naturally' led to take in the presence of imputed data (the naive calculation of the variance s^2 with all the data of which we have available is effectively quite natural). Unfortunately, this term underestimates the true accuracy, and it is therefore necessary to correct by multiplying it with the β term, greater than 1. The β term is for that matter pretty much equal to the inverse of the response rate. In effect, the mean imputation $\widehat{\overline{Y}}_r$ creates a lot of equal y_k^* (to $\widehat{\overline{Y}}_r$ in this case), and that artificially reduces the variance.

7. *Overview:*
   a) In I), we count on the validity of the response model (Bernoulli model here): it is necessary to specify the way in which we go from S to r by modelling the response probabilities. This is somewhat risky.
   b) In II), it is not necessary to know precisely how we go from S to r by modelling the response probabilities. However, a rather strong first assumption requires that the fact of whether or not to respond does not depend on the distribution producing y. A second rather strong assumption is the model of behaviour directly dealing with the variables y_k. Therefore:
   - If we favour the model on the response probabilities, use Approach I.
   - If we favour the model on the values y_k themselves, use Approach II.

   It is necessary to prejudge the model in which we have the most confidence, the one that seems most reliable.

III) *Imputation by sampling of individuals*

1. This method is intuitively justified if we consider that the respondents and the non-respondents have the same behaviour y 'on average', that is to say if we believe in the superpopulation model

$$y_k = a + z_k, \text{ for all } k = 1, 2, \ldots, N.$$

The natural mean estimator is:

$$\widehat{\overline{Y}}_I = \frac{\sum_{k \in r} y_k + \sum_{k \in S - r} y_k^*}{n},$$

where the y_k^* are in fact $y_j (j \in r)$. Since $\mathcal{E}(y_k) = a$ for all $k = 1, 2, \ldots, N$,

$$\mathcal{E}(\widehat{\overline{Y}}_I) = a = \mathcal{E}(\overline{Y}).$$

Thus
$$\mathcal{E}\mathrm{E}(\widehat{\overline{Y}}_I - \overline{Y}) = \mathrm{E}\mathcal{E}(\widehat{\overline{Y}}_I - \overline{Y}) = \mathrm{E}\left[\mathcal{E}(\widehat{\overline{Y}}_I) - \mathcal{E}(\overline{Y})\right] = 0.$$

The estimator $\widehat{\overline{Y}}_I$ is unbiased under the conditions of II.4. Thanks to the assumption that allows to interchange $\mathcal{E}$ and E, it is not worthwhile to specify how we go from S to r. Furthermore, the method of selecting the $(n - m)$ individuals in r (therefore the drawing of S^*) is 'without effect' on the calculation of the bias.

2. We have
$$\widehat{\overline{Y}}_I = \frac{\sum_{k\in r} y_k + \sum_{k\in S-r} y_k^*}{n}.$$

Indeed, y_k^*, for all $k \in S - r$, is in reality one of the y_j with $j \in r$. More precisely, we can say, by definition of S^*, that $j \in S^*$. We can therefore write, for all $k \in S - r$, that there exists $j \in S^*$ such that $y_k^* = y_j$, and

$$\widehat{\overline{Y}}_I = \frac{\sum_{k\in r} y_k + \sum_{j\in S^*} y_j}{n} = \sum_{k\in r}\left(\frac{1 + I\{k \in S^*\}}{n}\right) y_k,$$

where $I\{k \in S^*\}$ refers to the indicator variable of the occurrence $k \in S^*$. For $k \in r$ fixed, the sampling weight is:

$$\frac{1 + I\{k \in S^*\}}{n}.$$

It is therefore random, being explicitly dependent on S^*. It can take two values:

$$\frac{1 + I\{k \in S^*\}}{n} = \begin{cases} \dfrac{1+1}{n} = \dfrac{2}{n} & \text{with probability } \Pr[k \in S^*] = \dfrac{n - m}{m} \\ \dfrac{1}{n} & \text{with probability } \Pr[k \notin S^*] = 1 - \dfrac{n - m}{m}, \end{cases}$$

(reminder: $m > n/2$).

3. In the first place, let us condition with respect to r (r fixed, we are only interested in the randomness that produces S^*):

$$\mathop{\mathrm{E}}_{S^*|r}(\widehat{\overline{Y}}_I) = \mathop{\mathrm{E}}_{S^*|r}\left[\sum_{k\in r}\frac{1 + I\{k \in S^*\}}{n} y_k\right] = \sum_{k\in r}\left[\frac{1 + \mathop{\mathrm{E}}_{S^*|r} I\{k \in S^*\}}{n}\right] y_k.$$

Since
$$\mathop{\mathrm{E}}_{S^*|r} I\{k \in S^*\} = \Pr[k \in S^* \mid k \in r] = \frac{n - m}{m},$$

we have
$$\mathop{\mathrm{E}}_{S^*|r}(\widehat{\overline{Y}}_I) = \frac{1 + \frac{n}{m} - 1}{n}\left(\sum_{k\in r} y_k\right) = \widehat{\overline{Y}}_r.$$

Now, we saw in I) that with the Bernoulli model, we have

$$\mathrm{E}(\widehat{\overline{Y}}_r \mid m > 0) = \overline{Y},$$

the randomness this time being the sampling of S, then the sampling of r in S. Provided that $m > 0$, the estimator $\widehat{\overline{Y}}_I$ is therefore unbiased in the classical sense, together for all randomness.

Note: In point of view III), we bring into play up to three types of randomness which occurs in sequence: S, then r, then S^*.

*   *   *

Table of Notations

#	cardinal (number of elements in a set)
$\ll$	much less than
$\backslash$	$A \backslash B$ complement of B in A
$'$	the vector $\mathbf{u}'$ is the transpose of the vector $\mathbf{u}$
!	factorial: $n! = n \times (n-1) \times \cdots \times 2 \times 1$
$\binom{N}{n}$	$\frac{N!}{n!(N-n)!}$ number of combinations of n individuals among N
$[a \pm b]$	interval $[a - b, a + b]$
$\approx$	is approximately equal to
$\sim$	follows a specified distribution (for a random variable)
b	slope of the regression line for y on x in the population
$\hat{b}$	estimator of the slope of the regression line for y on x
$\text{CI}(1 - \alpha)$	confidence interval of probability level $1 - \alpha$
$\text{cov}(X, Y)$	covariance between random variables X and Y
CV	coefficient of variation
d_k	$d_k = 1/\pi_k$ natural sampling weight
DEFF	design effect
D	domain of U
$\text{E}(Y)$	mathematical expected value of random variable Y
$\text{E}(Y\|A)$	mathematical expected value of random variable Y given that event A occurs
$\mathcal{E}(Y)$	expected value of Y with respect to the randomness of a model
f	sampling rate $f = n/N$
$G_k(.)$	pseudo-distance
h	indicator of the stratum or post-stratum

I_k	is 1 if unit k is in the sample and 0 otherwise
$I\{A\}$	is 1 if A is true and 0 otherwise
k or i	generally indicates a statistical unit, $k \in U$ (identifying) or $i \in U$
m	number of clusters or primary units in the sample of primary units, or the sample size with replacement
M	number of clusters or primary units in the population
MSE	mean square error
n	sample size (without replacement)
$\bar{n}$	average sample size of SU in the PU
n_D	sample size intersecting domain D
n_i	number of sampled SU in PU i
n_r or m	number of respondents in the sample
n_S	sample size within S
N	population size
$\overline{N}$	average size of the PU in the population
n_h	sample size in stratum or post-stratum U_h
N_h	number of statistical units in stratum or post-stratum U_h
N_i	number of SU in PU i
N_{ij}	population size in case (i, j) of a contingency table
$p(s)$	probability associated with sample s
p_i	elementary sampling probability of unit i in a drawing with replacement
P or P_D	proportion of individuals belonging to a domain D
$\Pr(A)$	probability that event A occurs
$\Pr(A|B)$	probability that event A occurs, given that event B occurs
PU	primary unit
r	sample of respondents
R_k	random variable equalling 1 if k responds and 0 otherwise
s	sample or subset of the population, $s \subset U$
s_y^2	corrected sample variance of variable y
s_{yh}^2	corrected sample variance of y in stratum or post-stratum h
s_T^2	corrected sample variance of totals estimated for PU in the sample of PU
$s_{2,i}^2$	corrected sample variance of y in the sample of SU within PU i
s_{xy}	corrected covariance between variables x and y for the sample
S	random sample such that $Pr(S = s) = p(s)$

S_y^2	corrected population variance of variable y for the population
S_{xy}	corrected covariance between variables x and y for the population
S_h	random sample selected in stratum or post-stratum h
S_{yh}^2	corrected population variance of y in stratum or post-stratum h
S_T^2	corrected population variance of totals for PU in the population of PU
$S_{2,i}^2$	corrected population variance of y within PU i
SU	secondary unit
U	finite population of size N
U_h	finite population consisting of stratum or post-stratum h, where $h = 1, \cdots, H$
v_k	linearised variable
$\mathrm{var}(Y)$	variance of random variable Y
$\mathcal{V}(Y)$	variance of Y with respect to the randomness of a model
$\widehat{\mathrm{var}}(Y)$	estimator of the variance of random variable Y
w_k	weight associated with individual k in the sample
x	real auxiliary variable
x_k	value taken by the real auxiliary variable for unit k
$\mathbf{x}_k$	vector of $\mathbb{R}^p$ corresponding to the values taken by the p auxiliary variables for unit k
X	total of values taken by the auxiliary variable for all units of U
$\widehat{X}$ or $\widehat{X}_\pi$	Horvitz-Thompson estimator of X
$\overline{X}$	mean of values taken by the auxiliary variable for all units of U
$\widehat{\overline{X}}$ or $\widehat{\overline{X}}_\pi$	Horvitz-Thompson estimator of $\overline{X}$
y	variable of interest
y_k	value taken by the variable of interest for unit k
y_k^*	value of y imputed for individual k (treatment of non-response)
$y_{i,k}$	value of the variable of interest y for SU k of PU i
Y	total of values taken by the variable of interest for all units of U
Y_h	total of values taken by the variable of interest for all units of stratum or post-stratum U_h
Y_i	total of $y_{i,k}$ in PU i
$\widehat{Y}$ or $\widehat{Y}_\pi$	Horvitz-Thompson estimator of Y
$\overline{Y}$	mean of values taken by the variable of interest for all units of U
$\overline{Y}_h$	mean of values taken by the variable of interest for all units of stratum or post-stratum U_h

$\widehat{\overline{Y}}_h$ mean estimator of the values taken by the variable of interest for all the units of stratum or post-stratum U_h

$\widehat{\overline{Y}}$ or $\widehat{\overline{Y}}_\pi$ Horvitz-Thompson estimator of $\overline{Y}$

$\widehat{Y}_{\text{post}}$ post-stratified estimator of the total

$\widehat{Y}_{\text{reg}}$ regression estimator of the total

$\widehat{Y}_D$ difference estimator of the total

$\widehat{Y}_h$ estimator of the total Y_h in stratum or post-stratum U_h

$\widehat{Y}_H$ Hájek ratio of the total

$\widehat{Y}_I$ estimator used in the case of imputation for non-response

$\widehat{Y}_R$ ratio estimator of the total

$\widehat{\overline{Y}}_r$ simple mean of y for the individual respondents of the sample

$\widehat{Y}_\phi$ estimator used in the case of reweighting for non-response

z_p p-quantile of the standard normal distribution

α probability that the function of interest is found outside of the confidence interval

$\Delta_{k\ell}$ $\pi_{k\ell} - \pi_k \pi_\ell$

π_k inclusion probability of unit k

$\pi_{k\ell}$ second-order inclusion probability for units k and ℓ

$\pi_{k\ell} = Pr(k \text{ and } \ell \in S)$

σ^2 variance of randomness in a superpopulation model

ϕ_k response probability of individual k

ρ linear correlation coefficient between x and y for the population, or cluster effect

σ_y^2 population variance of variable y for the population

Normal Distribution Tables

Table 10.1. Table of quantiles of a standard normal variable

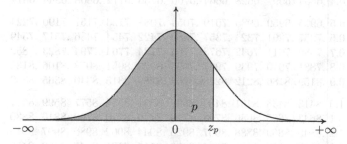

Order of quantile (p)	Quantile (z_p)	Order of quantile (p)	Quantile (z_p)
0.500	0.0000	0.975	1.9600
0.550	0.1257	0.976	1.9774
0.600	0.2533	0.977	1.9954
0.650	0.3853	0.978	2.0141
0.700	0.5244	0.979	2.0335
0.750	0.6745	0.990	2.3263
0.800	0.8416	0.991	2.3656
0.850	1.0364	0.992	2.4089
0.900	1.2816	0.993	2.4573
0.950	1.6449	0.994	2.5121
0.970	1.8808	0.995	2.5758
0.971	1.8957	0.996	2.6521
0.972	1.9110	0.997	2.7478
0.973	1.9268	0.998	2.8782
0.974	1.9431	0.999	3.0902

Table 10.2. Cumulative distribution function of the standard normal distribution

(Probability of finding a value less than u)

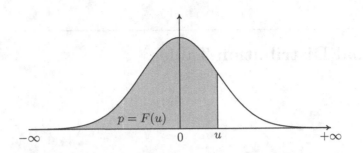

u	0.0	.01	.02	.03	.04	.05	.06	.07	.08	.09
0.0	.5000	.5040	.5080	.5120	.5160	.5199	.5239	.5279	.5319	.5359
0.1	.5398	.5438	.5478	.5517	.5557	.5596	.5636	.5675	.5714	.5753
0.2	.5793	.5832	.5871	.5910	.5948	.5987	.6026	.6064	.6103	.6141
0.3	.6179	.6217	.6255	.6293	.6331	.6368	.6406	.6443	.6480	.6517
0.4	.6554	.6591	.6628	.6664	.6700	.6736	.6772	.6808	.6844	.6879
0.5	.6915	.6950	.6985	.7019	.7054	.7088	.7123	.7157	.7190	.7224
0.6	.7257	.7291	.7324	.7357	.7389	.7422	.7454	.7486	.7517	.7549
0.7	.7580	.7611	.7642	.7673	.7704	.7734	.7764	.7794	.7823	.7852
0.8	.7881	.7910	.7939	.7967	.7995	.8023	.8051	.8078	.8106	.8133
0.9	.8159	.8186	.8212	.8238	.8264	.8289	.8315	.8340	.8365	.8389
1.0	.8413	.8438	.8461	.8485	.8508	.8531	.8554	.8577	.8599	.8621
1.1	.8643	.8665	.8686	.8708	.8729	.8749	.8770	.8790	.8810	.8830
1.2	.8849	.8869	.8888	.8907	.8925	.8944	.8962	.8980	.8997	.9015
1.3	.9032	.9049	.9066	.9082	.9099	.9115	.9131	.9147	.9162	.9177
1.4	.9192	.9207	.9222	.9236	.9251	.9265	.9279	.9292	.9306	.9319
1.5	.9332	.9345	.9357	.9370	.9382	.9394	.9406	.9418	.9429	.9441
1.6	.9452	.9463	.9474	.9484	.9495	.9505	.9515	.9525	.9535	.9545
1.7	.9554	.9564	.9573	.9582	.9591	.9599	.9608	.9616	.9625	.9633
1.8	.9641	.9649	.9656	.9664	.9671	.9678	.9686	.9693	.9699	.9706
1.9	.9713	.9719	.9726	.9732	.9738	.9744	.9750	.9756	.9761	.9767
2.0	.9772	.9778	.9783	.9788	.9793	.9798	.9803	.9808	.9812	.9817
2.1	.9821	.9826	.9830	.9834	.9838	.9842	.9846	.9850	.9854	.9857
2.2	.9861	.9864	.9868	.9871	.9875	.4878	.9881	.9884	.9887	.9890
2.3	.9893	.9896	.9898	.9901	.9904	.9906	.9909	.9911	.9913	.9916
2.4	.9918	.9920	.9922	.9925	.9927	.9929	.9931	.9932	.9934	.9936
2.5	.9938	.9940	.9941	.9943	.9945	.9946	.9948	.9949	.9951	.9952
2.6	.9953	.9955	.9956	.9957	.9959	.9960	.9961	.9962	.9963	.9964
2.7	.9965	.9966	.9967	.9968	.9969	.9970	.9971	.9972	.9973	.9974
2.8	.9974	.9975	.9976	.9977	.9977	.9978	.9979	.9979	.9980	.9981
2.9	.9981	.9982	.9982	.9983	.9984	.9984	.9985	.9985	.9986	.9986
3.0	.9987	.9987	.9987	.9988	.9988	.9989	.9989	.9989	.9990	.9990
3.1	.9990	.9991	.9991	.9991	.9992	.9992	.9992	.9992	.9993	.9993
3.2	.9993	.9993	.9994	.9994	.9994	.9994	.9994	.9995	.9995	.9995
3.3	.9995	.9995	.9995	.9996	.9996	.9996	.9996	.9996	.9996	.9997
3.4	.9997	.9997	.9997	.9997	.9997	.9997	.9997	.9997	.9997	.9998

Table 10.3. Quantiles of the standard normal distribution

(u: value having the probability α of being surpassed in absolute value)

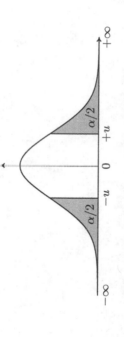

α	0	0.01	0.02	0.03	0.04	0.05	0.06	0.07	0.08	0.09
0	∞	2.5758	2.3263	2.1701	2.0537	1.9600	1.8808	1.8119	1.7507	1.6954
0.1	1.6449	1.5982	1.5548	1.5141	1.4758	1.4395	1.4051	1.3722	1.3408	1.3106
0.2	1.2816	1.2536	1.2265	1.2004	1.1750	1.1503	1.1264	1.1031	1.0803	1.0581
0.3	1.0364	1.0152	0.9945	0.9741	0.9542	0.9346	0.9154	0.8965	0.8779	0.8596
0.4	0.8416	0.8239	0.8064	0.7892	0.7722	0.7554	0.7388	0.7225	0.7063	0.6903
0.5	0.6745	0.6588	0.6433	0.6280	0.6128	0.5978	0.5828	0.5681	0.5534	0.5388
0.6	0.5244	0.5101	0.4958	0.4817	0.4677	0.4538	0.4399	0.4261	0.4125	0.3989
0.7	0.3853	0.3719	0.3585	0.3451	0.3319	0.3186	0.3055	0.2924	0.2793	0.2663
0.8	0.2533	0.2404	0.2275	0.2147	0.2019	0.1891	0.1764	0.1637	0.1510	0.1383
0.9	0.1257	0.1130	0.1004	0.0878	0.0753	0.0627	0.0502	0.0376	0.0251	0.0125

List of Tables

List of Figures

References

Ardilly, P. (1994). *Les Techniques de Sondage*. Technip, Paris.

Brewer, K. (1975). A simple procedure for πpswor. *Australian Journal of Statistics*, 17:166–172.

Brewer, K. and Hanif, M. (1983). *Sampling with Unequal Probabilities*. Springer-Verlag, New York.

Caron, N. (1999). Le logiciel POULPE, aspects méthodologiques. In *Actes des Journées de Méthodologie Statistique*, volume 84, pages 173–200.

Cassel, C.-M., Särndal, C.-E., and Wretman, J. (1993). *Foundations of Inference in Survey Sampling*. Wiley, New York.

Cicchitelli, G., Herzel, A., and Montanari, E. (1992). *Il Campionamento Statistico*. Il Mulino, Bologna.

Cochran, W. (1977). *Sampling Techniques*. Wiley, New York.

Deming, W. (1950). *Some Theory of Sampling*. Dover Publications, New York.

Deming, W. (1960). *Sample Design in Business Research*. Wiley, New York.

Deroo, M. and Dussaix, A.-M. (1980). *Pratique et analyse des enquêtes par sondage*. P.U.F., Paris.

Desabie, J. (1966). *Théorie et pratique des sondages*. Dunod, Paris.

Deville, J.-C. (1999). Variance estimation for complex statistics and estimators: linearization and residual techniques. *Survey Methodology*, 25:193–204.

Deville, J.-C. and Särndal, C.-E. (1992). Calibration estimators in survey sampling. *Journal of the American Statistical Association*, 87:376–382.

Deville, J.-C. and Tillé, Y. (1998). Unequal probability sampling without replacement through a splitting method. *Biometrika*, 85:89–101.

Deville, J.-C. and Tillé, Y. (2004). Efficient balanced sampling: the cube method. *Biometrika*, 91:893–912.

Droesbeke, J.-J., Fichet, B., and Tassi, P. (1987). *Les Sondages*. Economica, Paris.

Dussaix, A.-M. and Grosbras, J.-M. (1992). *Exercices de sondages*. Economica, Paris.

Dussaix, A.-M. and Grosbras, J.-M. (1996). *Les sondages : principes et méthodes*. P.U.F. (Que sais-je ?), Paris.

Efron, B. and Tibshirani, R. (1993). *An Introduction to the Bootstrap*. Chapman & Hall Ltd.

Fan, C., Muller, M., and Rezucha, I. (1962). Development of sampling plans by using sequential (item by item) selection techniques and digital computer. *Journal of the American Statistical Association*, 57:387–402.

Gabler, S. (1990). *Minimax Solutions in Sampling from Finite Populations*, volume 64. Springer-Verlag, Berlin.

Gouriéroux, C. (1981). *Théorie des sondages*. Economica, Paris.

Grosbras, J.-M. (1987). *Méthodes statistiques des sondages*. Economica, Paris.

Hájek, J. (1981). *Sampling from a Finite Population*. Marcel Dekker, New York.

Hansen, M. and Hurwitz, W. (1943). On the theory of sampling from finite populations. *Annals of Mathematical Statistics*, 14:333–362.

Hansen, M., Hurwitz, W., and Madow, W. (1953 reprint in 1993a). *Sample Survey Methods and Theory, I*. Wiley, New York.

Hansen, M., Hurwitz, W., and Madow, W. (1953 reprint in 1993b). *Sample Survey Methods and Theory, II*. Wiley, New York.

Hedayat, A. and Sinha, B. (1991). *Design and Inference Finite Population Sampling*. Wiley, New York.

Jessen, R. (1978). *Statistical Survey Techniques*. Wiley, New York.

Kish, L. (1965). *Survey Sampling*. Wiley, New York.

Kish, L. (1989). *Sampling method for agricultural surveys*. FAO Statistical Development Series.

Konijn, H. (1973). *Statistical theory of sample survey design and analysis*. North-Holland, Amsterdam.

Krishnaiah, P. and Rao, C. (1994). *Handbook of Statistics, Vol 6 (Sampling)*. Elsevier Science Publishers, New York.

Lohr, S. (1999). *Sampling: Design and Analysis*. Duxbury Press.

Madow, W. (1948). On the limiting distribution based on samples from finite universes. *Annals of Mathematical Statistics*, 19:535–545.

McLeod, A. and Bellhouse, D. (1983). A convenient algorithm for drawing a simple random sampling. *Applied Statistics*, 32:182–184.

Midzuno, H. (1952). On the sampling system with probability proportional to sum of size. *Annals of the Institute of Statistical Mathematics*, 3:99–107.

Morin, H. (1993). *Théorie de l'échantillonnage*. Les Presses de L'Université Laval, Sainte-Foy.

Raj, D. (1968). *Sampling Theory*. McGraw-Hill, New York.

Rao, J. and Sitter, R. (1995). Variance estimation under two-phase sampling with application to imputation for missing data. *Biometrika*, 82:453–460.

Ren, R. and Ma, X. (1996). *Sampling Survey Theory and Its Applications*. Henan University Press, Kaifeng, China. In Chinese.

Sen, A. (1953). On the estimate of the variance in sampling with varying probabilities. *Journal of Indian Society for Agricultural Statistics*, 5:119–127.

Singh, R. (1975). A note on the efficiency of ratio estimate with Midzuno's scheme of sampling. *Sankhyā*, C37:211–214.

Särndal, C.-E., Swensson, B., and Wretman, J. (1992). *Model Assisted Survey Sampling*. Springer Verlag, New York.

Stanley, R. (1997). *Enumerative Combinatorics*. Cambrige University Press, Cambrige.

Sukhatme, P. and Sukhatme, B. (1970). *Sampling Theory of Surveys with Applications*. Asian Publishing House, Calcutta, India.

Sunter, A. (1977). List sequential sampling with equal or unequal probabilities without replacement. *Applied Statistics*, 26:261–268.

Sunter, A. (1986). Solutions to the problem of unequal probability sampling without replacement. *International Statistical Review*, 54:33–50.

Thionet, P. (1953). *La théorie des sondages*. INSEE, Imprimerie nationale, Paris.

Thompson, S. (1992). *Sampling*. Wiley, New York.

Tillé, Y. (2001). *Théorie des sondages: échantillonnage et estimation en populations finies*. Dunod, Paris.

Valliant, R., Dorfman, A., and Royall, R. (2000). *Finite Population Sampling and Inference: A Prediction Approach*. Wiley Series in Probability and Statistics: Survey Methodology Section, New York.

Wolter, K. (1985). *Introduction to Variance Estimation*. Springer-Verlag, New York.

Yates, F. (1949). *Sampling Methods for Censuses and Surveys*. Griffin, London.

Yates, F. and Grundy, P. (1953). Selection without replacement from within strata with probability proportional to size. *Journal of the Royal Statistical Society*, B15:235–261.

Zarkovich, S. (1966). *Sondages et recensements*. F.A.O, Rome.

Author Index

Index

Matrix Algebra: Exercises and Solutions

D.A. Harville

This book contains over 300 exercises and solutions covering a wide variety of topics in matrix algebra. They can be used for independent study or in creating a challenging and stimulating environment that encourages active engagement in the learning process. Thus, the book can be of value to both teachers and students. The requisite background is some previous exposure to matrix algebra of the kind obtained in a first course. The exercises are those from an earlier book by the same author entitled Matrix Algebra From a Statistician's Perspective (ISBN 0-387-94978-X). They have been restated (as necessary) to stand alone, and the book includes extensive and detailed summaries of all relevant terminology and notation. The coverage includes topics of special interest and relevance in statistics and related disciplines, as well as standard topics.

2001. 271 p. Softcover ISBN 0-387-95318-3

Model Assisted Survey Sampling

C-E. Särndal, B. Swensson, and J. Wretman

This book provides a comprehensive account of survey sampling theory and methodology which will be suitable for students and researchers across a variety of disciplines. A central theme is to show how statistical modeling is a vital component of the sampling process and in the choice of estimation technique. Statistical modeling has strongly influenced sampling theory in recent years and has clarified many issues related to the uses of auxiliary information in surveys. This is the first textbook that systematically extends traditional sampling theory with the aid of a modern model assisted outlook. The central ideas of sampling theory are developed from the unifying perspective of unequal probability sampling. The book covers classical topics as well as areas where significant new developments have taken place notably domain estimation, variance estimation, methods for handling nonresponse, models for measurement error, and the analysis of survey data.

1991. 800 p. (Springer Series in Statistics) Softcover ISBN 0-387-40620-4